T0202864

PROBLEMS AND SOLUTIONS FOR
UNDERGRADUATE ANALYSIS

Rami Shakarchi

PROBLEMS AND SOLUTIONS FOR UNDERGRADUATE ANALYSIS

With 40 illustrations

Springer

Rami Shakarchi
Department of Mathematics
Princeton University
Fine Hall, Washington Road
Princeton, NJ 08544-1000
USA

Mathematics Subject Classification Codes: 26-00, 46-01

Printed on acid-free paper.

Production managed by Anthony K. Guardiola; manufacturing supervised by Jacqui Ashri.
Photocomposed pages prepared from the author's TeX files.
Printed and bound by Braun-Brumfield, Inc., Ann Arbor, MI.
Printed in the United States of America.

9 8 7 6 5 4 3 2 1

ISBN 0-387-98235-3 Springer-Verlag New York Berlin Heidelberg   SPIN 10574572

To my mother,
in appreciation for her dedication
and love for her children

Preface

The present volume contains all the exercises and their solutions for Lang's second edition of Undergraduate Analysis. The wide variety of exercises, which range from computational to more conceptual and which are of varying difficulty, cover the following subjects and more: real numbers, limits, continuous functions, differentiation and elementary integration, normed vector spaces, compactness, series, integration in one variable, improper integrals, convolutions, Fourier series and the Fourier integral, functions in n-space, derivatives in vector spaces, the inverse and implicit mapping theorem, ordinary differential equations, multiple integrals, and differential forms. My objective is to offer those learning and teaching analysis at the undergraduate level a large number of completed exercises and I hope that this book, which contains over 600 exercises covering the topics mentioned above, will achieve my goal.

The exercises are an integral part of Lang's book and I encourage the reader to work through all of them. In some cases, the problems in the beginning chapters are used in later ones, for example, in Chapter IV when one constructs-bump functions, which are used to smooth out singularities, and prove that the space of functions is dense in the space of regulated maps. The numbering of the problems is as follows. Exercise IX.5.7 indicates Exercise 7, §5, of Chapter IX.

Acknowledgments

I am grateful to Serge Lang for his help and enthusiasm in this project, as well as for teaching me mathematics (and much more) with so much generosity and patience.

My family's support proved invaluable. I thank my parents and brother for always being there when needed.

I thank the faculty at the Yale Mathematics Department for their encouragements and help in the past four years, as well as for the time they have spent teaching me mathematics.

Finally, I am indebted to Steven Joel Miller for his sense of humor and undying support and assistance when we do mathematics together.

Yale, 1997   *Rami Shakarchi*

Contents

0

Sets and Mappings

0.2 Mappings

Exercise 0.2.1 *Let S, T, T' be sets. Show that*

$$S \cap (T \cup T') = (S \cap T) \cup (S \cap T').$$

If $T_1, \ldots, T_n$ are sets, show that

$$S \cap (T_1 \cup \cdots \cup T_n) = (S \cap T_1) \cup \cdots \cup (S \cap T_n).$$

Solution. Assume that S, T, and T' are all non-empty (if not the equality is trivial). Suppose that $x \in S \cap (T \cup T')$. Then x belongs necessarily to S and to at least one of the sets T or T'. Thus x belongs to at least one of the sets $S \cap T$ or $S \cap T'$. Hence

$$S \cap (T \cup T') \subset (S \cap T) \cup (S \cap T').$$

To get the reverse inclusion, note that $T \subset (T \cup T')$ so $(S \cap T) \subset S \cap (T \cup T')$ and similarly, $(S \cap T') \subset S \cap (T \cup T')$. Therefore

$$(S \cap T) \cup (S \cap T') \subset S \cap (T \cup T').$$

For $1 \leq j \leq n$, let $V_j = T_1 \cup \cdots \cup T_j$. Then by our previous argument

$$S \cap V_n = S \cap (V_{n-1} \cup T_n) = (S \cap V_{n-1}) \cup (S \cap T_n).$$

Repeating this process $n - 1$ times we find

$$S \cap (T_1 \cup \cdots \cup T_n) = (S \cap V_1) \cup (S \cap T_2) \cup \cdots \cup (S \cap T_n).$$

But $V_1 = T_1$ so this proves the equality.

Exercise 0.2.2 *Show that the equalities of Exercise 1 remain true if the intersection and union signs $\cap$ and $\cup$ are interchanged.*

Solution. We want to show that $S \cup (T \cap T') = (S \cup T) \cap (S \cup T')$. Suppose $x \in S \cup (T \cap T')$, then x belongs to S or T and T'. But since $S \subset (S \cup T) \cap (S \cup T')$ and $(T \cap T') \subset (S \cup T) \cap (S \cup T')$ we must have $x \in (S \cup T) \cap (S \cup T')$. Conversely, if $x \in (S \cup T) \cap (S \cup T')$, then x belongs to $(S \cup T)$ and $(S \cup T')$. If x does not belong to S, then it must lie in T and T', thus lies in $S \cup (T \cap T')$ as was to be shown. The same argument as in Exercise 1 with union and intersection signs interchanged shows that if $T_1, \ldots, T_n$ are sets, then

$$S \cup (T_1 \cap \cdots \cap T_n) = (S \cup T_1) \cap \cdots \cap (S \cup T_n).$$

Exercise 0.2.3 *Let A, B be subsets of a set S. Denote by A^c the complement of A in S. Show that the complement of the intersection is the union of the complements, i.e.*

$$(A \cap B)^c = A^c \cup B^c \quad and \quad (A \cup B)^c = A^c \cap B^c.$$

Solution. Suppose $x \in (A \cap B)^c$, so x is not in both A and B, that is $x \notin A$ or $x \notin B$, thus

$$(A \cap B)^c \subset (A^c \cup B^c).$$

Conversely, if $x \in (A^c \cup B^c)$, then $x \notin A$ or $x \notin B$ so certainly, $x \notin A \cap B$, thus $x \in (A \cap B)^c$. Hence

$$(A^c \cup B^c) \subset (A \cap B)^c.$$

For the second formula, suppose $x \in (A \cup B)^c$, then $x \notin A \cup B$, so $x \notin A$ and $x \notin B$, thus $x \in A^c \cap B^c$. Conversely, if $x \notin A$ and $x \notin B$, then $x \notin A \cup B$ so $x \in (A \cup B)^c$.

Exercise 0.2.4 *If X, Y, Z are sets, show that*

$$(X \cup Y) \times Z = (X \times Z) \cup (Y \times Z),$$

$$(X \cap Y) \times Z = (X \times Z) \cap (Y \times Z),$$

Solution. Suppose $(a, b) \in (X \cup Y) \times Z$. Then $b \in Z$ and $a \in X$ or $a \in Y$, so $(a, b) \in (X \times Z)$ or $(a, b) \in (Y \times Z)$, thus $(a, b) \in (X \times Z) \cup (Y \times Z)$. Conversely, if $(a, b) \in (X \times Z) \cup (Y \times Z)$, then $b \in Z$ and $a \in X$ or $a \in Y$. Therefore $(a, b) \in (X \cup Y) \times Z$. This proves the first formula.

For the proof of the second formula, suppose that $(a, b) \in (X \cap Y) \times Z$, then $a \in X \cap Y$ and $b \in Z$. Hence $a \in X, a \in Y$ and $b \in Z$, thus $(a, b) \in (X \times Z)$ and $(a, b) \in (Y \times Z)$. This implies that $(a, b) \in (X \times Z) \cap (Y \times Z)$. Conversely, if $(a, b) \in (X \times Z) \cap (Y \times Z)$, then $(a, b) \in (X \times Z)$ and $(a, b) \in (Y \times Z)$, which implies that $a \in X$, $a \in Y$, and $b \in Z$. Thus $a \in (X \cap Y)$ and $b \in Z$. This implies that $(a, b) \in (X \cap Y) \times Z$ as was to be shown.

Exercise 0.2.5 *Let* $f: S \to T$ *be a mapping, and let* Y, Z *be subsets of* T. *Show that*

$$f^{-1}(Y \cap Z) = f^{-1}(Y) \cap f^{-1}(Z),$$
$$f^{-1}(Y \cup Z) = f^{-1}(Y) \cup f^{-1}(Z).$$

Solution. If $x \in f^{-1}(Y \cap Z)$, then $f(x) \in Y$ and $f(x) \in Z$, so $x \in f^{-1}(Y) \cap f^{-1}(Z)$. Conversely, if $x \in f^{-1}(Y) \cap f^{-1}(Z)$, then $f(x) \in Y$ and $f(x) \in Z$, so $f(x) \in Y \cap Z$ and therefore $x \in f^{-1}(Y \cap Z)$. This proves the first equality.

For the second equality suppose that $x \in f^{-1}(Y \cup Z)$, then $f(x) \in (Y \cup Z)$, so $f(x) \in Y$ or $f(x) \in Z$ which implies that $x \in f^{-1}(Y) \cup f^{-1}(Z)$. Conversely, if $x \in f^{-1}(Y) \cup f^{-1}(Z)$, then $f(x) \in Y$ or $f(x) \in Z$ which implies that $x \in f^{-1}(Y \cup Z)$.

Exercise 0.2.6 *Let* S, T, U *be sets, and let* $f: S \to T$ *and* $g: T \to U$ *be mappings. (a) If* g, f *are injective, show that* $g \circ f$ *is injective. (b) If* f, g *are surjective, show that* $g \circ f$ *is surjective.*

Solution. (a) Suppose that $x, y \in S$ and $x \neq y$. Since f and g are injective we have $f(x) \neq f(y)$ and therefore $g(f(x)) \neq g(f(y))$, thus $g \circ f$ is injective. (b) Since g is surjective, given $y \in U$ there exists $z \in T$ such that $g(z) = y$. Since f is surjective, there exists $x \in S$ such that $f(x) = z$. Then $g(f(x)) = y$, so $g \circ f$ is surjective.

Exercise 0.2.7 *Let* S, T *be sets and let* $f: S \to T$ *be a mapping. Show that* f *is bijective if and only if* f *has an inverse mapping.*

Solution. Suppose that f is bijective. Given any $y \in T$ there exists $x \in S$ such that $f(x) = y$ because f is surjective, and this x is unique because f is injective. Define a mapping $g: T \to S$ by $g(y) = x$, where x is the unique element of S such that $f(x) = y$. Then by construction we have $f \circ g = \mathrm{id}_T$ and $g \circ f = \mathrm{id}_S$.

Conversely, suppose that f has an inverse mapping g. Then given any $y \in T$ we have $f(g(y)) = y$ so f is surjective. If $x, x' \in S$ and $x \neq x'$, then $g(f(x)) \neq g(f(x'))$ which implies that $f(x) \neq f(x')$. Thus f is injective.

0.3   Natural Numbers and Induction

(In the exercises you may use the standard properties of numbers concerning addition, multiplication, and division.)

Exercise 0.3.1 *Prove the following statements for all positive integers.*
   (a) $1 + 3 + 5 + \cdots + (2n - 1) = n^2$.
   (b) $1^2 + 2^2 + 3^2 + \cdots + n^2 = n(n+1)(2n+1)/6$.
   (c) $1^3 + 2^3 + 3^3 + \cdots + n^3 = [n(n+1)/2]^2$.

Solution. (a) For $n = 1$ we certainly have $1 = 1$. Assume the formula is true for an integer $n \geq 1$. Then

$$1+3+5+\cdots+(2n-1)+(2(n+1)-1) = n^2+2(n+1)-1 = n^2+2n+1 = (n+1)^2.$$

(b) For $n = 1$ we certainly have $1^2 = (1 \cdot 2 \cdot 3)/6$. Assume the formula is true for some integer $n \geq 1$. Then

$$
\begin{aligned}
1^2 + 2^2 + 3^2 + \cdots + n^2 + (n+1)^2 &= \frac{n(n+1)(2n+1)}{6} + \frac{6(n+1)^2}{6} \\
&= \frac{(n+1)(2n^2+7n+6)}{6} \\
&= \frac{(n+1)((n+1)+1)(2(n+1)+1)}{6}.
\end{aligned}
$$

(c) For $n = 1$ we have $1^3 = (2/2)^3$. Assume the formula is true for an integer $n \geq 1$. Then

$$
\begin{aligned}
1^3 + 2^3 + 3^3 + \cdots + n^3 + (n+1)^3 &= \frac{n^2(n+1)^2}{4} + \frac{4(n+1)^3}{4} \\
&= \frac{(n+1)^2(n^2+4n+4)}{4} \\
&= \frac{(n+1)^2(n+2)^2}{4} \\
&= \left[\frac{(n+1)(n+2)}{2}\right]^2.
\end{aligned}
$$

Exercise 0.3.2 *Prove that for all numbers $x \neq 1$,*

$$(1+x)(1+x^2)(1+x^4)\cdots(1+x^{2^n}) = \frac{1-x^{2^{n+1}}}{1-x}.$$

Solution. The formula is true for $n = 0$ because $(1+x)(1-x) = 1-x^2$. Assume the formula is true for an integer $n \geq 0$, then

$$
\begin{aligned}
(1+x)(1+x^2)(1+x^4)\cdots(1+x^{2^n})(1+x^{2^{n+1}}) &= \frac{1-x^{2^{n+1}}}{1-x}(1+x^{2^{n+1}}) \\
&= \frac{1-\left(x^{2^{n+1}}\right)^2}{1-x} \\
&= \frac{1-x^{2^{n+2}}}{1-x}.
\end{aligned}
$$

Exercise 0.3.3 *Let $f: \mathbf{N} \to \mathbf{N}$ be a mapping such that $f(xy) = f(x)+f(y)$ for all x, y. Show that $f(a^n) = nf(a)$ for all $n \in \mathbf{N}$.*

Solution. The formula is true for $n = 1$ because $f(a^1) = 1 \cdot f(a)$. Suppose the formula is true for an integer $n \geq 1$. Then

$$f(a^{n+1}) = f(a^n a) = f(a^n) + f(a) = nf(a) + f(a) = (n+1)f(a),$$

as was to be shown.

Exercise 0.3.4 *Let* $\begin{pmatrix} n \\ k \end{pmatrix}$ *denote the binomial coefficient,*

$$\begin{pmatrix} n \\ k \end{pmatrix} = \frac{n!}{k!(n-k)!},$$

where n, k *are integers* ≥ 0, $0 \leq k \leq n$, *and* $0!$ *is defined to be 1. Also* $n!$ *is defined to be the product* $1 \cdot 2 \cdot 3 \cdots n$. *Prove the following assertions.*

(a) $\begin{pmatrix} n \\ k \end{pmatrix} = \begin{pmatrix} n \\ n-k \end{pmatrix}$ (b) $\begin{pmatrix} n \\ k-1 \end{pmatrix} + \begin{pmatrix} n \\ k \end{pmatrix} = \begin{pmatrix} n+1 \\ k \end{pmatrix}$ (*for* $k > 0$)

Solution. (a) This result follows from

$$\begin{pmatrix} n \\ n-k \end{pmatrix} = \frac{n!}{(n-k)!(n-(n-k))!} = \frac{n!}{(n-k)!k!} = \begin{pmatrix} n \\ k \end{pmatrix}.$$

(b) We simply have

$$
\begin{aligned}
\begin{pmatrix} n \\ k-1 \end{pmatrix} + \begin{pmatrix} n \\ k \end{pmatrix} &= \frac{n!}{(k-1)!(n-k+1)!} + \frac{n!}{(n-k)!k!} \\
&= \frac{n!k + n!(n-k+1)}{k!(n-k+1)!} = \frac{n!(n+1)}{k!(n+1-k)!} \\
&= \begin{pmatrix} n+1 \\ k \end{pmatrix}.
\end{aligned}
$$

Exercise 0.3.5 *Prove by induction that*

$$(x+y)^n = \sum_{k=0}^{n} \begin{pmatrix} n \\ k \end{pmatrix} x^k y^{n-k}.$$

Solution. For $n = 1$ the formula is true. Suppose that the formula is true for an integer $n \geq 1$. Then

$$
\begin{aligned}
(x+y)^{n+1} &= (x+y) \left[\sum_{k=0}^{n} \begin{pmatrix} n \\ k \end{pmatrix} x^k y^{n-k} \right] \\
&= \sum_{k=0}^{n} \begin{pmatrix} n \\ k \end{pmatrix} x^{k+1} y^{n-k} + \sum_{k=0}^{n} \begin{pmatrix} n \\ k \end{pmatrix} x^k y^{n+1-k} \\
&= \sum_{k=1}^{n+1} \begin{pmatrix} n \\ k-1 \end{pmatrix} x^k y^{n+1-k} + \sum_{k=0}^{n} \begin{pmatrix} n \\ k \end{pmatrix} x^k y^{n+1-k}
\end{aligned}
$$

$$= \sum_{k=1}^{n} \binom{n+1}{k} x^k y^{n+1-k} + \binom{n}{n} x^{n+1} + y^{n+1}$$

$$= \sum_{k=0}^{n+1} \binom{n+1}{k} x^k y^{n+1-k}$$

as was to be shown. The second to last identity follows from the previous exercise.

Exercise 0.3.6 *Prove that*

$$\left(1 + \frac{1}{1}\right)^1 \left(1 + \frac{1}{2}\right)^2 \cdots \left(1 + \frac{1}{n-1}\right)^{n-1} = \frac{n^{n-1}}{(n-1)!}.$$

Find and prove a similar formula for the product of the terms $(1+1/k)^{k+1}$ taken for $k = 1, \ldots, n-1$.

Solution. For $n = 2$ the formula is true because $2 = 2^{2-1}/(2-1)!$. Suppose the formula is true for an integer $n \geq 2$, then

$$\left(1 + \frac{1}{1}\right)^1 \cdots \left(1 + \frac{1}{n-1}\right)^{n-1} \left(1 + \frac{1}{n}\right)^n = \frac{n^{n-1}}{(n-1)!} \left(\frac{n+1}{n}\right)^n = \frac{(n+1)^n}{n!}$$

as was to be shown. The product of the terms $(1 + 1/k)^{k+1}$ taken for $k = 1, \ldots, n-1$ is given by the formula

$$\left(1 + \frac{1}{1}\right)^2 \cdots \left(1 + \frac{1}{n-1}\right)^n = \frac{n^n}{(n-1)!}.$$

The proof is also by induction.

0.4   Denumerable Sets

Exercise 0.4.1 *Let F be a finite non-empty set. Show that there is a surjective mapping of $\mathbf{Z}^+$ onto F.*

Solution. By definition the set F has n elements for some integer $n \geq 1$. There exists a bijection g between F and J_n. Define $f : \mathbf{Z}^+ \to F$ by

$$f(k) = \begin{cases} g(k) & \text{if } k \in J_n, \\ g(n) & \text{if } k > n. \end{cases}$$

The mapping f is surjective.

Exercise 0.4.2 *How many maps are there which are defined on a set of numbers $\{1, 2, 3\}$ and whose values are in the set of integers n with $1 \leq n \leq 10$?*

Solution. There are 10^3 maps. See the next exercise.

Exercise 0.4.3 *Let E be a set with m elements and F a set with n elements. How many maps are there defined on E and with values in F? [Hint: Suppose first that E has one element. Next use induction on m, keeping n fixed.]*

Solution. We prove by induction that there are n^m maps defined on E with values in F

Suppose $m = 1$. To the single element in E we can assign n values in F. Suppose the induction statement is true for an integer $m \geq 1$. Suppose that E has $m + 1$ elements. Choose $x \in E$. To x we can associate n elements of F. For each such association there is n^m maps defined on $E - \{x\}$ with values in F. So there is a total of $n \times n^m = n^{m+1}$ maps defined on E with values in F.

Exercise 0.4.4 *If S, T, S', T' are sets, and there is a bijection between S and S', T, and T', describe a natural bijection between $S \times T$ and $S' \times T'$. Such a bijection has been used implicitly in some proofs.*

Solution. Given the bijections $f : S \to S'$ and $g : T \to T'$ define a mapping $h : S \times T \to S' \times T'$ by

$$h(x, y) = (f(x), g(y)).$$

Given any $(x', y') \in S' \times T'$ there exists $x \in S$ and $y \in T$ such that $f(x) = x'$ and $g(y) = y'$. Then $h(x, y) = (x', y')$, so h is surjective. The map h is injective because if $h(x_1, y_1) = h(x_2, y_2)$, then $f(x_1) = f(x_2)$ and $g(y_1) = g(y_2)$, so $x_1 = x_2$ and $y_1 = y_2$ because both f and g are injective.

0.5   Equivalence Relations

Exercise 0.5.1 *Let T be a subset of $\mathbf{Z}$ having the property that if $m, n \in T$, then $m + n$ and $-n$ are in T. For $x, y \in \mathbf{Z}$ define $x \equiv y$ if $x - y \in T$. Show that this is an equivalence relation.*

Solution. Suppose that T is non-empty, otherwise there is nothing to prove. The element 0 belongs to T because if $n \in T$, then $-n$ and $0 = n - n$ belongs to T. Therefore $x \equiv x$ for all x. Since $y - x = -(x - y)$ we see that $x \equiv y$ implies $y \equiv x$. Finally, if $x \equiv y$ and $y \equiv z$, then $x \equiv z$ because $x - z = (x - y) + (y - z)$ so $x - z \in T$.

Exercise 0.5.2 *Let $S = \mathbf{Z}$ be the set of integers. Define the relation $x \equiv y$ for $x, y \in \mathbf{Z}$ to mean that $x - y$ is divisible by 3. Show that this is an equivalence relation.*

Solution. Since 0 is divisible by 3 we have $x \equiv x$ for all $x \in S$. If $x - y$ is divisible by 3, so is $y - x$ hence $y \equiv x$ whenever $x \equiv y$. Finally $x \equiv z$ whenever $x \equiv y$ and $y \equiv z$ because $x - z = (x - y) + (y - z)$.

I
Real Numbers

I.1 Algebraic Axioms

Exercise I.1.1 *Let x, y be numbers $\neq 0$. Show that $xy \neq 0$.*

Solution. Suppose that $xy = 0$ and both x and y are non-zero. Then y^{-1} exists and $xyy^{-1} = 0y^{-1} = 0$ so $x = 0$ which is a contradiction.

Exercise I.1.2 *Prove by induction that if $x_1, \ldots, x_n \neq 0$, then $x_1 \cdots x_n \neq 0$.*

Solution. The assertion is obviously true when $n = 1$. Suppose that the assertion is true for some integer $n \geq 1$. We have $x_1 \cdots x_n \neq 0$ and $x_{n+1} \neq 0$ so by Exercise 1 we have

$$(x_1 \cdots x_n)x_{n+1} \neq 0$$

thus $x_1 \cdots x_n x_{n+1} \neq 0$, as was to be shown.

Exercise I.1.3 *If $x, y, z \in \mathbf{R}$ and $x \neq 0$, and if $xy = xz$, prove that $y = z$.*

Solution. The number x^{-1} exists because x is non-zero. So $x^{-1}(xy) = x^{-1}(xz)$ whence $(x^{-1}x)y = (x^{-1}x)z$, and therefore $y = z$.

Exercise I.1.4 *Using the axioms, verify that*

$$(x + y)^2 = x^2 + 2xy + y^2 \quad and \quad (x + y)(x - y) = x^2 - y^2.$$

Solution. Distributivity and commutativity imply

$$(x+y)(x+y) = (x+y)x + (x+y)y$$
$$= x^2 + yx + xy + y^2$$
$$= x^2 + 2xy + y^2.$$

Similarly,

$$(x+y)(x-y) = (x+y)x - (x+y)y$$
$$= x^2 + yx - xy - y^2$$
$$= x^2 - y^2.$$

I.2   Ordering Axioms

Exercise I.2.1 *If $0 < a < b$, show that $a^2 < b^2$. Prove by induction that $a^n < b^n$ for all positive integers n.*

Solution. The axioms imply $aa < ba$ and $ab < bb$, so by transitivity (IN 1.) we have $a^2 < b^2$.

The general inequality is true when $n = 1$. Suppose the inequality is true for some integer $n \geq 1$. Then by assumption $a^n < b^n$ and since a and b are both > 0 with $a < b$ we find that

$$a^n a < a^n b \quad \text{and} \quad a^n b < b^n b.$$

Therefore $a^{n+1} < b^{n+1}$, as was to be shown.

Exercise I.2.2 *(a) Prove that $x \leq |x|$ for all real x. (b) If $a, b \geq 0$ and $a \leq b$, and if $\sqrt{a}, \sqrt{b}$ exist, show that $\sqrt{a} \leq \sqrt{b}$.*

Solution. (a) If $x \geq 0$, then $|x| = x$ so $x \leq |x|$. If $x \leq 0$, then $x \leq 0 \leq |x|$ and we get $x \leq |x|$.
(b) Suppose that $\sqrt{b} < \sqrt{a}$. Then by Exercise 1 we know that $(\sqrt{b})^2 < (\sqrt{a})^2$, whence $b < a$, which contradicts the assumption that $a \leq b$.

Exercise I.2.3 *Let $a \geq 0$. For each positive integer n, define $a^{1/n}$ to be a number x such that $x^n = a$, and $x \geq 0$. Show that such a number x, if it exists, is uniquely determined. Show that if $0 < a < b$, then $a^{1/n} < b^{1/n}$ (assuming that the n-th roots exist).*

Solution. If $a = 0$ we must have $a^{1/n} = 0$ for otherwise we get a contradiction with Exercise 2 of §1. Suppose that $a > 0$ and that x exists. Then we must have $x \neq 0$. If $x^n = y^n = a$ and $x, y > 0$, then $x = y$. Indeed, if $x < y$, then $x^n < y^n$ (Exercise 1) which is a contradiction. Similarly we cannot have $y < x$.

Now suppose that $0 < a < b$ and $b^{1/n} \leq a^{1/n}$. Then arguing like in Exercise 1 we find that $(b^{1/n})^n \leq (a^{1/n})^n$ so $b \leq a$ which is a contradiction. So $a^{1/n} < b^{1/n}$ and we are done.

Exercise I.2.4 *Prove the following inequalities for* $x, y \in \mathbf{R}$.

$$|x - y| \geq |x| - |y|,$$
$$|x - y| \geq |y| - |x|,$$
$$|x| \leq |x + y| + |y|.$$

Solution. All three inequalities are simple consequences of the triangle inequality. For the first we have

$$|x| = |x - y + y| \leq |x - y| + |y|.$$

The second inequality follows from

$$|y| = |y - x + x| \leq |y - x| + |x|.$$

Finally, the third inequality follows from

$$|x| = |x + y - y| \leq |x + y| + |y|.$$

Exercise I.2.5 *If* x, y *are numbers* ≥ 0 *show that*

$$\sqrt{xy} \leq \frac{x + y}{2}.$$

Solution. The inequality follows from the fact that

$$0 \leq (\sqrt{x} - \sqrt{y})^2 = x - 2\sqrt{xy} + y.$$

Exercise I.2.6 *Let* b, ϵ *be numbers and* $\epsilon > 0$. *Show that a number* x *satisfies the condition* $|x - b| < \epsilon$ *if and only if*

$$b - \epsilon < x < b + \epsilon.$$

Solution. Suppose that $|x - b| < \epsilon$. If $b \leq x$, then $0 \leq |x - b| = x - b < \epsilon$, so $x < b + \epsilon$. If $x \leq b$, then $0 \leq |x - b| = b - x < \epsilon$, so $b - \epsilon < x$.

Conversely, if $b - \epsilon < x < b + \epsilon$, then $-\epsilon < x - b < \epsilon$ so $|x - b| < \epsilon$.

Exercise I.2.7 *Notation as in Exercise 6, show that there are precisely two numbers* x *satisfying the condition* $|x - b| = \epsilon$.

Solution. Since $\epsilon > 0$ we must have $x \neq b$. If $x > b$, the equation $|x - b| = \epsilon$ is equivalent to $x - b = \epsilon$ which has a unique solution namely, $x = b + \epsilon$. If we have $x < b$, then $|x - b| = \epsilon$ is equivalent to $b - x = \epsilon$ which has a unique solution, namely $x = b - \epsilon$. So $|x - b| = \epsilon$ has exactly two solutions, $b + \epsilon$ and $b - \epsilon$.

Exercise I.2.8 *Determine all intervals of numbers satisfying the following equalities and inequalities:*
(a) $x + |x - 2| = 1 + |x|$.   *(b)* $|x - 3| + |x - 1| < 4$.

Solution. (a) Suppose $x \geq 2$. Then the equation is equivalent to $x + x - 2 = 1 + x$ which has a unique solution $x = 3$. If $0 \leq x \leq 2$ we are reduced to $x + 2 - x = 1 + x$ which has only one solution given by $x = 1$. If $x \leq 0$, then $x + 2 - x = 1 - x$ has a unique solution $x = -1$. So the set of solution to $x + |x - 2| = 1 + |x|$ is $S = \{-1, 1, 3\}$.
(b) Separating the cases, $3 \leq x$, $1 \leq x \leq 3$, and $x \leq 1$ we find that the interval solution is $S = (0, 4)$.

Exercise I.2.9 *Prove: If x, y, ϵ are numbers and $\epsilon > 0$, and if $|x - y| < \epsilon$, then*
$$|x| < |y| + \epsilon, \quad and \quad |y| < |x| + \epsilon.$$
Also,
$$|x| > |y| - \epsilon, \quad and \quad |y| > |x| - \epsilon.$$

Solution. Using the first inequality of Exercise 4 we get
$$|x| \leq |x - y| + |y| < \epsilon + |y|.$$

By the second inequality of Exercise 4 we get
$$|y| \leq |x - y| + |x| < \epsilon + |x|,$$
so $|y| < \epsilon + |x|$.

Exercise I.2.10 *Define the **distance** $d(x, y)$ between two numbers x, y to be $|x - y|$. Show that the distance satisfies the following properties: $d(x, y) = d(y, x)$; $d(x, y) = 0$ if and only if $x = y$; and for all x, y, z we have*
$$d(x, y) \leq d(x, z) + d(z, y).$$

Solution. We have
$$d(x, y) = |x - y| = |-(y - x)| = |y - x| = d(y, x).$$

Cleary, $x = y$ implies $d(x, y) = 0$ conversely, if $d(x, y) = 0$, then $|x - y| = 0$ so by the standard property of the absolute value (i.e. $|a| = 0$ if and only if $a = 0$) we conclude that $x - y = 0$, thus $x = y$. The last property follows from the triangle inequality for the absolute value:
$$d(x, y) = |x - y| = |x - z + z - y| \leq |x - z| + |z - y| = d(x, z) + d(z, y).$$

Exercise I.2.11 *Prove by induction that if $x_1, \ldots, x_n$ are numbers, then*
$$|x_1 + \cdots + x_n| \leq |x_1| + \cdots + |x_n|.$$

Solution. If $n = 1$ the inequality is obviously true. Suppose that the inequality is true for some integer $n \geq 1$. Then, by the triangle inequality and the induction hypothesis we obtain

$$|x_1 + \cdots + x_n + x_{n+1}| \leq |x_1 + \cdots + x_n| + |x_{n+1}| \leq |x_1| + \cdots + |x_n| + |x_{n+1}|,$$

as was to be shown.

I.3   Integers and Rational Numbers

Exercise I.3.1 *Prove that the sum of a rational number and an irrational number is always irrational.*

Solution. If not, then for some rational numbers x, y and some $\alpha \notin \mathbf{Q}$ we have $x + \alpha = y$. Then $\alpha = y - x$, but the difference of two rational numbers is rational, so $\alpha \in \mathbf{Q}$, which is a contradiction.

Exercise I.3.2 *Assume that $\sqrt{2}$ exists, and let $\alpha = \sqrt{2}$. Prove that there exists a number $c > 0$ such that for all integers q, p, and $q \neq 0$ we have*

$$|q\alpha - p| > \frac{c}{q}.$$

[Note: The same c should work for all q, p. Try rationalizing $q\alpha - p$, i.e. take the product $(q\alpha - p)(-q\alpha - p)$, show that it is an integer, so that its absolute value is ≥ 1. Estimate $q\alpha + p$.]

Solution. We may assume without loss of generality that $q > 0$. Let $a = 2$ in the solution of Exercise 4.

Exercise I.3.3 *Prove that $\sqrt{3}$ is irrational.*

Solution. Suppose that $\sqrt{3}$ is rational and write $\sqrt{3} = p/q$, where the fraction is in lowest form. Then $3q^2 = p^2$. If q is even, then $3q^2$ is even, which implies that p is even. This is a contradiction because the fraction p/q is in lowest form.

If q is odd, then $3q^2$ is odd, thus p must be odd. Suppose $q = 2n + 1$ and $p = 2m + 1$. Then we can rewrite $3q^2 = p^2$ as

$$12n^2 + 12n + 3 = 4m^2 + 4m + 1$$

which is equivalent to

$$6n^2 + 6n + 1 = 2m^2 + 2m.$$

The left-hand side of the above equality is odd and the right hand side is even. This contradiction shows that q cannot be odd and concludes the proof that $\sqrt{3}$ is not rational.

Exercise I.3.4 *Let a be a positive integer such that $\sqrt{a}$ is irrational. Let $\alpha = \sqrt{a}$. Show that there exists a number $c > 0$ such that for all integers p, q with $q > 0$ we have*

$$|q\alpha - p| > c/q.$$

Solution. We follow the suggestion given in Exercise 2. We have

$$(q\alpha - p)(-q\alpha - p) = -q^2\alpha^2 + p^2 = -q^2a + p^2 \in \mathbf{Z}^* = \mathbf{Z} - \{0\},$$

because α is irrational and $\alpha^2 = a$ is an integer. So the absolute value of the left-hand side is ≥ 1 which gives

$$|q\alpha - p| \geq \frac{1}{|q\alpha + p|}.$$

Let c be a number such that $0 < c < \min\{|\alpha|, 1/(3|\alpha|)\}$. We consider two cases.

Suppose that $|\alpha - p/q| < |\alpha|$, then

$$\left|\alpha + \frac{p}{q}\right| \leq |2\alpha| + \left|-\alpha + \frac{p}{q}\right| < 3|\alpha|.$$

Therefore

$$|q\alpha - p| \geq \frac{1}{|q\alpha + p|} > \frac{1}{3|\alpha|q} > \frac{c}{q}.$$

If $|\alpha - p/q| \geq |\alpha|$, then

$$|q\alpha - p| \geq q|\alpha| > \frac{c}{q}.$$

This concludes the exercise.

Exercise I.3.5 *Prove: Given a non-empty set of integers S which is bounded from below (i.e. there is some integer m such that $m < x$ for all $x \in S$), then S has a least element, that is an integer n such that $n \in S$ and $n \leq x$ for all $x \in S$. [Hint: Consider the set of all integers $x - m$ with $x \in S$, this being a set of positive integers. Show that if k is its least element, then $m + k$ is the least element of S.]*

Solution. Let $T = \{y \in \mathbf{Z} : y = x - m \text{ for some } x \in S\}$. The set T is non-empty and $T \subset \mathbf{Z}^+$. The well-ordering axiom implies that T has a least element k. Then for some $x_0 \in S$ we have $k = x_0 - m$ so $x_0 = k + m$. Clearly for all $x \in S$ we have

$$x - x_0 = x - m - (x_0 - m) = x - m - k \geq 0.$$

I.4 The Completeness Axiom

Exercise I.4.1 *In Proposition 4.3, show that one can always select the rational number a such that a ≠ z (in case z itself is rational). [Hint: If z is rational, consider z + 1/n.]*

Solution. If z is irrational, then there is no problem. If z is rational, let $a = z + 1/n \in \mathbf{Q}$, where $1/n < \epsilon$. Then $|z - a| \leq 1/n < \epsilon$.

Exercise I.4.2 *Prove: Let w be a rational number. Given ε > 0, there exists an irrational number y such that |y − w| < ε.*

Solution. Choose $z \in \mathbf{Q}$ such that $|(w/\sqrt{2}) - z| < \epsilon/\sqrt{2}$. Then $y = z\sqrt{2} \notin \mathbf{Q}$, and $|y - w| < \epsilon$.

Exercise I.4.3 *Prove: Given a number z, there exists an integer n such that n ≤ z < n + 1. This integer is usually denoted by [z].*

Solution. Let $S = \{n \in \mathbf{Z}$ such that $z - 1 < n\}$ which is non-empty. Then $n_0 = \inf(S)$ exists by Exercise 5 of the preceding section and $n_0 \in S$. Hence $z - 1 < n_0$. We cannot have $z - 1 < n_0 - 1$ because $n_0 = \inf(S)$, thus $z - 1 \geq n_0 - 1$, which implies $z \geq n_0$. Putting everything together we see that $n_0 \leq z < n_0 + 1$.

Exercise I.4.4 *Let x, y ∈ R. Define x ≡ y if x − y is an integer. Prove:*
(a) This defines an equivalence relation in R.
(b) If x ≡ y and k is an integer, then kx ≡ ky.
(c) If $x_1 ≡ y_1$ and $x_2 ≡ y_2$, then $x_1 + x_2 ≡ y_1 + y_2$.
(d) Given a number x ∈ R, there exists a unique number x̄ such that 0 ≤ x̄ < 1 and such that x̄ ≡ x (in other words, x − x̄ is an integer). Show that x̄ = x − [x], where the bracket is that of Exercise 3.

Solution. (a) Since 0 is an integer, $x \equiv x$ for all x. If $x \equiv y$, then $y \equiv x$ because $y - x$ is an integer whenever $x - y$ is an integer. Finally if $x \equiv y$ and $y \equiv x$, then $x \equiv z$ because $x - z = x - y + y - z$.
(b) The result follows from the fact that $kx - ky = k(x - y)$.
(c) Immediate from the fact that $(x_1 + x_2) - (y_1 + y_2) = (x_1 - y_1) + (x_2 - y_2)$.
(d) By Exercise 3, we know that given $x \in \mathbf{R}$ there exists an integer n such that $n \leq x < n + 1$. Let $\bar{x} = x - n = x - [x]$. Then $0 \leq \bar{x} < 1$, thereby proving existence. For uniqueness suppose that there exists two numbers a and b such that $0 \leq a, b < 1$ and $a \equiv x$ and $b \equiv x$. Then by (b) and (c) $a - b \equiv 0$ so $a - b$ is an integer. But $0 \leq a, b < 1$, hence $-1 < a - b < 1$ which implies that $a - b = 0$ as was to be shown.

Exercise I.4.5 *Denote the number x̄ of Exercise 4 by R(x). Show that if x, y are numbers, and R(x) + R(y) < 1, then R(x + y) = R(x) + R(y). In general, show that*
$$R(x + y) \leq R(x) + R(y).$$

Show that $R(x) + R(y) - R(x + y)$ is an integer, i.e.

$$R(x + y) \equiv R(x) + R(y).$$

Solution. We have the following unique expressions, $x = R(x) + n_x$ and $y = R(y) + n_y$, with $0 \le R(x), R(y) < 1$, and $n_x, n_y \in \mathbf{Z}$. Then

$$x + y = R(x) + R(y) + n_x + n_y.$$

By assumption, $R(x) + R(y) < 1$ and since $n_x + n_y \in \mathbf{Z}$ we conclude that $R(x + y) = R(x) + R(y)$.

If $R(x) + R(y) \ge 1$, then $R(x + y) < 1 \le R(x) + R(y)$, so in general we have $R(x + y) \le R(x) + R(y)$. Finally, $R(x + y) \equiv x + y$, $R(x) \equiv x$, and $R(y) \equiv y$ so by part (c) of the previous exercise we find that

$$R(x + y) - R(x) - R(y) \equiv x + y - x - y \equiv 0$$

as was to be shown.

Exercise I.4.6 *(a) Let α be an irrational number. Let $\epsilon > 0$. Show that there exist integers m, n with $n > 0$ such that $|m\alpha - n| < \epsilon$.*
(b) In fact, given a positive integer N, show that there exist integers m, n, and $0 < m \le N$ such that $|m\alpha - n| < 1/N$.
(c) Let w be any number and $\epsilon > 0$. Show that there exist integers q, p such that

$$|q\alpha - p - w| < \epsilon.$$

[In other words, the numbers of type $q\alpha - p$ come arbitrarily close to w. Use part (a), and multiply $m\alpha - n$ by a suitable integer.]

Solution. Since (b) implies (a), it suffices to prove (b). Let n_z be the unique integer such that $0 \le z - n_z = \overline{z} < 1$. Divide the interval $[0, 1]$ in N intervals $[j/N, (j + 1)/N]$, $j = 0, \dots, N - 1$. Then consider $\overline{k\alpha}$ for $k = 0, 1, \dots, N$. The number α is irrational, there are N intervals and $N + 1$ numbers $\overline{k\alpha}$, therefore for some k_1, k_2 the numbers $\overline{k_1\alpha}$ and $\overline{k_2\alpha}$ belong to the same interval. Thus

$$|\overline{k_1\alpha} - \overline{k_1\alpha}| = |(k_1 - k_2)\alpha - n_{k_1\alpha} + n_{k_2\alpha}| < \frac{1}{N}.$$

(c) Let $\epsilon > 0$. Select integers m, n such that

$$0 < |m\alpha - n| < \epsilon.$$

Assume without loss of generality that $0 < m\alpha - n < \epsilon$. Let k_0 be the greatest lower bound of the set of integers k such that $w \le k(m\alpha - n)$. Then

$$(k_0 - 1)(m\alpha - n) < w \le k_0(m\alpha - n),$$

which implies that

$$|k_0 m\alpha - k_0 n - w| < \epsilon.$$

Exercise I.4.7 *Let S be a non-empty set of real numbers, and let b be a least upper bound for S. Let $-S$ denote the set of all numbers $-x$, with $x \in S$. Show that $-b$ is a greatest lower bound for $-S$. Show that one-half of the completeness axiom implies the other half.*

Solution. Since $x \leq b$ we have $-b \leq -x$ for all $x \in S$ so $-b$ is a lower bound for $-S$. Suppose there exists $c \in \mathbf{R}$ such that $-b < c \leq y$ for all $y \in -S$. Then $b > -c \geq x$ for all $x \in S$. This contradicts the fact that b is a least upper bound for S.

The first half of the completeness axiom implies the other half. Indeed, suppose S is bounded from below. Then $-S$ is bounded from above, thus $-S$ has a least upper bound. Hence S has a greatest lower bound.

Use the same kind of argument to prove that the second half of the completeness axiom implies the first half.

Exercise I.4.8 *Given any real number ≥ 0, show that it has a square root.*

Solution. Assume that $a > 0$ (the case $a = 0$ is obvious). Let $S = \{x \in \mathbf{R}$ such that $0 \leq x$ and $x^2 \leq a\}$. Since S is non-empty, $\sup(S)$ exists, call it b. Then, proceed as in the proof of Proposition 4.2. Suppose $b^2 < a$. If $n > (2b + 1)/(a - b^2)$, then

$$\left(b + \frac{1}{n}\right)^2 = b^2 + \frac{2b}{n} + \frac{1}{n^2} \leq b^2 + \frac{2b}{n} + \frac{1}{n} < a$$

because of our choice for n. This is a contradiction. If $a < b^2$, select n such that $1/n < (b^2 - a)/(2b)$. Then

$$\left(b - \frac{1}{n}\right)^2 = b^2 - \frac{2b}{n} + \frac{1}{n^2} > b^2 - \frac{2b}{n} > a$$

because of our choice for n. This contradiction concludes the exercise.

Exercise I.4.9 *Let $x_1, \ldots, x_n$ be real numbers. Show that $x_1^2 + \cdots + x_n^2$ is a square.*

Solution. Since $x_i^2 \geq 0$ for all $i = 1, \ldots, n$, an easy induction argument shows that $x_1^2 + \cdots + x_n^2 \geq 0$. The previous exercise implies that $x_1^2 + \cdots + x_n^2$ has a square root.

II
Limits and Continuous Functions

II.1 Sequences of Numbers

Determine in each case whether the given sequence has a limit, and if it does, prove that your stated value is a limit.

Exercise II.1.1 $x_n = \frac{1}{n}$.

Solution. Given $\epsilon > 0$ choose N such that $N > 1/\epsilon$. Then for all $n \geq N$, $|1/n| < \epsilon$, so $\{x_n\}$ converges to 0.

Exercise II.1.2 $x_n = \frac{(-1)^n}{n}$.

Solution. Given $\epsilon > 0$ choose N such that $N > 1/\epsilon$. Then for all $n \geq N$, we have $|x_n| = |1/n| < \epsilon$, so $\{x_n\}$ converges to 0.

Exercise II.1.3 $x_n = (-1)^n \left(1 - \frac{1}{n}\right)$.

Solution. The sequence $\{x_n\}$ has at least two distinct points of accumulation, 1 and -1. Indeed, $|x_n - 1| = 1/n$ for even n, and $|x_n - (-1)| = 1/n$ for odd n. This shows that the sequence $\{x_n\}$ does not have a limit.

Exercise II.1.4 $x_n = \frac{1 + (-1)^n}{n}$.

Solution. By the triangle inequality we have

$$\left| \frac{1 + (-1)^n}{n} \right| \leq \frac{2}{n}.$$

Given $\epsilon > 0$ choose N so that $N > 2/\epsilon$. Then for all $n \geq N$, we have $|x_n| < \epsilon$. Hence the sequence $\{x_n\}$ converges to 0.

Exercise II.1.5 $x_n = \sin n\pi$.

Solution. For all positive integers n, $x_n = \sin n\pi = 0$, thus $\{x_n\}$ converges to 0.

Exercise II.1.6 $x_n = \sin\left(\frac{n\pi}{2}\right) + \cos n\pi$.

Solution. Let n be a positive integer. We have

$$x_{4n} = \sin\left(\frac{4n\pi}{2}\right) + \cos 4\pi = 1,$$

and

$$x_{4n+1} = \sin\left(2n\pi + \frac{\pi}{2}\right) + \cos(4n\pi + \pi) = 1 - 1 = 0.$$

Thus $\{x_n\}$ has at least two distinct points of accumulation (actually it has exactly three, -2, 0, and 1) and therefore the sequence $\{x_n\}$ does not have a limit.

Exercise II.1.7 $x_n = \frac{n}{n^2+1}$.

Solution. Since $n^2 + 1 > n^2$ for all n we have

$$|x_n| < \frac{n}{n^2} = \frac{1}{n},$$

so $\{x_n\}$ converges to 0.

Exercise II.1.8 $x_n = \frac{n^2}{n^2+1}$.

Solution. For all $n \geq 1$ we have

$$|x_n - 1| = \left|\frac{n^2}{n^2+1} - \frac{n^2+1}{n^2+1}\right| = \left|\frac{-1}{n^2+1}\right| = \frac{1}{n^2+1} \leq \frac{1}{n^2} \leq \frac{1}{n},$$

so $\{x_n\}$ converges to 1.

Exercise II.1.9 $x_n = \frac{n^3}{n^2+1}$.

Solution. For all $n \geq 1$, we have $1 + 1/n^2 \leq 2$. Thus

$$x_n = \frac{n^3}{n^2\left(1 + \frac{1}{n^2}\right)} = \frac{n}{1 + \frac{1}{n^2}} \geq \frac{n}{2}.$$

Given any positive real number M, $x_n \geq M$ whenever $n > 2M$, thus $\{x_n\}$ does not have a limit.

Exercise II.1.10 $x_n = \frac{n^2-n}{n^3+1}$.

Solution. For all $n \geq 1$, we have $0 \leq n^2 - n \leq n^2$ and $n^3 \leq n^3 + 1$, thus

$$|x_n| \leq \frac{n^2}{n^3} = \frac{1}{n}.$$

This implies that $\{x_n\}$ converges to 0.

Exercise II.1.11 *Let S be a bounded set of real numbers. Let A be the set of its points of accumulation. That is, A consists of all numbers $a \in \mathbf{R}$ such that a is the point of accumulation of an infinite subset of S. Assume that A is not empty. Let b be its least upper bound.*
(a) Show that b is a point of accumulation of S. Usually, b is called the **limit superior** *of S, and is denoted by $\limsup S$.*
(b) Let c be a real number. Prove that c is the limit superior of S if and only if c satisfies the following property. For every ϵ there exists only a finite number of elements $x \in S$ such that $x > c + \epsilon$, and there exists infinitely many elements x of S such that $x > c - \epsilon$.

Solution. (a) There exists a point of accumulation d of S at distance less than $\epsilon/2$ of b. The open ball of radius $\epsilon/2$ centered at d contains infinitely many elements of S. Hence the open ball of radius ϵ centered at b contains infinitely many elements of S. This proves that b is a point of accumulation of S.

(b) Suppose that $c = \limsup S$. Then given any $\epsilon > 0$, part (a) implies that there exists infinitely many $x \in S$ such that $x > c - \epsilon$. If for some ϵ_0 there exists infinitely many $x \in S$ such that $x > c + \epsilon_0$, then the Weierstrass-Bolzano theorem implies that S has a point of accumulation strictly greater than c, which is a contradiction.

Conversely, suppose that c is a number that satisfies both properties. Given $\epsilon > 0$, there exists infinitely many $x \in S$ such that $x > c - \epsilon$ and there exists only finitely many $x \in S$ such that $x > c + \epsilon$ so there are infinitely many elements of S in the open ball centered at c of radius 2ϵ. Hence c is a point of accumulation of S. Now suppose that b is a point of accumulation of S such that $c < b$. Then if ϵ is small, say $\epsilon < (b - c)/2$ we know that there exists infinitely many elements in S at distance $< \epsilon$ from b. This implies that there exists infinitely many elements in S that are $> x + \epsilon$, which contradicts the second property of c.

Exercise II.1.12 *Let $\{a_n\}$ be a bounded sequence of real numbers. Let A be the set of its point of accumulation in $\mathbf{R}$. Assume that A is not empty. Let b be its least upper bound.*
(a) Show that b is a point of accumulation of the sequence. We call b the **limit superior** *of the sequence, denoted by $\limsup a_n$.*
(b) Let c be a real number. Show that c is the $\limsup$ of the sequence $\{a_n\}$ if and only if c has the following property. For every ϵ, there exists only a finite number of n such that $a_n > c + \epsilon$, and there exist infinitely many n such that $a_n > c - \epsilon$.

(c) If $\{a_n\}$ and $\{b_n\}$ are two sequences of numbers, show that

$$\limsup(a_n + b_n) \leq \limsup a_n + \limsup b_n$$

provided the limsups on the right exist.

Solution. The proofs of (a) and (b) are analoguous to the proofs given in the previous exercise.

(c) Let $\epsilon > 0$, $a = \limsup a_n$ and $b = \limsup b_n$. If $x \leq a + \epsilon$ and $y \leq b + \epsilon$, then $x + y \leq a + b + 2\epsilon$. By (b) we conclude that there exists only finitely many n such that $a + b + 2\epsilon < a_n + b_n$. This implies that

$$\limsup(a_n + b_n) \leq \limsup a_n + \limsup b_n.$$

Exercise II.1.13 *Define the* **limit inferior** *(lim inf). State and prove the properties analoguous to those in Exercise 12.*

Solution. (a) Let S be a bounded set of real numbers. Let A be the set of its points of accumulation. Assume that A is non-empty. Then the greatest lower bound of A is called the **limit inferior** of S. If $b = \liminf S$, then b is also a point of accumulation of S because given $\epsilon > 0$ there exists a point of accumulation of S at distance $\epsilon/2$ hence there are infinitely many elements of S at distance $< \epsilon$ of b.

(b) We prove that a real number c is the limit inferior of S if and only if given $\epsilon > 0$ there exists only a finite number of x in S such that $x < c - \epsilon$ and there exists infinitely many x in S such that $x < c + \epsilon$. If $c = \liminf S$, then the second property holds because c is a point of accumulation of S and if the first property does not hold, then the Weierstrass-Bolzano theorem implies that there exists a point of accumulation b of S such that $b < c$ which is a contradiction. Conversely, suppose that c satisfies both properties. Then any ball of positive radius centered at c contains infinitely many points of S, so c is a point of accumulation of S. If there were a point of accumulation b of S with $b < c$, then the first property would be violated. So $c = \liminf S$, as was to be shown.

II.2   Functions and Limits

Exercise II.2.1 *Let $d > 1$. Prove: Given $B > 1$, there exists N such that if $n > N$, then $d^n > B$. [Hint: Write $d = 1 + b$ with $b > 0$. Then*

$$d^n = 1 + nb + \cdots \geq 1 + nb.]$$

Solution. Write $d = 1 + b$ with $b > 0$. By the binomial formula we get

$$d^n = (1 + b)^n = \sum_{k=0}^{n} \binom{n}{k} b^k = 1 + nb + \cdots \geq 1 + nb.$$

So given $B > 1$ choose N such that $N > (B-1)/b$. Then for all $n > N$, we have $d^n \geq 1 + nb > B$, as was to be shown.

Exercise II.2.2 *Prove that if $0 < c < 1$, then*

$$\lim_{n \to \infty} c^n = 0.$$

What if $-1 < c \leq 0$? [Hint: Write $c = -1/d$ with $d > 1$.]

Solution. Write $c = 1/d$ with $d > 1$. Exercise 1 implies that given $\epsilon > 0$ there exists N so that for all $n > N$ we have $d^n > 1/\epsilon$. Then for all $n > N$, we get $c^n < \epsilon$. Hence $\lim_{n \to \infty} c^n = 0$.

If $c = 0$ the result is trivial. If $-1 < c < 0$, then $0 < |c| < 1$ and $\lim_{n \to \infty} |c|^n = 0$, so $\lim_{n \to \infty} c^n = 0$.

Exercise II.2.3 *Show that for any number $x \neq 1$ we have*

$$1 + x + \cdots + x^n = \frac{x^{n+1} - 1}{x - 1}.$$

If $|c| < 1$, show that

$$\lim_{n \to \infty} (1 + c + \cdots + c^n) = \frac{1}{1 - c}.$$

Solution. We simply expand

$$(x-1)(x^n + x^{n-1} + \cdots + x + 1) = x^{n+1} + x^n + \cdots + x - x^n - \cdots - x - 1$$
$$= x^{n+1} - 1.$$

When $|c| < 1$, consider the difference

$$1 + c + \cdots + c^n - \frac{1}{1-c} = \frac{1 - c^{n+1} - 1}{1 - c} = \frac{-c^{n+1}}{1 - c}.$$

So

$$\left| 1 + c + \cdots + c^n - \frac{1}{1-c} \right| \leq \frac{|c|^{n+1}}{1 - |c|}.$$

But $\lim_{n \to \infty} |c|^n = 0$, so the desired limit follows.

Exercise II.2.4 *Let a be a number. Let f be a function defined for all numbers $x < a$. Assume that when $x < y < a$ we have $f(x) \leq f(y)$ and also that f is bounded from above. Prove that $\lim_{x \to a} f(x)$ exists.*

Solution. Let $a_n = a - 1/n$, defined for all large n and consider the sequence whose general term is given by $b_n = \{f(a_n)\}$. Then $\{b_n\}$ is an increasing sequence of real numbers and is bounded, so $\lim_{n \to \infty} b_n$ exists. Denote this limit by b. We contend that $\lim_{x \to a} f(x) = b$. Clearly, for each x, we have $f(x) \leq b$ because there exists a_n (depending on x) such that

$x \leq a_n$ and therefore $f(x) \leq f(a_n) \leq b$. Given $\epsilon > 0$ choose N such that $b - b^n < \epsilon$ whenever $n \geq N$. If $a_N \leq x \leq a$, then

$$b_N = f(a_N) \leq f(x) \leq b$$

so $0 \leq b - f(x) \leq b - b_N < \epsilon$, which proves our contention.

Exercise II.2.5 *Let $x > 0$. Assume that the n-th root $x^{1/n}$ exists for all positive integers n. Find $\lim_{n \to \infty} x^{1/n}$.*

Solution. If $x = 1$ the result is trivial. Suppose $x > 1$ and write $x^{1/n} = 1 + h_n$ with $h_n > 0$. Then

$$x = (1 + h_n)^n \geq 1 + nh_n.$$

This implies that

$$0 \leq h_n \leq \frac{x - 1}{n}.$$

Therefore $\lim_{n \to \infty} h_n = 0$ and hence $\lim_{n \to \infty} x^{1/n} = 1$.

If $0 < x < 1$, then $1 < 1/x$ and $\lim_{n \to \infty} (1/x)^{1/n} = 1$. Hence $\lim_{n \to \infty} x^{1/n} = 1$.

Exercise II.2.6 *Let f be the function defined by*

$$f(x) = \lim_{n \to \infty} \frac{1}{1 + n^2 x}.$$

Show that f is the characteristic function of the set $\{0\}$, that is $f(0) = 1$ and $f(x) = 0$ if $x \neq 0$.

Solution. We have

$$f(0) = \lim_{n \to \infty} \frac{1}{1 + 0} = 1,$$

so $f(0) = 1$. If $x \neq 0$, choose N so that $N^2 |x| - 1 > 1/\epsilon$. Then for all $n \geq N$ we have

$$\left| \frac{1}{1 + n^2 x} \right| \leq \frac{1}{n^2 |x| - 1} < \epsilon,$$

so $f(x) = 0$.

II.3   Limits with Infinity

Exercise II.3.1 *Formulate completely the rules for limits of products, sums, and quotients when $L = -\infty$. Prove explicitly as many of these as are needed to make you feel comfortable with them.*

Solution. If M is a number > 0, then $\lim_{n \to \infty} f(x)g(x) = -\infty$ because given any $B > 0$ we can find numbers C_1 and C_2 such that for all $x > C_1$ we have $g(x) > M/2$ and such that for all $x > C_2$ we have $f(x) < -2B/M$. Then for all $x > \max(C_1, C_2)$ we have $B < -f(x)g(x)$.

If $M = \infty$, then $\lim_{n \to \infty} f(x)g(x) = -\infty$ because given any $B > 0$ we can find numbers C_1 and C_2 such that for all $x > C_1$ we have $g(x) > 1$ and such that for all $x > C_2$ we have $f(x) < -B$. Then $x > \max(C_1, C_2)$ implies $f(x)g(x) < -B$.

If M is a number, then $\lim_{n \to \infty} f(x) + g(x) = -\infty$. Choose C_1 such that $x > C_1$ implies $g(x) < M + 1$. Choose C_2 such that $x > C_2$ implies $f(x) < -B - M - 1$. Then for all $x > \max(C_1, C_2)$ we have $f(x) + g(x) < -B$.

If M is a number $\neq 0$, then $\lim_{n \to \infty} g(x)/f(x) = 0$. Indeed, there exists a number K (fixed) such that for all large x we have $|g(x)| < K$. Given ϵ there exists a number C_2 such that for all $x > C_2$ we have $|f(x)| > K/\epsilon$ and $|g(x)| < K$. Then for all $x > C_2$ we have $|g(x)/f(x)| < \epsilon$.

Exercise II.3.2 *Let $f(x) = a_d x^d + \cdots + a_0$ be a polynomial of degree d. Describe the behavior of $f(x)$ as $x \to \infty$ depending on whether $a_d > 0$ or $a_d < 0$. (Of course the case $a_d > 0$ has already been treated in the text.) Similarly, describe the behavior of $f(x)$ as $x \to -\infty$ depending on whether $a_d > 0$, $a_d < 0$, d is even, or d is odd.*

Solution. We can write

$$f(x) = a_d x^d \left(1 + \frac{a_{d-1}}{a_d x} + \cdots + \frac{a_0}{a_d x^d} \right)$$

for all large $|x|$. Since the expression in parentheses $\to 1$ as $|x| \to \infty$ we conclude that

$$\lim_{x \to \infty} f(x) = \infty$$

if $a_d > 0$ and

$$\lim_{x \to \infty} f(x) = -\infty$$

if $a_d < 0$.

Similarly we have

$$\lim_{x \to -\infty} f(x) = \infty$$

if $a_d > 0$ and d is even or $a_d < 0$ and d is odd. Also

$$\lim_{x \to -\infty} f(x) = -\infty$$

if $a_d > 0$ and d is odd or $a_d < 0$ and d is even.

Exercise II.3.3 *Let $f(x) = x^n + a_{n-1}x^{n-1} + \cdots + a_0$ be a polynomial. A root of f is a number c such that $f(c) = 0$. Show that any root satisfies the condition*

$$|c| \leq 1 + |a_{n-1}| + \cdots + |a_0|.$$

[Hint: Consider $|c| \leq 1$ and $|c| > 1$ separately.]

Solution. If $|c| \leq 1$, the result is trivial. Suppose $|c| > 1$. Since c is a root of f we have $-c^n = a_{n-1}c^{n-1} + \cdots + a_0$, thus

$$|c|^n \leq |a_{n-1}||c|^{n-1} + \cdots + |a_0|.$$

Dividing by $|c|^{n-1}$ implies

$$|c| \leq |a_{n-1}| + \cdots + \frac{|a_0|}{|c|^{n-1}},$$

but since $0 < 1/|c| < 1$, we get

$$|c| \leq |a_{n-1}| + \cdots + |a_0| \leq 1 + |a_{n-1}| + \cdots + |a_0|,$$

as was to be shown.

Exercise II.3.4 *Prove: Let f, g be functions defined for all sufficiently large numbers. Assume that there exists a number $c > 0$ such that $f(x) \geq c$ for all sufficiently large x, and that $g(x) \to \infty$ as $x \to \infty$. Show that $f(x)g(x) \to \infty$ as $x \to \infty$.*

Solution. There exists $C_1 > 0$ such that for all $x > C_1$, $f(x)$ is defined and $f(x) \geq c$. Let $B > 0$. There is a $C_2 > 0$ such that whenever $x > C_2$ we have $g(x) > B/c$. Then whenever $x > \max(C_1, C_2)$ we have $f(x)g(x) > B$ and therefore $f(x)g(x) \to \infty$ as $x \to \infty$.

Exercise II.3.5 *Give an example of two sequences $\{x_n\}$ and $\{y_n\}$ such that*

$$\lim_{n \to \infty} x_n = 0, \quad \lim_{n \to \infty} y_n = \infty,$$

and

$$\lim_{n \to \infty} (x_n y_n) = 1.$$

Solution. Take, for example, $x_n = 1/n$ and $y_n = n$ defined for $n \geq 1$.

Exercise II.3.6 *Give an example of two sequences $\{x_n\}$ and $\{y_n\}$ such that*

$$\lim_{n \to \infty} x_n = 0, \quad \lim_{n \to \infty} y_n = \infty,$$

and $\lim_{n \to \infty}(x_n y_n)$ does not exists, and such that $|x_n y_n|$ is bounded, i.e. there exists $C > 0$ such that $|x_n y_n| < C$ for all n.

Solution. Let $x_n = (-1)^n/n$ and $y_n = n$. Then $x_n y_n = (-1)^n$ and $|x_n y_n| \leq 1$ for all $n \geq 1$.

Exercise II.3.7 *Let*

$$f(x) = a_n x^n + \cdots + a_0$$
$$g(x) = b_m x^m + \cdots + b_0$$

be polynomials, with $a_n, b_m \neq 0$, so of degree n, m respectively. Assume that $a_n, b_m > 0$. Investigate the limit

$$\lim_{n \to \infty} \frac{f(x)}{g(x)},$$

distinguishing the cases $n > m$, $n = m$, and $n < m$.

Solution. For large values of x we can write

$$\frac{f(x)}{g(x)} = \frac{a_n x^n \left(1 + \cdots + \frac{a_0}{a_n x^n}\right)}{b_m x^m \left(1 + \cdots + \frac{b_0}{b_m x^m}\right)},$$

where $(1 + \cdots + a_0/a_n x^n)$ and $(1 + \cdots + b_0/b_m x^m) \to 1$ as $x \to \infty$. Thus we have the three cases

$$\begin{cases} n > m & \Rightarrow \lim_{x \to \infty} \frac{f(x)}{g(x)} = \infty, \\ n = m & \Rightarrow \lim_{x \to \infty} \frac{f(x)}{g(x)} = a_n/b_m, \\ n < m & \Rightarrow \lim_{x \to \infty} \frac{f(x)}{g(x)} = 0. \end{cases}$$

Exercise II.3.8 *Prove in detail: Let f be defined for all numbers $>$ some number a, let g be defined for all numbers $>$ some number b, and assume that $f(x) > b$ for all $x > a$. Suppose that*

$$\lim_{x \to \infty} f(x) = \infty \quad and \quad \lim_{x \to \infty} g(x) = \infty.$$

Show that

$$\lim_{x \to \infty} g(f(x)) = \infty.$$

Solution. For all $x > a$, $g(f(x))$ is defined. Let $B > 0$. Choose $M_g > b$ such that for all $x > M_g$ we have $g(x) > B$. Choose $M_f > a$ such that for all $x > M_f$ we have $f(x) > M_g$. Then for all $x > M_f$ we have $g(f(x)) > B$.

Exercise II.3.9 *Prove: Let S be a set of numbers, and let a be adherent to S. Let f be defined on S and assume*

$$\lim_{x \to a} f(x) = \infty.$$

Let g be defined for all sufficiently large numbers, and assume

$$\lim_{x \to \infty} g(x) = L,$$

where L is a number. Show that

$$\lim_{x \to \infty} g(f(x)) = L.$$

Solution. Let $\epsilon > 0$. Choose M such that if $y > M$, then $|g(y) - L| < \epsilon$. Select $\delta > 0$ such that whenever $x \in S$ and $|x - a| < \delta$ we have $f(x) > M$. Clearly for all $x \in S$ and $|x - a| < \delta$ we have $|g(f(x)) - L| < \epsilon$.

Exercise II.3.10 *Let the assumptions be as in Exercise 9, except that L now stands for the symbol ∞. Show that*

$$\lim_{x \to \infty} g(f(x)) = \infty.$$

Solution. Let $B > 0$. Choose M such that for all $y > M$ we have $g(y) > B$. Choose δ as in Exercise 9. Then for all $x \in S$ and $|x - a| < \delta$ we have $g(f(x)) > B$.

Exercise II.3.11 *State and prove the results analogous to Exercises 9 and 10 for the cases when $a = \infty$ and L is a number or ∞.*

Solution. Suppose L is a number. Given $\epsilon > 0$ choose A and B such that $y > B$ implies $|g(y) - L| < \epsilon$ and such that $x > A$ implies $f(x) > B$ $(x \in S)$. Then for all $x > A$ we have $|g(f(x)) - L| < \epsilon$.

Now suppose $L = \infty$. Given $M > 0$, choose A and B such that for all $y > B$ we have $g(y) > M$ and such that $x > A$ implies $f(x) > B$. Then $g(f(x)) > M$ whenever $x > A$.

Exercise II.3.12 *Find the following limits as $n \to \infty$:*
(a) $\frac{1+n}{n^2}$. (b) $\sqrt{n} - \sqrt{n+1}$. (c) $\frac{\sqrt{n}}{\sqrt{n+1}}$.
(d) $\frac{1}{1+nx}$ *if $x \neq 0$.* (e) $\sqrt{n} - \sqrt{n+10}$.

Solution. (a) The limit is 0. To see this write

$$\frac{1+n}{n^2} = \frac{1}{n^2} + \frac{1}{n}.$$

Since $1/n^2 \to 0$ and $1/n \to 0$ as $n \to \infty$ we have $\lim_{n \to \infty} (1+n)/n^2 = 0$.
(b) The limit is 0. To see this, write

$$|\sqrt{n} - \sqrt{n+1}| = \left| \frac{n - (n+1)}{\sqrt{n} + \sqrt{n+1}} \right| \leq \frac{1}{2\sqrt{n}}.$$

(c) The limit is 1. Indeed, we can write

$$\frac{\sqrt{n}}{\sqrt{n+1}} = \frac{1}{\sqrt{1 + 1/n}}.$$

(d) The limit is 0. For n large, we have

$$\left| \frac{1}{1+nx} \right| \leq \frac{1}{n|x| - 1}$$

and $n|x| - 1 \to \infty$ as $n \to \infty$.
(e) The limit is 0 because we have the bound

$$|\sqrt{n} - \sqrt{n+10}| = \left| \frac{n - (n+10)}{\sqrt{n} + \sqrt{n+10}} \right| \leq \frac{10}{2\sqrt{n}}.$$

II.4   Continuous Functions

Exercise II.4.1 *Let $f: \mathbf{R} \to \mathbf{R}$ be a function such that $f(tx) = tf(x)$ for all $x, t \in \mathbf{R}$. Show that f is continuous. In fact, describe all such functions.*

Solution. The only functions verifying the given property are the linear functions with 0 constant term. Indeed, $f(x) = f(x \cdot 1) = xf(1)$, thus f is a linear function with $f(0) = 0$.

Conversely, suppose f is a linear function with zero constant term, $f(x) = ax$ for some real number a. Then for all $t, x \in \mathbf{R}$ we have

$$f(tx) = atx = tax = tf(x)$$

as was to be shown.

It is now sufficient to prove the continuity of a linear function with zero constant term. Suppose $f(x) = ax$ for some $a \in \mathbf{R}$. Let $\epsilon > 0$ and let $\delta = \epsilon/(|a| + 1)$. If $|x - x_0| < \delta$, then

$$|f(x) - f(x_0)| < \epsilon \frac{|a|}{|a| + 1} < \epsilon.$$

Hence f is continuous.

Exercise II.4.2 *Let $f(x) = [x]$ be the greatest integer $\leq x$ and let $g(x) = x - [x]$. Sketch the graphs of f and g. Determine the points at which f and g are continuous.*

Solution. If $n \leq x < n + 1$ with $n \in \mathbf{Z}$, then $[x] = n$ and $x - [x] = x - n$ so $0 \leq g \leq 1$.

From the definitions of f and g we see that these functions are continuous on $\mathbf{R} - \mathbf{Z}$.

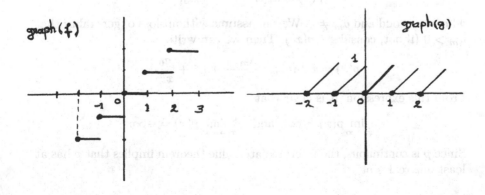

graph(f)   graph(g)

Exercise II.4.3 *Let f be the function such that $f(x) = 0$ if x is irrational and $f(p/q) = 1/q$ if p/q is a rational number, $q > 0$, and the fraction is in reduced form. Show that f is continuous at irrational numbers and not continuous at rational numbers. [Hint: For a fixed denominator q, consider all fractions m/q. If x is irrational, such fractions must be at distance $> \delta$ from x. Why?]*

Solution. Suppose $x_0 = p_0/q_0$ is a rational number such that the fraction is in lowest form and $q_0 > 0$. Then $f(x_0) = 1/q_0$. Every non-trivial open interval contains an irrational number, therefore $f(x) = 0$ for x arbitrarily close to x_0. Thus f is not continuous at x_0, thereby proving that f is not continuous at rational numbers.

Let $\epsilon > 0$ and suppose x_0 is irrational. Let $q_0 \in \mathbf{Z}^+$ such that $1/q_0 < \epsilon$. For each $q \in \mathbf{Z}^+$ with $q \leq q_0$ let S_q be the set of $p \in \mathbf{Z}$ such that

$$\left| \frac{p}{q} - x_0 \right| < 1.$$

The set S_q has finitely many elements. So there are only finitely many rationals with denominator $\leq q_0$ which are at distance less than 1 from x_0. So we can find δ such that all rationals in $(x_0 - \delta, x_0 + \delta)$ have denominator $> q_0$. To be precise we let

$$\text{dist}(x_0, S_q) = \min_{p \in S_q} \left\{ \text{dist}\left(x_0, \frac{p}{q} \right) \right\}.$$

Then $\text{dist}(x_0, S_q) > 0$ because x_0 is irrational, so select δ such that

$$0 < \delta < \min\{1, \min_{1 \leq q \leq q_0} \text{dist}(x_0, S_q)\}.$$

Then $|x - x_0| < \delta$ implies $|f(x) - f(x_0)| < \epsilon$.

Exercise II.4.4 *Show that a polynomial of odd degree with real coefficients has a root.*

Solution. Suppose we have a polynomial

$$p(x) = a_m x^m + \cdots + a_0,$$

where m is odd and $a_m \neq 0$. We can assume without loss of generality that $a_m > 0$ (if not, consider $-p(x)$). Then we can write

$$p(x) = x^m \left[a_m + \frac{a_{m-1}}{x} + \cdots + \frac{a_0}{x^m} \right].$$

From this expression, it is clear that

$$\lim_{x \to \infty} p(x) = \infty \quad \text{and} \quad \lim_{x \to -\infty} p(x) = -\infty.$$

Since p is continuous, the intermediate value theorem implies that p has at least one real root.

Exercise II.4.5 *For $x \neq -1$ show that the following limit exists:*

$$f(x) = \lim_{n \to \infty} \left(\frac{x^n - 1}{x^n + 1} \right)^2 .$$

(a) What are $f(1), f(\frac{1}{2}), f(2)$?
(b) What is $\lim_{x \to 1} f(x)$?
(c) What is $\lim_{x \to -1} f(x)$?
(d) For which values of $x \neq -1$ is f continuous? Is is possible to define $f(-1)$ in such a way that f is continuous at -1.

Solution. If $x = 1$, then clearly, $f(x) = 0$. If $|x| > 1$, then

$$\left(\frac{x^n - 1}{x^n + 1} \right)^2 = \left(\frac{1 - 1/x^n}{1 + 1/x^n} \right)^2$$

so $f(x) = 1$. If $|x| < 1$, then

$$\lim_{n \to \infty} \left(\frac{x^n - 1}{x^n + 1} \right)^2 = \left(\frac{-1}{1} \right)^2 = 1.$$

(a) The above argument shows that $f(1) = 0, f(1/2) = 1$, and $f(2) = 1$.
(b) Note that $f(x) = 1$ for all x such that $|x| \neq 1$, but $f(1)$ is defined and $f(1) = 0$. So $\lim_{x \to 1} f(x)$ does not exist, but

$$\lim_{x \to 1, x \neq 1} f(x) = 1.$$

(c) Similarly, $\lim_{x \to -1} f(x)$ does not exist, but

$$\lim_{x \to -1, x \neq -1} f(x) = 1.$$

(d) The function f is continuous at all $x \neq 1, -1$. However, f can be extended continuously as -1 by defining $f(-1) = 1$.

Exercise II.4.6 *Let*

$$f(x) = \lim_{n \to \infty} \frac{x^n}{1 + x^n} .$$

(a) What is the domain of definition of f, i.e. for which numbers x does the limit exist?
(b) Give explicitly the values $f(x)$ of f for the various x in the domain of f.
(c) For which x in the domain is f continuous at x?

Solution. (a) The domain of f is $\mathbf{R} - \{-1\}$.
(b) We have

$$\begin{cases} |x| < 1 \ \Rightarrow \ f(x) = 0, \\ |x| > 1 \ \Rightarrow \ f(x) = 1, \\ x = 1 \ \Rightarrow \ f(x) = 1/2. \end{cases}$$

(c) The function f is continuous for x such that $|x| < 1$ or $|x| > 1$.

Exercise II.4.7 *Let f be a function on an interval I. The equation of a line being given as usual by the formula $y = sx + c$ where s is the slope, write down the equation of the line segment between two points $(a, f(a))$ and $(b, f(b))$ of the graph of f, if $a < b$ are elements of the interval I.*

We define the function f above to be **convex upward** *if*

$$f((1-t)a + tb) \le (1-t)f(a) + tf(b) \tag{II.1}$$

for all a, b in the interval, $a \le b$ and $0 \le t \le 1$. Equivalently, we can write the condition as

$$f(ua + tb) \le uf(a) + tf(b)$$

for $t, u \ge 0$ and $t + u = 1$. Show that the definition of convex upward means that the line segment between $(a, f(a))$ and $(b, f(b))$ lies above the graph of the curve $y = f(x)$.

Solution. The equation of the line passing through $(a, f(a))$ and $(b, f(b))$ is given by

$$y - f(a) = \frac{f(b) - f(a)}{b - a}(x - a)$$

thus

$$s = \frac{f(b) - f(a)}{b - a} \quad \text{and} \quad c = \frac{bf(a) - af(b)}{b - a}.$$

Let $x \in [a, b]$ and let $t = (x - a)/(b - a)$. Then $t \in [0, 1]$, $(1 - t)a + tb = x$ and $(1 - t)f(a) + tf(b) = sx + c$. The definition of convexity implies

$$f(x) \le sx + c.$$

Exercise II.4.8 *A function f is said to be* **convex downward** *if the inequality (II.1) holds when $\le$ is replaced by $\ge$. Interpret this definition in terms of the line segment being below the curve $y = f(x)$.*

Solution. Let $x \in [a, b]$. Using the same argument and the same notation as in Exercise 7 we get

$$f(x) \ge (1 - t)f(a) + tf(b) = sx + c.$$

Exercise II.4.9 *Let f be convex upward on an open interval I. Show that f is continuous. [Hint: Suppose we want to show continuity at a point $c \in I$. Let $a < c$ and $a \in I$. For $a < x < c$, by Exercise 7 the convexity condition gives*

$$f(x) \le \frac{f(c) - f(a)}{c - a}(x - a) + f(a).$$

Given ϵ, for x sufficiently close to c and $x < c$, this shows that

$$f(x) \le f(c) + \epsilon.$$

For the reverse inequality, fix a point $b \in I$ with $c < b$ and use

$$f(c) \le \frac{f(b) - f(x)}{b - x}(c - x) + f(x). \,]$$

If the interval is not open, show that the function need not be continuous.

Solution. The inequality

$$f(x) \le \frac{f(c) - f(a)}{c - a}(x - a) + f(a)$$

follows from the equation we gave in Exercise 7 of the line segment between $(a, f(a))$ and $(c, f(c))$ and the fact that the curve $y = f(x)$ lies below this line. Write $x = c - \delta$. Then we have

$$\begin{aligned}
f(x) \ &\le \ \frac{f(c) - f(a)}{c - a}(x - a) + f(a) \\
&= \ \frac{f(c) - f(a)}{c - a}(c - a - \delta) + f(a) \\
&= \ f(c) - \frac{f(c) - f(a)}{c - a}\delta \\
&\le \ f(c) + \epsilon,
\end{aligned}$$

whenever

$$\left| \frac{f(c) - f(a)}{c - a}\delta \right| \le \epsilon.$$

This happens for all small δ, hence for all x close to c and $x < c$ we have $f(x) \le f(c) + \epsilon$.

Now we prove the reverse inequality. Fix a point $b \in I$ such that $c < b$. Then since the line between $(x, f(x))$ and $(b, f(b))$ lies above the curve $y = f(x)$ we see that

$$f(c) \le \frac{f(b) - f(x)}{b - x}(c - x) + f(x).$$

Now we claim that

$$\frac{f(b) - f(x)}{b - x} \le \frac{f(b) - f(c)}{b - c}.$$

This inequality is equivalent to proving that

$$(b - c)(f(b) - f(x)) \le (b - x)(f(b) - f(c))(b - x)$$

which in turn is equivalent to

$$f(c)(b - x) \le f(b)(c - x) + f(x)(b - c)$$

which we can rewrite as

$$f(c) \le \frac{c - x}{b - x}f(b) + \frac{b - c}{b - x}f(x).$$

Let $u = (c - x)/(b - x)$ and let $t = (b - c)/(b - x)$. Then $u, t \geq 0$ and $u + t = 1$. Moreover $ub + tx = c$, so by the convexity assumption we find that our desired inequality holds. This implies that

$$f(x) \leq \frac{f(b) - f(x)}{b - x}(c - x) + f(x) \leq \frac{f(b) - f(c)}{b - c}(c - x) + f(x),$$

so if $c - x > 0$ is small we conclude that $f(c) \leq f(x) + \epsilon$ as was to be shown. The same argument applies when $c < x$.

Exercise II.4.10 *Let f, g be convex upward and assume that the image of f is contained in the interval of definition of g. Assume that g is an increasing function, that is if $x < y$, then $g(x) \leq g(y)$. Show that $g \circ f$ is convex upward.*

Solution. We have the following

$$g \circ f((1 - t)a + tb) \leq g((1 - t)f(a) + tf(b)) \leq (1 - t)g \circ f(a) + tg \circ f(b).$$

Exercise II.4.11 *Let f, g be functions defined on the same set S. Define $\max(f, g)$ to be the function h such that*

$$h(x) = \max(f(x), g(x))$$

and similarly, define the minimum of the two functions, $\min(f, g)$. Let f, g be defined on a set of numbers. Show that if f, g are continuous, then $\max(f, g)$ and $\min(f, g)$ are continuous.

Solution. Since

$$\max(f, g) = \frac{1}{2}(f + g + |f - g|) \quad \text{and} \quad \min(f, g) = \frac{1}{2}(f + g - |f - g|),$$

Exercise 12 implies the continuity of $\max(f, g)$ and $\min(f, g)$.

Exercise II.4.12 *Let f be defined on a set of numbers, and let $|f|$ be the funcion whose value at x is $|f(x)|$. If f is continuous, show that $|f|$ is continuous.*

Solution. The result follows from the inequality

$$||f(x)| - |f(x_0)|| \leq |f(x) - f(x_0)|.$$

III
Differentiation

III.1 Properties of the Derivative

Exercise III.1.1 *Let α be an irrational number having the following property. There exists a number $c > 0$ such that for any rational number p/q (in lowest form) with $q > 0$ we have*

$$\left| \alpha - \frac{p}{q} \right| > \frac{c}{q^2},$$

or equivalently,

$$|q\alpha - p| > \frac{c}{q}.$$

(a) Let f be the function defined for all numbers as follows. If x is not a rational number, then $f(x) = 0$. If x is a rational number, which can be written as a fraction p/q, with integers q, p and if this fraction is in lowest form, $q > 0$, then $f(x) = 1/q^3$. Show that f is differentiable at α.
(b) Let g be the function defined for all numbers as follows. If x is irrational, then $g(x) = 0$. If x is rational, written as a fraction p/q in lowest form, $q > 0$, then $g(x) = 1/q$. Investigate the differentiability of g at the number α as above.

Solution. (a) The Newton quotient of f at α is

$$\frac{f(x) - f(\alpha)}{x - \alpha} = \frac{f(x)}{x - \alpha}.$$

Given $\epsilon > 0$ select $\delta > 0$ such that if $|x - \alpha| < \delta$ and $p/q = x \in \mathbf{Q}$, then $1/q < c\epsilon$. Then

$$\left| \frac{f(x)}{x - \alpha} \right| \leq \frac{q^2}{c} \cdot \frac{1}{q^3} < \epsilon,$$

thus

$$\lim_{\mathbf{Q} \ni x \to \alpha} \frac{f(x) - f(\alpha)}{x - \alpha} = 0.$$

If x is irrational, then the Newton quotient is 0, so f is differentiable at α and $f'(\alpha) = 0$.

(b) For $p/q = x \in \mathbf{Q}$, the Newton quotient of g at α becomes

$$\left| \frac{g(\alpha) - g(\alpha)}{x - \alpha} \right| = \frac{1}{|p - q\alpha|}.$$

By Exercise 6, §4, of Chapter 1 we know that given $N > 0$ there exists integers p_N, q_N such that

$$\left| \frac{1}{p_N - q_N\alpha} \right| \geq N \quad \text{and} \quad \left| \frac{p_N}{q_N} - \alpha \right| \leq \frac{1}{N},$$

thus g is not differentiable at α.

Exercise III.1.2 *(a) Show that the function $f(x) = |x|$ is not differentiable at 0. (b) Show that the function $g(x) = x|x|$ is differentiable for all x. What is its derivative?*

Solution. (a) For $h > 0$ we have

$$\lim_{h \to 0} \frac{f(0 + h) - f(0)}{h} = \lim_{h \to 0} \frac{h}{h} = 1,$$

and if $h < 0$, then

$$\lim_{h \to 0} \frac{f(0 + h) - f(0)}{h} = \lim_{h \to 0} \frac{-h}{h} = -1,$$

whence f is not differentiable at 0.

(b) If $x > 0$, then $f(x) = x^2$ and if $h > 0$ we get

$$\lim_{h \to 0} \frac{f(0 + h) - f(0)}{h} = \lim_{h \to 0} h = 0.$$

If $x < 0$, then $f(x) = -x^2$ and the Newton quotient at 0 tends to 0 as $h \to 0$ with $h < 0$. Thus f is differentiable for all x and for $x > 0$ its derivative is $2x$ and for $x < 0$ its derivative is $-2x$.

Exercise III.1.3 *For a positive integer k, let $f^{(k)}$ denote the k-th derivative of f. Let $P(x) = a_0 + a_1 x + \cdots + a_n x^n$ be a polynomial. Show that for all k,*

$$P^{(k)}(0) = k!a_k.$$

Solution. We prove by induction that for $0 \leq k \leq n$ we have the formula

$$P^{(k)}(x) = k!a_k + \frac{(k+1)!}{1!}a_{k+1}x + \frac{(k+2)!}{2!}a_{k+2}x^2 + \cdots + \frac{(k+n-k)!}{(n-k)!}a_n x^{n-k}.$$

When $k = 0$ the formula holds. Differentiating the above expression we get

$$(k+1)!a_{k+1} + \frac{(k+2)!}{1!}a_{k+2}x + \cdots + \frac{(k+n-k)!}{(n-k-1)!}a_n x^{n-k-1}$$

which is equal to $P^{(k+1)}(x)$, thereby concluding the proof by induction. We immediadely get that $P^{(k)}(0) = k!a_k$ whenever $0 \leq k \leq n$. If $k > n$, then $P^{(k)}$ is identically 0.

Exercise III.1.4 *By induction, obtain a formula for the n-th derivative of a product, i.e.* $(fg)^{(n)}$, *in terms of lower derivatives* $f^{(k)}, g^{(j)}$.

Solution. We prove by induction that

$$(fg)^{(n)} = \sum_{k=0}^{n} \binom{n}{k} (f)^{(k)}(g)^{(n-k)}.$$

When $n = 1$ the formula yields $(fg)' = f'g + fg'$ which holds. Differentiating the above formula using the product rule and splitting the sum in two we get

$$(fg)^{(n+1)} = \sum_{k=0}^{n} \binom{n}{k} (f)^{(k+1)}(g)^{(n-k)} + \sum_{k=0}^{n} \binom{n}{k} (f)^{(k)}(g)^{(n+1-k)}.$$

The change of index $j = k + 1$ in the first sum shows that $(fg)^{(n+1)}$ is

$$= \sum_{j=1}^{n+1} \binom{n}{j-1} (f)^{(j)}(g)^{(n+1-j)} + \sum_{k=0}^{n} \binom{n}{k} (f)^{(k)}(g)^{(n+1-k)}$$

$$= (f)^{(0)}(g)^{(n+1)} + (f)^{(n+1)}(g)^{(0)} + \sum_{k=1}^{n} \left[\binom{n}{k-1} + \binom{n}{k} \right] f^{(k)} g^{(n+1-k)}$$

$$= (f)^{(0)}(g)^{(n+1)} + (f)^{(n+1)}(g)^{(0)} + \sum_{k=1}^{n} \binom{n+1}{k} f^{(k)} g^{(n+1-k)}$$

$$= \sum_{k=0}^{n+1} \binom{n+1}{k} (f)^{(k)}(g)^{(n+1-k)}.$$

The second to last equality follows from Exercise 4, §3, of Chapter 0.

III.2   Mean Value Theorem

Exercise III.2.1 *Let $f(x) = a_n x^n + \cdots + a_0$ be a polynomial with $a_n \neq 0$. Let $c_1 < c_2 < \cdots < c_r$ be numbers such that $f(c_i) = 0$ for $i = 1, \ldots, r$. Show that $r \leq n$. [Hint: Show that f' has at least $r - 1$ roots, continue to take the derivatives, and use induction.]*

Solution. Suppose $r > n$. By Lemma 2.2, f' has at least one root in (c_j, c_{j+1}) for all $1 \leq j \leq r - 1$. Therefore f' has at least $r - 1$ distinct roots. Suppose that for some $1 \leq k \leq n - 1$, the function $f^{(k)}$ has at least $r - k$ distinct roots, $c_{k,1} < c_{k,2} < \cdots < c_{k,r-k}$. Then by Lemma 2.2, $f^{(k+1)}$ has at least one root in $(c_{k,j}, c_{k,j+1})$ for all $1 \leq j \leq r - k - 1$. Thus $f^{(k+1)}$ has at least $r - (k + 1)$ distinct roots. Therefore $f^{(n)}$ has at least $r - n$ roots. But $f^{(n)} = a_n n!$, so $f^{(n)}$ has no roots. This contradiction shows that $r \leq n$.

Exercise III.2.2 *Let f be a function which is twice differentiable. Let $c_1 < c_2 < \cdots < c_r$ be numbers such that $f(c_i) = 0$ for all i. Show that f' has at least $r - 1$ zeros (i.e. numbers b such that $f'(b) = 0$).*

Solution. Lemma 2.2 implies that for each $1 \leq j \leq r - 1$ there exists numbers d_j such that $c_j < d_j < c_{j+1}$ and $f'(d_j) = 0$. So f' has at least $r - 1$ roots.

Exercise III.2.3 *Let $a_1, \ldots, a_n$ be numbers. Determine x so that $\sum_{i=1}^{n}(a_i - x)^2$ is a minimum.*

Solution. Let $f(x) = \sum_{i=1}^{n}(a_i - x)^2$. The limits $\lim_{x \to \infty} f(x) = \infty$ and $\lim_{x \to -\infty} f(x) = \infty$ imply that f has a minimum. The minimum verifies $f'(x) = 0$, which is equivalent to

$$-\sum_{i=1}^{n} 2(a_i - x) = 0.$$

We conclude that f is at a minimum at $x = \sum a_i / n$.

Exercise III.2.4 *Let $f(x) = x^3 + ax^2 + bx + c$ where a, b, c are numbers. Show that there is a number d such that f is convex downward if $x \leq d$ and convex upward if $x \geq d$.*

Solution. The function f'' exists and $f''(x) = 6x + 2a$. Then for all $x \leq d = -a/3$, the function f is convex downward, and for all $x \geq d$, f is convex upward.

Exercise III.2.5 *A function f on an interval is said to satisfy a **Lipschitz condition** with **Lipschitz constant** C if for all x, y in the interval, we have*

$$|f(x) - f(y)| \leq C|x - y|.$$

Prove that a function whose derivative is bounded on an interval is Lipschitz. In particular, a C^1 function on a closed interval is Lipschitz. Also, note that a Lipschitz function is uniformly continuous. However, the converse if not necessarily true. See Exercise 5 of Chapter IV, §3.

Solution. Let M be a bound for the derivative. Given x and y in the interval, there exists c in (x, y) such that $f(x) - f(y) = f'(c)(x - y)$ which implies
$$|f(x) - f(y)| = |f'(c)||x - y| \leq M|x - y|.$$

Exercise III.2.6 *Let f be a C^1 function on an open interval, but such that its derivative is not bounded. Prove that f is not Lipschitz. Give an example of such a function.*

Solution. Assume that f is Lipschitz with Lipschitz constant C. By assumption there exists x_0 in the interval such that $|f'(x_0)| > 2C$. By continuity, we have $|f'(x)| > 2C$ for all x in a small open interval I centered at x_0. Choose $x_1 \in I$ with $x_1 \neq x_0$. By the mean value theorem, there exists $c \in I$ such that
$$f(x_1) - f(x_0) = f'(c)(x_1 - x_0)$$
so, since we assumed f Lipschitz, we get
$$2C|x_1 - x_0| < |f'(c)||x_1 - x_0| = |f(x_1) - f(x_0)| \leq C|x_1 - x_0|.$$

Thus $2C < C$ which is a contradiction. For an example of such a function, consider $x \mapsto 1/x$ on $(0, 1)$ or $x \mapsto x \sin(1/x)$ also on $(0, 1)$. See Exercise 5, §3, Chapter IV.

Exercise III.2.7 *Let f, g be functions defined on an interval $[a, b]$, continuous on this interval, differentiable on $a < x < b$. Assume that $f(a) \leq g(a)$, and $f'(x) < g'(x)$ on $a < x < b$. Show that $f(x) < g(x)$ if $a < x \leq b$.*

Solution. Let $h(x) = g(x) - f(x)$. The function h verifies $h'(x) > 0$ and $h(a) \geq 0$. Thus $h(x) > 0$ for $a < x \leq b$.

III.3   Inverse Functions

For each one of the following functions f restrict f to an interval so that the inverse function g is defined in an interval containing the indicated point, and find the derivative of the inverse function at that point.

Exercise III.3.1 $f(x) = x^3 + 1$; *find* $g'(2)$.

Solution. Restrict f to $[0, 2]$ because for all $x \in [0, 2]$, $f'(x) = 3x^2 \geq 0$. Then $f(0) = 1$ and $f(2) = 9$. Thus the inverse function $g \colon [1, 9] \to [0, 2]$ of f is well defined and
$$g'(2) = \frac{1}{f'(1)} = \frac{1}{3}.$$

Exercise III.3.2 $f(x) = x^2 - x + 5$; find $g'(7)$.

Solution. Restrict f to $[1,3]$ because for all $x \in [1,3]$, $f'(x) = 2x - 1 > 0$. Then $f(1) = 5$ and $f(3) = 11$, $f(2) = 7$. Thus the inverse function g: $[5,11] \to [1,3]$ of f is well defined and

$$g'(7) = \frac{1}{f'(2)} = \frac{1}{3}.$$

Exercise III.3.3 $f(x) = x^4 - 3x^2 + 1$; find $g'(-1)$.

Solution. Restrict f to $[0, \sqrt{3/2}]$ because for all $x \in [0, \sqrt{3/2}]$, $f'(x) = 2x(2x^2 - 3) \geq 0$. Furthermore, $f(0) = 1$ and $f(\sqrt{3/2}) = -5/4$, $f(1) = -1$. Thus the inverse function g: $[0, -5/4] \to [0, \sqrt{3/2}]$ of f is well defined and

$$g'(-1) = \frac{1}{f'(1)} = -\frac{1}{2}.$$

Exercise III.3.4 $f(x) = -x^3 + 2x + 1$; find $g'(2)$.

Solution. Restrict f to $[\sqrt{2/3}, 2]$ because for all x in this interval, $f'(x) = -3x^2 + 2 \leq 0$. Furthermore, $f(\sqrt{2/3}) = \alpha > 2$, $f(2) = -3$, and $f(1) = 2$. Thus the inverse function g: $[-3, \alpha] \to [\sqrt{2/3}, 2]$ of f is well defined and

$$g'(2) = \frac{1}{f'(1)} = -1.$$

Exercise III.3.5 $f(x) = 2x^3 + 1$; find $g'(21)$.

Solution. Restrict f to $[0,3]$ because for all x in this interval, $f'(x) = 6x^2 \geq 0$. Furthermore, $f(0) = 1$, $f(3) = 55$, and $f(\sqrt[3]{10}) = 21$. Then the inverse function $g : [1, 55] \to [0, 3]$ of f is well defined and

$$g'(21) = \frac{1}{f'(\sqrt[3]{10})} = \frac{1}{6 \cdot 10^{2/3}}.$$

Exercise III.3.6 *Let f be a continuous function on the interval $[a, b]$. Assume that f is twice differentiable on the open interval $a < x < b$, and that $f'(x) > 0$ and $f''(x) > 0$ on this interval. Let g be the inverse function of f.*

(a) Find an expression for the second derivative of g.

(b) Show that $g''(y) < 0$ on its interval of definition. Thus g is convex in the opposite direction of f.

Solution. (a) Since $g'(x) = 1/f'(g(x))$, the chain rule and the rule for differentiating quotients apply, leading to the following expression for g'':

$$g''(x) = \frac{-f''(g(x))g'(x)}{[f'(g(x))]^2}.$$

(b) For $a < x < b$ we have $f'(x) > 0$ and $f''(x) > 0$. Therefore we see from the above expression of $g''(x)$, that $g''(x) < 0$ on its interval of definition.

Exercise III.3.7 *In Theorem 3.2, prove that if f is of class C^p with $p \geq 1$, then its inverse function g is also of class C^p.*

Solution. We prove the result by induction on p. Assume that f is of class C^p, then f' is of class C^{p-1}. By induction the inverse g of f is of class C^{p-1} and the map $x \mapsto 1/x$ is C^∞ so the formula of Theorem 3.2 implies that g' is of class C^{p-1}, hence g is of class C^p.

IV

Elementary Functions

IV.1 Exponential

Exercise IV.1.1 *Let f be a differentiable function such that*

$$f'(x) = -2xf(x).$$

Show that there is some constant C such that $f(x) = Ce^{-x^2}$.

Solution. We simply have

$$\frac{d}{dx}\left(\frac{f(x)}{e^{-x^2}}\right) = \frac{f'(x)e^{-x^2} + 2xf(x)e^{-x^2}}{(e^{-x^2})^2} = \frac{-2xf(x) + 2xf(x)}{e^{-x^2}} = 0.$$

Exercise IV.1.2 *(a) Prove by induction that for any positive integer n, and $x \geq 0$,*

$$1 + x + \frac{x^2}{2!} + \cdots + \frac{x^n}{n!} \leq e^x.$$

[Hint: Let $f(x) = 1 + x + \cdots + x^n/n!$ and $g(x) = e^x$.]
(b) Prove that for $x \geq 0$,

$$e^{-x} \geq 1 - x + \frac{x^2}{2!} - \frac{x^3}{3!}.$$

(c) Show that $2.7 < e < 3$.

Solution. (a) The inequality is true for $n = 1$. Indeed,

$$\frac{d}{dx}\left(e^x - (1+x)\right) = e^x - 1 \geq 0$$

whenever $x \geq 0$ and $e^0 - (1+0) = 0$, so $e^x - (1+x) \geq 0$ for all $x \geq 0$. Suppose the inequality is true for some integer n. Let $f(x) = 1 + x + \cdots + \frac{x^{n+1}}{(n+1)!}$ and $g(x) = e^x$. Then, the induction hypothesis implies that

$$\frac{d}{dx}(g(x) - f(x)) = e^x - \left(1 + x + \cdots + \frac{x^n}{n!}\right) \geq 0.$$

Since $g(0) = f(0) = 1$, the desired inequality follows.
(b) Let

$$f(x) = e^{-x} - 1 + x - \frac{x^2}{2!} + \frac{x^3}{3!}.$$

Then $f'''(x) = -e^{-x} + 1 \geq 0$ for all $x \geq 0$. Since $f(0) = f'(0) = f''(0) = f'''(0) = 0$, we conclude that $f(x) \geq 0$ for all $x \geq 0$ and if $x > 0$ we have $f(x) > 0$.
(c) Let $n = 4$ and $x = 1$ in (a), so that

$$2.7 < 1 + 1 + \frac{1}{2} + \frac{1}{6} + \frac{1}{24} \leq e.$$

Let $x = 1$ in (b) to get

$$e^{-1} > \frac{1}{2} - \frac{1}{6} = \frac{1}{3}.$$

Exercise IV.1.3 *Sketch the graph of the following functions:*
(a) xe^x;
(b) xe^{-x}
(c) $x^2 e^x$; and (c) $x^2 e^{-x}$.

Solution. (a) Let $f(x) = xe^x$. Then $f'(x) = (1+x)e^x$, so f is increasing on $[-1, \infty)$ and decreasing on $(-\infty, -1]$. The function f is positive for $x > 0$ and negative for $x < 0$, and $f(0) = 0$. Clearly, $\lim_{x \to \infty} = xe^x = \infty$. Moreover since $\lim_{u \to \infty} u/e^u = 0$ letting $x = -u$ we get $\lim_{x \to -\infty} f(x) = 0$.

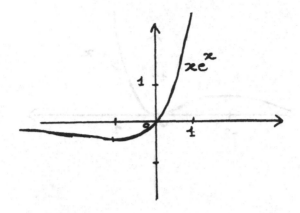

(b) Let $f(x) = xe^{-x}$. Note that $f(-x) = -xe^x$ so the graph of f is the image of the graph obtained in (a) in the symmetry with respect to the origin.

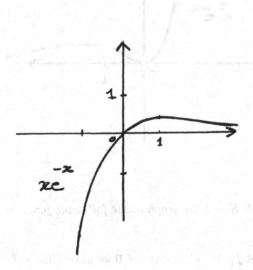

(c) Let $f(x) = x^2 e^x$. Then $f'(x) = xe^x(2+x)$ so f is decreasing on $[-2, 0]$ and increasing on $\mathbf{R}-[-2, 0]$. For all x we have $f(x) \geq 0$ and $\lim_{x \to \infty} f(x) = \infty$. We also have $\lim_{x \to -\infty} f(x) = \lim_{u \to \infty} u^2/e^u = 0$.

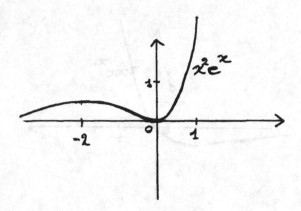

(d) Let $f(x) = x^2 e^{-x}$. Since $f(-x) = x^2 e^x$, the graph of f is the reflection across the y-axis of the graph obtained in (c).

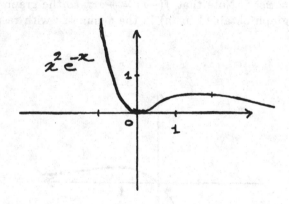

Exercise IV.1.4 *Sketch the graph of the following functions: (a) $e^{1/x}$; and (b) $e^{-1/x}$.*

Solution. (a) Let $f(x) = e^{1/x}$. For $x \neq 0$ we have $f'(x) = (-1/x^2)e^{1/x} < 0$. For $x < 0$, letting $u = -1/x$, we get $\lim_{x \to 0^-} f'(x) = \lim_{u \to \infty} -u^2 e^{-u} = 0$. The behavior of f at the boundary of its domain of definition is described by the following limits:

$$\lim_{x \to -\infty} f(x) = e^0 = 1, \quad \lim_{x \to 0^-} f(x) = 0, \quad \lim_{x \to 0^+} f(x) = \infty, \quad \lim_{x \to \infty} f(x) = 1.$$

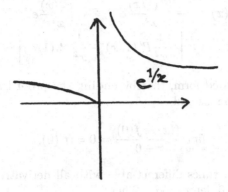

(b) Let $f(x) = e^{-1/x}$. Notice that $f(-x) = e^{1/x}$, so the graph of f is the reflection across the y-axis of the graph obtained in (a).

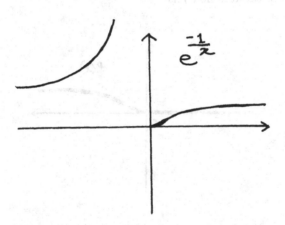

Exercise IV.1.5 *(a) Let f be the function such that $f(x) = 0$ if $x \leq 0$ and $f(x) = e^{-1/x}$ if $x > 0$. Show that f is infinitely differentiable at 0, and that all its derivatives at 0 are equal to 0. [Hint: Use induction to show that the n-th derivative of f for $x > 0$ is of type $P_n(1/x)e^{-1/x}$ where P_n is a polynomial.]*
(b) Sketch the graph of f.

Solution. (a) For $x > 0$ there exists a sequence of polynomials $\{P_n\}$ such that $f^{(n)}(x) = P_n(1/x)e^{-1/x}$. Indeed, we have $f'(x) = (1/x^2)e^{-1/x}$ which is of the desired form. We assume that the assertion is true for some integer n. Differentiating we obtain

$$f^{(n+1)}(x) = -\frac{P_n'(1/x)}{x^2}e^{-1/x} + \frac{P_n(x)}{x^2}e^{-1/x}$$

$$= \left[-\frac{1}{x^2}P_n'(1/x) + \frac{1}{x^2}P_n(1/x)\right]e^{-1/x},$$

which is of the desired form, thereby ending the proof by induction. The function f is continuous and

$$\lim_{x\to 0}\frac{f(x) - f(0)}{x - 0} = 0 = f'(0).$$

Assume that f is n times differentiable with all derivatives equal to 0 at the origin. For $x > 0$, let $x = 1/u$. Then

$$\lim_{x\to 0^+}\frac{f^{(n)}(x) - f^{(n)}(0)}{x - 0} = \lim_{u\to\infty} uP_n(u)e^{-u} = 0$$

so $f^{(n+1)}(0) = 0$. By induction we conclude that the function f is infinitely differentiable at 0 and $f^{(n)}(0) = 0$ for all n.
(b) The graph of the function f is

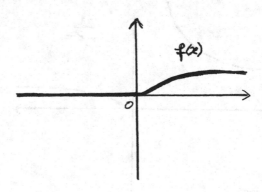

Exercise IV.1.6 *(a)* **(Bump Functions).** *Let a, b be numbers, $a < b$. Let f be the function such that $f(x) = 0$ if $x \le a$ or $x \ge b$, and*

(a) $f(x) = e^{-1/(x-a)(b-x)}$ or   (b) $f(x) = e^{-1/(x-a)}e^{-1/(b-x)}$

if $a < x < b$. Sketch the graph of f. Show that f is infinitely differentiable at both a and b.
(b) We assume you know about the elementary integral. Show that there exists a C^∞ function F such that $F(x) = 0$ if $x \le a$, $F(x) = 1$ if $x \ge b$, and F is strictly increasing on $[a, b]$.

(c) Let $\delta > 0$ be so small that $a + \delta < b - \delta$. Show that there exists a C^∞ function g such that:

$g(x) = 0$ if $x \le a$, and $g(x) = 0$ if $x \ge b$;
$g(x) = 1$ on $[a + \delta, b - \delta]$; and
g is strictly increasing on $[a, a + \delta]$, and strictly decreasing on $[b - \delta, b]$.
Sketch the graphs of F and g.

Solution. (a) We take $f(x) = e^{-1/(x-a)(b-x)}$. For $a < x < b$ we have

$$f'(x) = \frac{(b - x) - (x - a)}{(x - a)^2 (x - b)^2} e^{-1/(x-a)(b-x)}.$$

Thus f is increasing on $(a, (a+b)/2)$ and decreasing on $((a+b)/2, b)$. Just as in Exercise 5, use induction to show that there exists a sequence of polynomials $\{P_n\}$ and a sequence of positive integers $\{k_n\}$ such that

$$f^{(n)}(x) = \frac{P_n(x)}{[(x - a)(b - x)]^{k_n}} e^{-1/(x-a)(b-x)}.$$

A linear change of variable and the limits computed in Exercise 5 prove that f is infinitely differentiable at both a and b.

(b) Let $I = \int_a^b f(t)dt = \int_{-\infty}^\infty f(t)dt$. Since f is continuous, non-negative, and not identically zero, we know that $I \ne 0$, so we can define

$$F(x) = \frac{1}{I} \int_{-\infty}^x f(t)dt.$$

Then $F(x) = 0$ if $x \le a$ and $F(x) = 1$ if $x \ge b$ and for all x we have

$$\frac{d}{dx}F(x) = f(x).$$

So F is strictly increasing on $[a, b]$ and F is C^∞.

(c) By (b) we can construct a function F_1 on $\mathbf{R}$ such that F_1 is C^∞, $F_1(x) = 0$ if $x \le a$, $F_1(x) = 1$ if $x \ge a + \delta$, and F is strictly increasing on $[a, a + \delta]$. Arguing as in (b) with the function

$$G(x) = \frac{1}{I} \int_x^\infty$$

we see that we can construct a function F_2 on $\mathbf{R}$ which is C^∞ such that $F_2(x) = 1$ if $x \le b - \delta$, $F_2(x) = 0$ if $x \ge b$, and F_2 is strictly decreasing on $[b - \delta, b]$. Define g as follows:

$$\begin{cases} F_1(x) & \text{if } x \le a + \delta, \\ 1 & \text{if } a + \delta \le x \le b - \delta, \\ F_2(x) & \text{if } b - \delta \le x. \end{cases}$$

The function g verifies the desired properties. The graph of F and g are

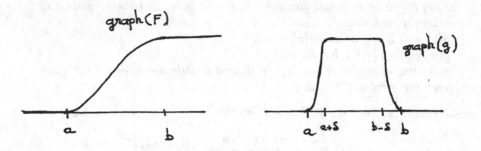

Exercise IV.1.7 *Let $f(x) = e^{-1/x^2}$ if $x \neq 0$ and $f(0) = 0$. Show that f is infinitely differentiable and that $f^{(n)}(0) = 0$ for all n. After you learn the terminology of Taylor's formula, you will see that the function provides an example of a C^∞ function which is not identically 0 but all its Taylor polynomials are identically 0.*

Solution. Since $\lim_{x\to 0} f(x) = 0$, the function f is continuous. There exists polynomials $\{P_n\}$ such that

$$f^{(n)}(x) = P_n(1/x)e^{-1/x^2}.$$

The proof is by induction and is similar to the one given in Exercise 5. Assuming that $f^{(n)}(0) = 0$ and letting $u = 1/x$ we see that

$$\lim_{x\to 0} \frac{f^{(n)}(x) - f^{(n)}(0)}{x - 0} = \lim_{u\to\pm\infty} uP_n(u)e^{-u^2} = 0$$

because for all integers m we have $u^m e^{-u^2} \to 0$ as $u \to \infty$ or $u \to -\infty$. By induction we conclude that f is infinitely differentiable on $\mathbf{R}$ and $f^{(n)}(0) = 0$ for all n.

Exercise IV.1.8 *Let n be an integer ≥ 1. Let $f_0, \ldots, f_n$ be polynomials such that*

$$f_n(x)e^{nx} + f_{n-1}(x)e^{(n-1)x} + \cdots + f_0(x) = 0$$

for arbitrarily large numbers x. Show that $f_0, \ldots, f_n$ are identically 0. [Hint: Divide by e^{nx} and let $x \to \infty$.]

Solution. Dividing the relation by e^{nx} we obtain

$$f_n(x) + f_{n-1}(x)e^{-x} + \cdots + f_0(x)e^{-nx} = 0.$$

By Theorem 1.1 and the fact that $f_0, \ldots, f_n$ are polynomials, we see that for all $1 \le k \le n$ we have

$$\lim_{x \to \infty} f_{n-k}(x)e^{-kx} = 0.$$

So $\lim_{x \to \infty} f_n(x) = 0$ which proves that f_n is identically 0. By induction we see that f_j is identically 0 for all $0 \le j \le n$.

IV.2   Logarithm

Exercise IV.2.1 *Let $f(x) = x^x$ for $x > 0$. Sketch the graph of f.*

Solution. Since $f(x) = e^{x \log x}$ the function f is positive for all $x > 0$. Furthermore, $f'(x) = (\log x + 1)e^{x \log x}$. Thus f is increasing on $(1/e, \infty)$ and decreasing on $(0, 1/e)$ and therefore attains a minimum at $(1/e, e^{-1/e})$. We also have $\lim_{x \to \infty} f(x) = \infty$, and since $\lim_{x \to 0+} x \log x = \lim_{u \to \infty} (\log u)/u = 0$, we have $\lim_{x \to 0+} f(x) = 1$.

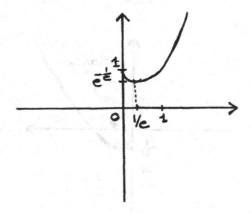

Exercise IV.2.2 *Let f be as in Exercise 1, except that we restrict f to the infinite interval $x > 1/e$. Show that the inverse function g exists. Show that one can write*

$$g(y) = \frac{\log y}{\log \log y} \psi(y),$$

where $\lim_{y \to \infty} \psi(y) = 1$.

Solution. Since f is continuous and increasing on $(1/e, \infty)$ the function g exists and is defined on $(e^{-1/e}, \infty)$. Put $y = f(x) = x^x$, so $\log y = x \log x$ and

$$x = \frac{\log y}{\log x} = \frac{\log y}{\log\log y - \log\log x} = \frac{\log y}{\log\log y \left(1 - \frac{\log\log x}{\log\log y}\right)}.$$

Put

$$\psi(y) = \frac{1}{1 - \frac{\log\log x}{\log\log y}} = \frac{1}{1 - h(x)}.$$

But $\log\log y = \log x + \log\log x$ so $h(x) \to 0$ as $x \to \infty$ and therefore $\psi(y) \to 1$ as $y \to \infty$.

Exercise IV.2.3 *Sketch the graph of: (a) $x\log x$; and (b) $x^2\log x$.*

Solution. (a) Since

$$\frac{d}{dx}(x\log x) = \log x + 1,$$

the function is increasing on $(1/e, \infty)$ and decreasing on $(0, 1/e)$. Furthermore,

$$\lim_{x\to\infty} x\log x = \infty \quad \text{and} \quad \lim_{x\to 0^+} x\log x = \lim_{u\to\infty}(\log u)/u = 0.$$

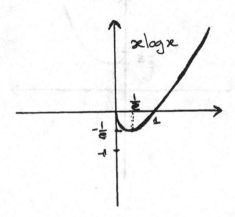

(b) Since

$$\frac{d}{dx}(x^2\log x) = x(2\log x + 1),$$

the given function is increasing on $(1/\sqrt{e}, \infty)$ and decreasing on $(0, 1/\sqrt{e})$. Futhermore, we have

$$\lim_{x\to\infty} x^2\log x = \infty \quad \text{and} \quad \lim_{x\to 0^+} xx\log x = 0.$$

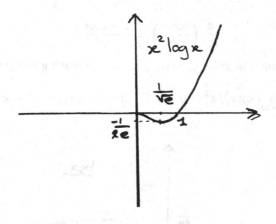

Exercise IV.2.4 *Sketch the graph of: (a) $(\log x)/x$; and (b) $(\log x)/x^2$.*

Solution. (a) We have

$$\frac{d}{dx}\left(\frac{\log x}{x}\right) = \frac{1 - \log x}{x^2},$$

thus the function is increasing on $(0, e)$ and decreasing on (e, ∞). Futhermore, we have

$$\lim_{x \to \infty} (\log x)/x = 0 \quad \text{and} \quad \lim_{x \to 0^+} (\log x)/x = -\infty.$$

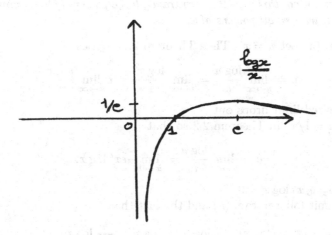

(b) We have
$$\frac{d}{dx}\left(\frac{\log x}{x^2}\right) = \frac{1 - 2\log x}{x^2},$$
thus the function is increasing on $(0, \sqrt{e})$ and decreasing on $(\sqrt{e}, \infty)$. Futhermore
$$\lim_{x\to\infty}(\log x)/x^2 = 0 \quad\text{and}\quad \lim_{x\to 0+}(\log x)/x^2 = -\infty.$$

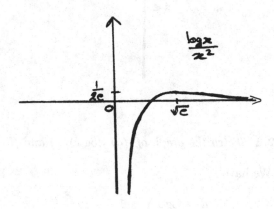

Exercise IV.2.5 *Let* $\epsilon > 0$. *Show:* (a) $\lim_{x\to\infty}(\log x)/x^\epsilon = 0$; (b) $\lim_{x\to 0} x^\epsilon \log x = 0$; *and* (c) *let* n *be a positive integer, and let* $\epsilon > 0$. *Show that*
$$\lim_{x\to\infty}\frac{(\log x)^n}{x^\epsilon} = 0.$$
Roughly speaking, this says that arbitrarily large powers of $\log x$ *grow slower than arbitrarily small powers of* x.

Solution. (a) Let $u = x^\epsilon$. Then Theorem 2.2 implies
$$0 = \lim_{u\to\infty}\frac{\log u}{u} = \lim_{x\to\infty}\frac{\log x^\epsilon}{x^\epsilon} = \epsilon \lim_{x\to\infty}\frac{\log x}{x^\epsilon},$$
so the desired limit drops out.

(b) Put $u = 1/x^\epsilon$ in Theorem 2.2 so that
$$0 = \lim_{u\to\infty}\frac{\log u}{u} = \lim_{x\to 0}-\epsilon x^\epsilon \log x,$$
hence $\lim_{x\to 0} x^\epsilon \log x = 0$.

(c) The limit follows from (a) and the fact that
$$\frac{(\log x)^n}{x^\epsilon} = \frac{\log x}{x^{\epsilon/n}}\cdots\frac{\log x}{x^{\epsilon/n}} = \prod_{i=1}^{n}\frac{\log x}{x^{\epsilon/n}}.$$

Exercise IV.2.6 *Let $f(x) = x \log x$ for $x > 0$, $x \neq 0$, and $f(0) = 0$.*
(a) Is f continuous on $[0,1]$? Is f uniformly continuous on $[0,1]$?
(b) Is f right differentiable at 0? Prove all your assertions.

Solution. (a) The function f is continuous at 0 because for $x > 0$ we have $\lim_{x \to 0} x \log x = 0$. Hence f is continuous on $[0,1]$. A function continuous on a compact set is uniformly continuous, so f is uniformly continuous on $[0,1]$.
(b) The Newton quotient of f at 0 is

$$\frac{f(x) - f(0)}{x - 0} = \frac{x \log x}{x} = \log x,$$

which tends to $-\infty$ as x approaches 0. So f is not right differentiable.

Exercise IV.2.7 *Let $f(x) = x^2 \log x$ for $x > 0$, $x \neq 0$, and $f(0) = 0$.*
Is f right differentiable at 0? Prove your assertion. Investigate the differentiability of $f(x) = x^k \log x$ for an integer $k > 0$, i.e. how many right derivatives does this function have at 0?

Solution. The function is right differentiable at 0 because for $x > 0$ we have

$$\lim_{x \to 0} \frac{f(x) - f(0)}{x - 0} = \lim_{x \to 0} \frac{x^2 \log x}{x} = \lim_{x \to 0} x \log x = 0.$$

Now let $f(x) = x^k \log x$. We claim that f has $k - 1$ right derivatives. By induction we see that there exists positive numbers a_n and b_n such that on $(0,1]$ we have

$$f^{(n)}(x) = a_n x^{k-n} \log x + b_n x^{k-n},$$

so by induction for $n = 0, 1, \ldots, k - 2$ we have $f^{(n+1)}(0) = 0$ because

$$\frac{f^{(n)}(x) - f^{(n)}(0)}{x - 0} = \frac{a_n x^{k-n} \log x + b_n x^{k-n}}{x} = a_n x^{k-n-1} \log x + b_n x^{k-n-1}$$

hence

$$\lim_{x \to 0} \frac{f^{(n)}(x) - f^{(n)}(0)}{x - 0} = 0.$$

But

$$\lim_{x \to 0} \frac{f^{(k-1)}(x) - f^{(k-1)}(0)}{x - 0} = \lim_{x \to 0} \frac{a_n x \log x + b_n x}{x} = -\infty,$$

which proves that f has $k - 1$ right derivatives.

Exercise IV.2.8 *Let n be an integer ≥ 1. Let $f_0, \ldots, f_n$ be polynomials such that*

$$f_n(x)(\log x)^n + f_{n-1}(x)(\log x)^{n-1} + \cdots + f_0(x) = 0$$

for all numbers $x > 0$. Show that $f_0, \ldots, f_n$ are identically 0. [Hint: Let $x = e^y$ and rewrite the above relation in the form

$$\sum a_{ij}(e^y)^i y^j,$$

where a_{ij} are numbers. Use Exercise 8 of the preceding section.]

Solution. Let d_j be the degree of f_j and write $f_j(x) = \sum_{i=0}^{d_j} a_{ij}x^i$. Then letting $x = e^y$ we get

$$\sum_{j=0}^{n}\left(\sum_{i=0}^{d_j} a_{ij}(e^y)^i\right) y^j = 0.$$

Collecting terms and factoring out the $(e^y)^i$'s we get an expression of the form

$$g_0(y) + g_1(y)(e^y)^1 + \cdots + g_m(y)(e^y)^m = \sum_{i=0}^{m} g_i(y)(e^y)^i = 0,$$

where $m = \max_{1 \le j \le n}\{d_j\}$ and where the g_i's are polynomials in y. Exercise 8, §1, of Chapter IV concludes the exercise.

Exercise IV.2.9 (a) Let $a > 1$ and $x > 0$. Show that

$$x^a - 1 \ge a(x - 1).$$

(b) Let p, q be numbers ≥ 1 such that $1/p + 1/q = 1$. If $x \ge 1$, show that

$$x^{1/p} \le \frac{x}{p} + \frac{1}{q}.$$

Solution. (a) The function h defined by $h(x) = x^a - 1 - a(x - 1)$ is continuous on $\mathbf{R}_{\ge 0}$ and differentiable on $\mathbf{R}_{>0}$ with $h'(x) = a(x^{a-1} - 1)$. Thus h is decreasing on $(0,1)$, increasing on $(1,\infty)$ and $h(1) = 0$. Hence $h(x) \ge 0$ for all $x > 0$.
(b) Let $f(x) = x^{1/p}$ and $g(x) = x/p + 1/q$. Then $f(1) = g(1) = 1$ and for $x > 1$ we have

$$f'(x) = \frac{1}{p}x^{(1-p)/p} \quad \text{and} \quad g'(x) = \frac{1}{p}.$$

But if $p > 1$ and $x > 1$, then $x^{(1-p)/p} < 1$ which implies that $f'(x) < g'(x)$ whenever $x > 1$. Hence $f(x) \le g(x)$ and $f(x) = g(x)$ if and only if $x = 1$, as was to be shown.

Exercise IV.2.10 (a) Let u, v be positive numbers, and let p, q be as in Exercise 8. Show that
$$u^{1/p}v^{1/q} \le \frac{u}{p} + \frac{v}{q}.$$

(b) Let u, v be positive numbers, and $0 < t < 1$. Show that

$$u^t v^{1-t} \le tu + (1 - t)v,$$

and that equality holds if and only if $u = v$.

Solution. (a) Let $x = u/v$ in the inequality of Exercise 9(b) and multiply both sides of the inequality by v.

(b) Fix t in $(0,1)$, and let $p = 1/t$ and $q = 1/(1-t)$. Then since $1/p + 1/q = 1$ and $p, q > 1$, the previous exercise implies that $u^t v^{1-t} \leq tu + (1-t)v$. Clearly, if $u = v$ the equality holds. Conversely, if the equality holds, then $u/v = 1$ because as we have seen in Exercise 9(b), $f(x) = g(x)$ if and only if $x = 1$.

Exercise IV.2.11 *Let a be a number > 0. Find the minimum and maximum of the function $f(x) = x^2/a^x$. Sketch the graph of $f(x)$.*

Solution. Since

$$\frac{d}{dx}\left(\frac{x^2}{a^x}\right) = \frac{2xa^x - x^2 a^x \log a}{a^{2x}}$$

the function attains its minimum 0 at 0. If $a > 1$, then $\lim_{x \to -\infty} f(x) = \infty$ and if $a < 1$, then $\lim_{x \to \infty} f(x) = \infty$. If $a = 1$, then $f(x) = x^2$.

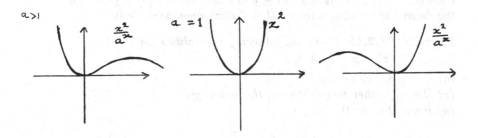

Exercise IV.2.12 *Using the mean value theorem, find the limit*

$$\lim_{n \to \infty} (n^{1/3} - (n+1)^{1/3}).$$

Generalize by replacing $\frac{1}{3}$ by $1/k$ for any integer $k \geq 2$.

Solution. Let $f_k(x) = x^{1/k}$. Then $f_k'(x) = 1/kx^{(1-k)/k}$, so by the mean value theorem we know that given n there exists a number $c_{n,k}$ such that $n \leq c_{n,k} \leq n+1$ and

$$f_k(n+1) - f_k(n) = f_k'(c_{n,k}) = \frac{1}{k}c_{n,k}^{(1-k)/k}.$$

Therefore

$$0 \le (n+1)^{1/k} - n^{1/k} \le \frac{1}{kn^{1-1/k}}.$$

Hence $\lim_{n \to \infty}(n+1)^{1/k} - n^{1/k} = 0$ for any integer $k \ge 2$.

Exercise IV.2.13 *Find the limit*

$$\lim_{h \to 0} \frac{(1+h)^{1/3} - 1}{h}.$$

Solution. We recognize the Newton quotient of $f(x) = x^{1/3}$ at 1, thus the desired limit is $f'(1) = \frac{1}{3}$.

Exercise IV.2.14 *Show that for $x \ge 0$ we have $\log(1 + x) \le x$.*

Solution. Let $f(x) = \log(1 + x)$ and $g(x) = x$. Then for $x \ge 0$ we have

$$f'(x) = \frac{1}{1+x} \quad \text{and} \quad g'(x) = 1.$$

Clearly, if $x \ge 0$ we have $f'(x) \le g'(x)$, and since $f(0) = g(0) = 0$ we get the desired inequality, namely $f(x) \le g(x)$ whenever $x \ge 0$.

Exercise IV.2.15 *Prove the following inequalities for $x \ge 0$:*
(a) $\log(1 + x) \le x - \frac{x^2}{2} + \frac{x^3}{3}$.
(b) $x - \frac{x^2}{2} \le \log(1 + x)$.
(c) Derive further inequalities of the same type.
(d) Prove that for $0 \le x \le 1$,

$$\log(1 + x) = \lim_{n \to \infty}\left(x - \frac{x^2}{2} + \cdots + (-1)^{n+1}\frac{x^n}{n}\right).$$

Solution. (a) Let $f(x) = x - x^2/2 + x^3/3 - \log(1 + x)$. Then

$$f'(x) = 1 - x + x^2 - \frac{1}{1+x} \ge 0$$

because $1 - x + x^2 \ge 1/(1 + x)$ as one sees by multiplying both sides by $1 + x$. Furthermore, $f(0) = 0$ which implies that $f(x) \ge 0$ for all $x \ge 0$.
(b) Let $f(x) = \log(1 + x) - x + x^2/2$. Then

$$f'(x) = \frac{1}{1+x} - 1 + x \ge 0$$

because $1/(1 + x) \ge 1 - x$ as one sees by multiplying both sides by $1 + x$. Since $f(0) = 0$ we see that $f(x) \ge 0$ for all $x \ge 0$.
(c) Let

$$P_{2n}(x) = x - \frac{x^2}{2} + \frac{x^3}{3} - \cdots - \frac{x^{2n}}{2n}$$

and

$$P_{2n+1}(x) = x - \frac{x^2}{2} + \frac{x^3}{3} - \cdots + \frac{x^{2n+1}}{2n+1}$$

and $f(x) = \log(1+x)$. Then

$$P_{2n}(x) \le f(x) \le P_{2n+1}(x).$$

Indeed, $P_{2n}(0) = f(0) = P_{2n}(0) = 0$, $f'(x) = 1/(1+x)$, and

$$P'_{2n}(x) = 1 - x + x^2 - \cdots - x^{2n-1} \quad \text{and} \quad P'_{2n+1}(x) = 1 - x + x^2 - \cdots + x^{2n}.$$

However we can sum the first m terms of a geometric series and obtain the formula

$$1 - x + x^2 - \cdots + (-1)^m x^m = \frac{1 - (-1)^{m+1} x^{m+1}}{1+x} = \frac{1}{1+x} - \frac{(-1)^{m+1} x^{m+1}}{1+x},$$

so that $P'_{2n}(x) \le f'(x) \le P'_{2n+1}(x)$ which implies that $P_{2n}(x) \le f(x) \le P_{2n+1}(x)$.

We also have the inequality $x/(1+x) \le \log(1+x)$ for all $x \ge 0$. Indeed, consider $f(x) = x/(1+x) - \log(1+x)$. Then

$$f'(x) = \frac{-x}{(1+x)^2} \le 0.$$

Since $f(0) = 0$, the inequality follows.

(d) Notation being as in part (c) we have $P_{2n+1}(x) - P_{2n}(x) \to 0$ as $n \to \infty$ for all $x \in [0,1]$, hence

$$\log(1+x) = \lim_{n \to \infty} P_{2n}(x) = \lim_{n \to \infty} P_{2n+1}(x)$$

for all $x \in [0,1]$. If $L_n(x) = x - x^2/2 + \cdots + (-1)^{n+1} x^n/n$, then for all positive integers n we have $L_{2n}(x) = P_{2n}(x)$ and $L_{2n+1}(x) = P_{2n+1}(x)$ which implies

$$\lim_{n \to \infty} L_n(x) = \log(1+x)$$

for all $x \in [0,1]$.

Exercise IV.2.16 *Show that for every positive integer k one has*

$$\left(1 + \frac{1}{k}\right)^k < e < \left(1 + \frac{1}{k}\right)^{k+1}.$$

Taking the product for $k = 1, 2, \ldots, n-1$, conclude by induction that

$$\frac{n^{n-1}}{(n-1)!} < e^{n-1} < \frac{n^n}{(n-1)!}$$

and consequently

$$en^n e^{-n} < n! < en^{n+1} e^{-n}.$$

For another way to get this inequality, see Exercise 20.

Solution. By Exercise 14 we have $\log(1+1/k) < 1/k$. Therefore $k\log(1+1/k) < 1$ which implies $(1+1/k)^k < e$. For the second inequality plug $x = 1/k$ in (c) of the previous exercise to get $1/(k+1) < \log(1+1/k)$ which implies $e < (1+1/k)^{k+1}$.

Since $e^n = e^{n-1}e$ we have the following inequalities

$$\frac{n^{n-1}}{(n-1)!}\left(\frac{n+1}{n}\right)^n < e^n < \frac{n^n}{(n-1)!}\left(\frac{n+1}{n}\right)^{n+1}$$

which implies

$$\frac{(n+1)^n}{n!} < e^n < \frac{(n+1)^{n+1}}{n!}.$$

By induction we conclude that the double inequality holds for all positive integers n. This double inequality is equivalent to $e^{-n+1}n^{n-1} < (n-1)! < e^{-n+1}n^n$ which implies $ee^{-n}n^n < n! < ee^{-n}n^{n+1}$, as was to be shown.

Exercise IV.2.17 *Show that*

$$\lim_{n\to\infty}\left(1+\frac{x}{n}\right)^n = e^x.$$

Solution. We first show that $\lim_{n\to\infty} n\log(1+x/n) = x$. Let $h = x/n$, then

$$\lim_{n\to\infty} n\log(1+x/n) = \lim_{h\to 0} x\frac{\log(1+h)}{h} = x.$$

Exponentiating yields the desired limit.

Exercise IV.2.18 *Let $\{a_n\}$, $\{b_n\}$ be sequences of positive numbers. Define these sequences to be **equivalent**, and write $a_n \equiv b_n$ for $n \to \infty$ to mean that there exists a sequence of positive numbers $\{u_n\}$ such that $b_n = u_n a_n$ and $\lim u_n^{1/n} = 1$. Alternatively, this amounts to the property that $\lim(a_n/b_n)^{1/n} = 1$.*
(a) Prove that the above relation is an equivalence relation for sequences.
(b) Show that $n! \equiv n^n e^{-n}$ for $n \to \infty$. Give a similar equivalence for $(3n)!$.
(c) Show that if $a_n \equiv a'_n$ and $b_n \equiv b'_n$, then $a_n b_n \equiv a'_n b'_n$ for $n \to \infty$.

Solution. (a) Clearly, $a_n \equiv a_n$ because $\lim 1^{1/n} = 1$. If $a_n \equiv b_n$, then $b_n \equiv a_n$ because $a_n = b_n(1/u_n)$ with $\lim(1/u_n)^{1/n} = 1$. Finally suppose that $a_n \equiv b_n$ and $b_n \equiv c_n$. Write $b_n = u_n a_n$ and $c_n = t_n b_n$, where $u^{1/n} \to 1$ and $t^{1/n} \to 1$ as $n \to \infty$. Then we have $c_n = u_n t_n a_n$ with $\lim(u_n t_n)^{1/n} = 1$ so $a_n \equiv c_n$, as was to be shown.
(b) The inequalities deduced in Exercise 16 imply $e < n!/(n^n e^{-n}) < en$ so

$$e^{1/n} < \left(\frac{n!}{n^n e^{-n}}\right)^{1/n} < e^{1/n}n^{1/n}.$$

But $\lim e^{1/n} = \lim n^{1/n} = 1$, hence $n! \equiv n^n e^{-n}$. Replacing n by $3n$ in the inequalities of Exercise 16 we find

$$e^{1/n} < \left(\frac{3n!}{(3n)^{3n}e^{-3n}} \right)^{1/n} < e^{1/n}3^{1/n}n^{1/n},$$

so $(3n!) \equiv (3n)^{3n}e^{-3n}$.

(c) We write $a_n = u_n a'_n$ and $b_n = t_n b'_n$ where $u^{1/n} \to 1$ and $t^{1/n} \to 1$ as $n \to \infty$. Then $a_n b_n = u_n t_n a'_n b'_n$, so we have $a_n b_n \equiv a'_n b'_n$ because $\lim(u_n t_n)^{1/n} = 1$.

Exercise IV.2.19 *Find the following limits as $n \to \infty$: (a)$\left(\frac{(3n!)}{n^{3n}} \right)^{1/n}$; (b) $\left(\frac{(n!)^3}{n^{3n}e^{-n}} \right)^{1/n}$; (c) $\left(\frac{(n!)^2}{n^{2n}} \right)^{1/n}$; and (d) $\left(\frac{n^{2n}}{(2n)!} \right)^{1/n}$*

Solution. (a) Since $n! \equiv n^n e^{-n}$ we have

$$\lim((3n!)/n^{3n})^{1/n} = \lim(3n^{-2n}e^{-n})^{1/n} = 0.$$

(b) We have $(n!)^3 \equiv n^{3n}e^{-3n}$ so

$$\lim \left(\frac{(n!)^3}{n^{3n}e^{-n}} \right)^{1/n} = \lim \left(\frac{n^{3n}e^{-3n}}{n^{3n}e^{-n}} \right)^{1/n} = e^{-2}.$$

(c) Since $(n!)^2 = n^{2n}e^{-2n}$ the desired limit is e^{-2}.

(d) We have $(2n!) \equiv (2n)^{2n}e^{-2n}$ so the desired limit is $e^{-2}/4$.

For the next exercises, which concern the logarithm, we assume that you know elementary integration and upper-lower sums associated with the integral. Some of the proofs are easiest using such sums.

Exercise IV.2.20 *We shall give here an alternate proof for the estimate of Exercise 16. Write down upper and lower sums for the integral of $\log x$ over the interval $[1, n]$ for each positive integer n. Use the partition of the interval at the integers k such that $1 \leq k \leq n$. Using the inequalities*

lower sum $\leq$ integral $\leq$ upper sum,

give a proof of the inequality

$$n \log n - n + 1 \leq \log(n!) \leq (n+1) \log n - n + 1.$$

Exponentiating, you have a proof of the inequality

$$en^n e^{-n} \leq n! \leq en^{n+1}e^{-n}.$$

Solution. The lower sum is $\log 2 + \log 3 + \cdots + \log(n-1) = \log(n-1)!$ and the upper sum is $\log 2 + \log 3 + \cdots + \log n = \log(n!)$. The integral of $\log x$ from 1 to n is

$$\int_1^n \log x\, dx = [x \log x - x]_1^n = n \log n - n + 1,$$

so we get

$$\log(n-1)! \le n\log n - n + 1 \le \log n!.$$

Therefore

$$n\log n - n + 1 \le \log(n!) \le (n+1)\log n - n + 1$$

as was to be shown.

Exercise IV.2.21 *(a) Using an upper and lower sum, prove that for every positive integer n, we have*

$$\frac{1}{n+1} < \log\left(1+\frac{1}{n}\right) < \frac{1}{n}.$$

(b) By the same technique, prove that

$$\frac{1}{2}+\cdots+\frac{1}{n} < \log n < 1 + \frac{1}{2}+\cdots+\frac{1}{n-1}.$$

Solution. (a) Consider the function $f(x) = 1/(1+x)$ for $0 \le x \le 1$. Noting that f is decreasing on the interval from $(0,1)$ and comparing areas we get

$$\frac{1}{n}\frac{1}{1+1/n} < \int_0^{1/n} f(x)dx < \frac{1}{n}.$$

But the integral is equal to $\log(1 + 1/n)$ so the desired inequality drops out.

(b) Consider the function $f(x) = 1/x$ from 1 to n. Then the lower sum is $\frac{1}{2}+\cdots+\frac{1}{n}$ and the upper sum is $1 + \frac{1}{2}+\cdots+\frac{1}{(n-1)}$ so we have the inequalities

$$\frac{1}{2}+\cdots+\frac{1}{n} < \int_1^n f(x)dx < 1 + \frac{1}{2}+\cdots+\frac{1}{n-1}.$$

But the integral is equal to $\log n$ so the desired inequalities follow.

Exercise IV.2.22 *(a) For each integer $n \ge 1$, let*

$$a_n = 1 + \frac{1}{2}+\cdots+\frac{1}{n} - \log n.$$

Show that $a_{n+1} < a_n$. [Hint: Consider $a_n - a_{n+1}$ and use Exercise 21.]
(b) Let $b_n = a_n - 1/n$. Show that $b_{n+1} > b_n$.
(c) Prove that the sequences $\{a_n\}$ and $\{b_n\}$ are Cauchy sequences. Their limit is called the **Euler number** γ.

Solution. (a) We have

$$a_n - a_{n+1} = -\log n - \frac{1}{n+1} + \log(n+1) = -\frac{1}{n+1} + \log(1+1/n) > 0$$

the last inequality coming from Exercise 21 (a).

(b) We simply take the difference and get

$$b_{n+1} - b_n = a_{n+1} - a_n - \frac{1}{n+1} + \frac{1}{n} = -\log\left(1 + \frac{1}{n}\right) + \frac{1}{n} > 0$$

the last inequality coming from Exercise 21(a).

(c) Let $\epsilon > 0$. Choose an integer $N > 1/\epsilon$. Then for all $m \geq n > N$ we have

$$0 \leq a_n - a_m \leq b_n - b_m + 1/n - 1/m \leq 1/n < \epsilon,$$

and

$$0 \leq b_m - a_n \leq a_m - a_n + 1/n - 1/m \leq 1/n < \epsilon.$$

So both $\{a_n\}$ and $\{b_n\}$ are Cauchy sequences and they converges to the same limit because $\lim_{n\to\infty}(a_n - b_n)$.

Exercise IV.2.23 *If $0 \leq x \leq 1/2$, show that $\log(1 - x) \geq -x - x^2$. [Note: When you have Taylor's formula and series later, you can see that*

$$\log(1 - x) = -x - \frac{x^2}{2} - \frac{x^3}{3} - \cdots.$$

The point is that $-x$ is a good approximation to $\log(1-x)$ when x is small.]

Solution. Let $f(x) = \log(1 - x)$ and let $g(x) = -x - x^2$. Then $f'(x) = -1/(1+x)$ and $g'(x) = -1 - 2x$. We wish to show that $f'(x) \geq g'(x)$. This inequality is equivalent to $1 \leq (1 - x)(1 + 2x)$ or $0 \leq x - 2x^2 = x(1 - 2x)$. Clearly, this last inequality is true whenever $0 \leq x \leq 1/2$ so we have $f'(x) \geq g'(x)$ on this interval. Finally, $f(0) = g(0) = 0$ so we must have $f(x) \geq g(x)$.

Exercise IV.2.24 *(a) Let s be a number. Define the **bimomial coefficients***

$$\binom{s}{1} = s \quad \text{and} \quad \binom{s}{n} = B(n, s) = s(s-1)(s-2) \cdots (s-n+1)/n! \quad \text{for } n \geq 2.$$

Prove the estimate $|B(n, s)| \leq |s|e^{|s|}(n - 1)^{|s|}/n$. In particular,

$$\limsup_{n\to\infty} |B(n, s)|^{1/n} \leq 1.$$

Note that the above estimate applies as well if s is complex.

(b) If s is not an integer ≥ 0, show that $\lim |B(n, s)|^{1/n} = 1$.

Solution. (a) Since

$$B(n, s) = \frac{1}{n}s(s - 1)\left(\frac{s}{2} - 1\right) \cdots \left(\frac{s}{n-1} - 1\right)$$

letting $\alpha = |s|$, we get

$$|B(n,s)| \leq \frac{1}{n}[\alpha(1+\alpha)(1+\alpha/2)\cdots(1+\alpha/(n-1))]$$

hence the estimate for $\log n|B(n,s)|$,

$$\log n|B(n,s)| \leq \log\alpha + \log(1+\alpha) + \cdots + \log(1+\alpha/(n-1)).$$

Exercises 14 and 21 (b) imply that

$$\log n|B(n,s)| \leq \log\alpha + \alpha + \alpha\log(n-1).$$

Exponentiating we get that $n|B(n,s)| \leq \alpha e^\alpha(n-1)^\alpha$. The lim sup statement follows from the fact that

$$\lim_{n\to\infty}(e^{|s|}|s|(n-1)^{|s|})^{1/n} = 1.$$

whenever $|s| \neq 0$. This limit follows from taking the logarithm.
(b) Because of part (a) it is sufficient to bound $|B(n,s)|^{1/n}$ from below. Let $\alpha = |s|$. Then the triangle inequality implies

$$|B(n,s)| \geq \frac{1}{n}\alpha|\alpha-1|\left|\frac{\alpha}{2}-1\right|\cdots\left|\frac{\alpha}{n-1}-1\right|.$$

Let n_0 be an integer such that for all $n \geq n_0$ we have $\alpha/n < 1/2$. Then, since $\lim c^{1/n} = \lim 1/n^{1/n} = 1$ for $c \neq 0$ we can forget finitely many of the beginning terms, namely those up to n_0 including $1/n$, so that after taking the n-th power and the log we see that is suffices to show that

$$A_n = \frac{1}{n}\left[\log\left(1-\frac{\alpha}{n_0}\right) + \cdots + \log\left(1-\frac{\alpha}{n-1}\right)\right] \to 0.$$

By assumption on n_0 we have $A_n \leq 0$, and Exercise 23 implies the following estimate

$$A_n \geq \frac{1}{n}\left[-\frac{\alpha}{n_0} - \frac{\alpha^2}{n_0^2} - \cdots - \frac{\alpha}{n-1} - \frac{\alpha^2}{n-1}\right]$$

$$\geq \frac{1}{n}\left[-\alpha\left(\frac{1}{n_0} + \cdots + \frac{1}{n-1}\right) - \alpha^2 C\right],$$

where C is a positive real number. The last inequality follows from the fact that $\sum 1/n^2$ converges. Exercise 21 implies (assuming that $n_0 \geq 2$ which we obviously can)

$$0 \geq A_n \geq \frac{-\alpha}{n}\log(n-1) - \frac{\alpha^2 C}{n},$$

so $\lim_{n\to\infty} A_n = 0$ as was to be shown.

Exercise IV.2.25 *Let α be a real number > 0. Let*

$$a_n = \frac{\alpha(\alpha+1)\cdots(\alpha+n)}{n!n^\alpha}.$$

Show that $\{a_n\}$ is monotonically decreasing for sufficiently large values of n, and hence approaches a limit. This limit is denoted by $1/\Gamma(\alpha)$, where Γ is called the **gamma function.**

Solution. For all $n \geq 1$, $a_n > 0$ and

$$\frac{a_{n+1}}{a_n} = \frac{\alpha+n+1}{n+1}\frac{n^\alpha}{(n+1)^\alpha} = \left(1+\frac{\alpha}{n+1}\right)\left(\frac{n}{n+1}\right)^\alpha.$$

The inequalities deduced in Exercises 14 and 21(a) imply that for sufficiently large n

$$\log\left(1+\frac{\alpha}{n+1}\right) \leq \frac{\alpha}{n+1} < \alpha\log\left(1+\frac{1}{n}\right).$$

Exponentiating, we find that $a_{n+1}/a_n < 1$ for all large n, and since $a_n > 0$ we see that $\{a_n\}$ is monotonically decreasing for sufficiently large values of n.

IV.3 Sine and Cosine

Exercise IV.3.1 *Define $\tan x = \sin x/\cos x$. Sketch the graph of $\tan x$. Find*

$$\lim_{h\to 0} \frac{\tan h}{h}.$$

Solution. The tangent is not defined for $x = \pi/2 + k\pi$ when $k \in \mathbf{Z}$, and is periodic of period π. We have

$$\frac{d}{dx}\tan x = \frac{1}{\cos^2 x} > 0$$

and $\tan x = 0$ is equivalent to $x = k\pi$, $k \in \mathbf{Z}$. The limit

$$\lim_{h\to 0} \frac{\tan h}{h} = 1$$

follows from the expression

$$\frac{\tan h}{h} = \frac{1}{\cos h}\frac{\sin h}{h},$$

and the limit $\lim_{h\to 0}(\sin h)/h = 1$. The graph of the tangent function is

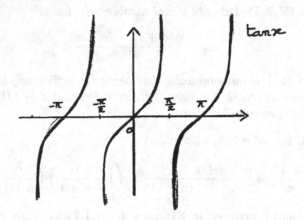

Exercise IV.3.2 *Restrict the sine function to the interval* $-\pi/2 \le x \le \pi/2$, *on which it is continuous, and such that its derivative is* > 0 *on* $-\pi/2 < x < \pi/2$. *Define the inverse function, called the* **arcsine**. *Sketch the graph, and show that the derivation of* $\arcsin x$ *is* $1/\sqrt{1-x^2}$.

Solution. By symmetry with respect to the line $y = x$ we get the graph of $\arcsin x$. To compute its derivative on the given interval, we simply use the formula for the derivative of the inverse map, so that

$$\frac{d}{dx}(\arcsin x) = \frac{1}{\cos(\arcsin x)} = \frac{1}{\sqrt{1-\sin^2(\arcsin x)}} = \frac{1}{\sqrt{1-x^2}}.$$

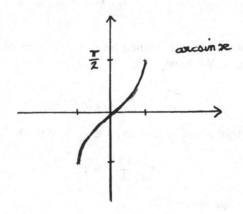

Exercise IV.3.3 *Restrict the cosine function to the interval* $0 \le x \le \pi$. *Show that the inverse function exists. It is called the* **arccosine**. *Sketch its*

graph, and show that the derivative of arccos x *is* $-1/\sqrt{1-x^2}$ *on* $0 < x < \pi$.

Solution. The inverse function exists because on the given interval, $\cos x$ is continuous and its derivative is < 0 on $0 < x < \pi$. Arguing as in the previous exercise we find

$$\frac{d}{dx}(\arccos x) = \frac{1}{-\sin(\arccos x)} = \frac{-1}{\sqrt{1 - \cos^2(\arccos x)}} = \frac{-1}{\sqrt{1 - x^2}}.$$

We can also graph the function arccos x using symmetry with respect to the diagonal $y = x$.

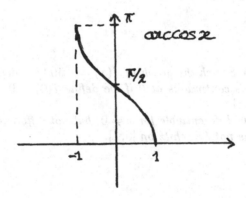

Exercise IV.3.4 *Restrict the tangent function to* $-\pi/2 < x < \pi/2$. *Show that its inverse function exists. It is called the* **arctangent**. *Show that arctan is defined for all numbers, sketch its graph, and show that the derivative of* arctan x *is* $1/(1 + x^2)$.

Solution. Since the tangent function is continuous, $(\arctan x)' = 1 + \tan^2 x > 0$ and $\lim_{x \to \pi/2} \tan x = \infty$ and $\lim_{x \to -\pi/2} \tan x = -\infty$, we conclude that the inverse function exists on $\mathbf{R}$, that it is continuous with derivative

$$\frac{d}{dx}(\arctan x) = \frac{1}{1 + \tan^2(\arctan x)} = \frac{1}{1 + x^2}.$$

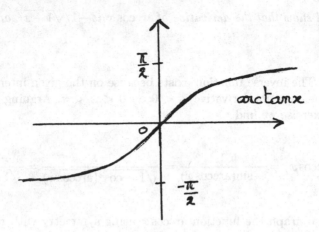

Exercise IV.3.5 *Sketch the graph of $f(x) = x \sin 1/x$, defined for $x \neq 0$.*
(a) Show that f is continuous at 0 if we define $f(0) = 0$. Is f uniformly continuous on $[0, 1]$?
(b) Show that f is differentiable for $x \neq 0$, but not differentiable at $x = 0$.
(c) Show that f is not Lipschitz on $[0, 1]$.

Solution. (a) Since $|\sin x| \leq 1$ we have the inequalities $-|x| \leq f(x) \leq |x|$ for all $x \neq 0$ so $f(x) \to 0$ as $x \to 0$, which proves that f is continuous at 0 if we define $f(0) = 0$. Since $[0, 1]$ is compact and f is continuous, we conclude that f is uniformly continuous on $[0, 1]$. The function f tends to 1 as $|x| \to \infty$ because $(\sin u)/u \to 1$ as $u \to 0$. Moreover,

$$f'(x) = \sin(1/x) - (1/x)\cos(1/x),$$

so there are oscillations near zero as the graph illustrates. In fact, if $x_n = 1/(n\pi)$, $n \in \mathbf{Z}$, and $n \neq 0$ we have $f(x_n) = 0$.

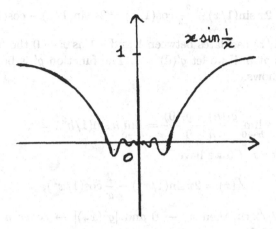

$x \sin \frac{1}{x}$

(b) The Newton quotient of f at 0 is

$$\frac{f(h) - f(0)}{h - 0} = \sin(1/h)$$

which does not tend to a limit as $h \to 0$, so f is not differentiable at 0.
(c) The function f' is not bounded on $(0, 1)$ because

$$\left| f'\left(\frac{1}{2\pi n}\right) \right| = 2\pi n,$$

so by Exercise 6, §2, of Chapter III, we conclude that f is not Lipschitz on $(0, 1)$, hence f is not Lipschitz on $[0, 1]$.

Exercise IV.3.6 *Let $g(x) = x^2 \sin(1/x)$ if $x \neq 0$ and $g(0) = 0$.*
(a) Show that g is differentiable at 0, and is thus differentiable on the closed interval $[0, 1]$.
(b) Show that g is Lipschitz on $[0, 1]$.
(c) Show that g' is not continuous at 0, but is continuous for all $x \neq 0$. Is g' bounded? Why?
(d) Let $g_1(x) = x^2 \sin(1/x^2)$ for $x \neq 0$ and $g_1(0) = 0$. Show that $g_1'(0) = 0$, but g_1' is not bounded on $(0, 1]$. Is g_1 Lipschitz?

Solution. (a) We form the Newton quotient of g at 0:

$$\frac{g(h) - g(0)}{h - 0} = h \sin h.$$

But $h \sin h \to 0$ as $h \to 0$, so g is differentiable at 0 and $g'(0) = 0$.
(b) The derivative of g is bounded (see (c)) on $[0, 1]$ so g is Lipschitz on $[0, 1]$. This is a consequence of the mean value theorem.
(c) For $x \neq 0$ we have

$$g'(x) = 2x \sin(1/x) - \frac{x^2}{x^2} \cos(1/x) = 2x \sin(1/x) - \cos(1/x),$$

and since $\cos(1/x)$ oscillates between 1 and -1 as $x \to 0$ the function g' is not continuous at 0, if we let $g'(0) = 0$. The function g' is bounded on $\mathbf{R}$ as Exercise 5 shows.

(d) We have

$$\lim_{h \to 0} \frac{g_1(h) - g_1(0)}{h - 0} = \lim_{h \to 0} h \sin(1/h^2) = 0,$$

so $g_1'(0) = 0$. For $x \neq 0$ we have

$$g_1'(x) = 2x \sin(1/x^2) - \frac{2}{x} \cos(1/x^2),$$

and if $x_n = 1/\sqrt{2\pi n}$, then $x_n \to 0$ and $|g_1'(x_n)| \to \infty$ as $n \to \infty$, g_1' is unbounded on $(0, 1]$.

Exercise IV.3.7 *Show that if $0 < x < \pi/2$, then $\sin x < x$ and $2/\pi < (\sin x)/x$.*

Solution. Let $f(x) = \sin x - x$. Then f is differentiable on $(0, \pi/2)$ with derivative $f'(x) = \cos x - 1 < 0$, so $f(x) < f(0)$ and since $f(0) = 0$ we get the first inequality.

For the second inequality, consider $f(x) = \sin x - 2x/\pi$. The derivative of f is $f'(x) = \cos x - 2/\pi$, so f is increasing on $(0, a)$ and decreasing on $(a, \pi/2)$, where a is the number such that $0 < a < \pi/2$ and $\cos a = 2/\pi$. Conclude using the fact that $f(0) = f(\pi/2) = 0$.

Exercise IV.3.8 *Let $0 \leq x$. (a) Show that $\sin x \leq x$. (b) Show that $\cos x \geq 1 - x^2/2$. (c) Show that $\sin x \geq x - x^3/3!$ (d) Give the general inequalities similar to the preceding ones, by induction.*

Solution. (a) This result follows from the first part of Exercise 7.

(b) Let $f_2(x) = \cos x - 1 + x^2/2$, then $f_2'(x) = -\sin x + x \geq 0$ and $f_2(0) = 0$ which proves the desired inequality.

(c) Let $f_3(x) = \sin x - x + x^3/3!$, then $f_3'(x) = f_2(x) \geq 0$ and $f_3(0) = 0$ which proves the desired inequality.

(d) Let $f_n(x) = \cos x - 1 + x^2/2! - x^4/4! + \cdots + (-1)^{n+1} x^{2n}/(2n)!$ and consider also $g_n(x) = \sin x - x + x^3/3! - \cdots + (-1)^{n+1} x^{2n-1}/(2n-1)!$. Then the general inequalities are

$$\begin{cases} f_n(x) \leq 0 \text{ for } n \text{ even}, \\ f_n(x) \geq 0 \text{ for } n \text{ odd}, \end{cases}$$

and

$$\begin{cases} g_n(x) \leq 0 \text{ for } n \text{ even}, \\ g_n(x) \geq 0 \text{ for } n \text{ odd}. \end{cases}$$

These inequalities follow by induction and the fact that $f_n(0) = g_n(0) = 0$ and $f_n'(x) = g_n(x)$ and $g_{n+1}'(x) = f_n(x)$.

IV.4   Complex Numbers

Exercise IV.4.1 *Let α be a complex number $\neq 0$. Show that there are two distinct complex numbers whose square is α.*

Solution. The only two numbers solution to $z^2 = \alpha$ are $\sqrt{|\alpha|}e^{i(\varphi/2)}$ and $\sqrt{|\alpha|}e^{i(\pi+\varphi/2)}$. See the next exercise.

Exercise IV.4.2 *Let α be a complex, $\neq 0$. Let n be a positive integer. Show that there are exactly n distinct complex numbers z such that $z^n = \alpha$. Write these complex numbers in polar form.*

Solution. Write z and α in polar form, that is $z = re^{i\theta}$ and $\alpha = r_0e^{i\varphi}$. Suppose $z^n = \alpha$. This equality is equivalent to $r^ne^{in\theta} = r_0e^{i\varphi}$. Hence

$$\frac{r^n}{r_0}e^{i(n\theta-\varphi)} = 1.$$

Taking the absolute value of both sides we see that $|z|^n = |\alpha|$. Moreover we must have $n\theta - \varphi = 0 \pmod{2\pi}$, so the set of solutions of the equation $z^n = \alpha$ is

$$S = \left\{|\alpha|^{1/n}e^{i\left(\frac{\varphi}{n}\right)}, |\alpha|^{1/n}e^{i\left(\frac{\varphi}{n}+\frac{2\pi}{n}\right)}, \ldots, |\alpha|^{1/n}e^{i\left(\frac{\varphi}{n}+(n-1)\frac{2\pi}{n}\right)}\right\}.$$

Exercise IV.4.3 *Let w be a complex number and suppose that z is a complex number such that $e^z = w$. Describe all complex numbers u such that $e^u = w$.*

Solution. We have $e^z = e^u$ so taking absolute values we see that u and z must have the same real part. Furthermore, their imaginary part must differ by an integer multiple of $2\pi i$. Conclude.

Exercise IV.4.4 *What are the complex numbers z such that $e^z = 1$?*

Solution. Write $z = x + iy$. Then $e^z = e^xe^{iy}$. So $e^z = 1$ implies that $x = 0$ and $y = 2k\pi$ for some integer k.

Exercise IV.4.5 *If θ is real, show that*

$$\cos\theta = \frac{e^{i\theta}+e^{-i\theta}}{2} \quad \text{and} \quad \sin\theta = \frac{e^{i\theta}-e^{-i\theta}}{2i}.$$

Solution. Since

$$e^{i\theta} + e^{-i\theta} = \cos\theta + i\sin\theta + \cos(-\theta) + i\sin(\theta) = 2\cos\theta$$

the first inequality drops out. For the second formula note that

$$e^{i\theta} - e^{-i\theta} = \cos\theta + i\sin\theta - \cos(-\theta) - i\sin(\theta) = 2i\sin\theta.$$

Exercise IV.4.6 *Let F be a differentiable complex valued function defined on some interval. Show that*

$$\frac{d(e^{F(t)})}{dt} = F'(t)e^{F(t)}.$$

Solution. Let $G(t) = e^{it} = \cos t + i \sin t$. Then $G'(t) = -\sin t + i \cos t = ie^{it}$. Now let $F(t) = x(t) + iy(t)$, so that $e^{F(t)} = e^{x(t)}e^{iy(t)}$. The rule for differentiating products and the chain rule imply

$$\frac{d(e^{F(t)})}{dt} = x'(t)e^{x(t)}e^{iy(t)} + e^{x(t)}ie^{iy(t)}y'(t) = F'(t)e^{F(t)}.$$

V

The Elementary Real Integral

V.2   Properties of the Integral

Exercise V.2.1 *(a) Let f, g be continuous functions on $[a, b]$ with $a < b$. Assume g positive. Show that there exists $c \in [a, b]$ such that*

$$\int_a^b f(x)g(x)dx = f(c)\int_a^b g(x)dx.$$

(b) **Bonnet Mean Value Theorem** *(1849). Let f, g be continuous real valued functions on $[a, b]$. Assume f positive monotone decreasing. Show that there exists a point $c \in [a, b]$ such that*

$$\int_a^b f(x)g(x)dx = f(a)\int_a^c g(x)dx.$$

This result is definitely harder. First assume that f is C^1, so $f' \leq 0$. Let $G(x)$ be the integral of g from a to x. Integrate by parts. Using the intermediate value theorem, show that there is some $c_1 \in [a, b]$ such that

$$\int_a^b f(x)g(x)dx = f(b)G(b) + G(c_1)(f(a) - f(b)).$$

Divide by $f(a)$, use the hypothesis that f is decreasing to conclude that the right side in on the segment between $G(b)$ and $G(c_1)$, so by the intermediate value theorem again, is equal to $G(c)$ for some c, thus proving the result in this case. In general, possibly wait until Chapter X, §3, Exercise 7. Show *that there exists a sequence $\{f_n\}$ of C^1 functions with $f_n(a) = f(a)$, $f_n(b) = f(b)$, each f_n is monotone decreasing, and f_n converges uniformly to f. Use bump functions to do this. The theorem is true for each f_n, with some c_n instead of c. By Weierstrass Bolzano, the sequence $\{c_n\}$ has a point of accumulation $c \in [a, b]$ which does what you want.*

Solution. (a) The function f is continuous on the compact interval $[a, b]$, so f attains its minimum m and its maximum M on $[a, b]$. For all $x \in [a, b]$ we have $m \le f(x) \le M$ and g is positive so $mg(x) \le f(x)g(x) \le Mg(x)$, which implies

$$m \int_a^b g(x)dx \le \int_a^b f(x)g(x)dx \le M \int_a^b g(x)dx.$$

But g is positive and continuous so $\int_a^b g(x)dx > 0$, and therefore

$$m \le \frac{\int_a^b f(x)g(x)dx}{\int_a^b g(x)dx} \le M.$$

The intermediate value theorem guarantees the existence of a number $c \in [a, b]$ such that

$$\frac{\int_a^b f(x)g(x)dx}{\int_a^b g(x)dx} = f(c),$$

concluding the proof of (a).

(b) We follow the hint. First assume that f is C^1 and let $G(x) = \int_a^x g(t)dt$. Integration by parts yields

$$\int_a^b f(x)g(x)dx = f(x)G(x)\big|_a^b - \int_a^b f'(x)G(x)dx. \tag{V.1}$$

The function G is continuous and therefore attains its maximum M and minimum m on $[a, b]$. Since $-f'(x)$ is positive, we have

$$m(-f'(x)) \le -f'(x)G(x) \le M(-f'(x)).$$

By the intermediate value theorem there is some $c_1 \in [a, b]$ such that

$$\int_a^b -f'(x)G(x)dx = G(c_1) \int_a^b -f'(x)dx = G(c_1)(f(a) - f(b)).$$

Therefore equation (V.1) becomes

$$\int_a^b f(x)g(x)dx = f(b)G(b) + G(c_1)(f(a) - f(b)).$$

Dividing by $f(a)$ and letting $u = f(b)/f(a)$ we obtain

$$\frac{1}{f(a)} \int_a^b f(x)g(x)dx = uG(b) + G(c_1)(1 - u).$$

Since G is continuous and $u \in [0, 1]$, there exists $c \in [a, b]$ such that $uG(b) + G(c_1)(1 - u) = G(c)$. This concludes the proof when $f \in C^1$.

In the general case we approximate f uniformly by a sequence $\{f_n\}$ of monotone decreasing C^1 functions. This can be done as follows. We first approximate f uniformly by step functions $\{\varphi_n\}$ such that each φ_n is decreasing. Indeed, given n choose δ such that $|f(x) - f(y)| \le 1/n$ whenever $|x - y| \le \delta$ (by the uniform continuity of f). Then select a partition $a = t_0 < t_1 < \cdots < t_k = b$ of $[a, b]$ of size $< \delta$ and define φ_n on $[t_j, t_{j+1})$ to be equal to $f(t_j)$. Then $|f(x) - \varphi_n(x)| \le 1/n$ for all $x \in [a, b]$, and φ_n is decreasing. Then using a bump function as in Exercise 6(b), §1, of Chapter IV and Exercise 7, §3, of Chapter X, we can approximate φ_n by a C^1 function f_n such that $|f_n(x) - \varphi_n(x)| \le 2/n$ for all $x \in [a, b]$ and f_n is decreasing. Then we are reduced to part (a), so we can find for each n a number c_n such that

$$\int_a^b f_n(x)g(x)dx = f_n(a) \int_a^{c_n} g(x)dx.$$

By Weierstrass Bolzano, $\{c_n\}$ has a point of accumulation $c \in [a, b]$. We may assume without loss of generality that $\lim_{n \to \infty} c_n = c$. Then letting $n \to \infty$ in the above equation we get

$$\int_a^b f(x)g(x)dx = f(a) \int_a^c g(x)dx$$

as was to be shown.

Exercise V.2.2 *Let*

$$P_n(x) = \frac{1}{2^n n!} \frac{d^n}{dx^n} ((x^2 - 1)^n).$$

Show that

$$\int_{-1}^1 P_n(x)P_m(x)dx = 0 \quad if \; m \ne n,$$

and that

$$\int_{-1}^1 P_n(x)^2 dx = \frac{2}{2n + 1}.$$

Solution. (i) For simplicity, let $g = g(x) = x^2 - 1$ and let $K_n = 1/(2^n n!)$. By induction one proves that for all $0 \le k \le n$, $(d^k/dx^k)g^n = g^{n-k}Q_k(x)$ where $Q_0(x), \ldots, Q_n(x)$ are polynomials. We have

$$\int_{-1}^1 P_n(x)P_m(x)dx = K_n K_m \int_{-1}^1 \frac{d^n}{dx^n} g^n \frac{d^m}{dx^m} g^m dx = K_n K_m J,$$

where

$$J = \int_{-1}^{1} \frac{d^n}{dx^n} g^n \frac{d^m}{dx^m} g^m \, dx.$$

Assume without loss of generality that $m > n$. Integrating by parts we obtain

$$J = \left[\frac{d^n}{dx^n} g^n \frac{d^{m-1}}{dx^{m-1}} g^m \right]_{-1}^{1} - \int_{-1}^{1} \frac{d^{n+1}}{dx^{n+1}} g^n \frac{d^{m-1}}{dx^{m-1}} g^m \, dx.$$

The expression in brackets is 0, so integrating by parts n successive times we get,

$$J = (-1)^n \int_{-1}^{1} \frac{d^{2n}}{dx^{2n}} g^n \frac{d^{m-n}}{dx^{m-n}} g^m \, dx = (-1)^n (2n)! \left[\frac{d^{m-n-1}}{dx^{m-n-1}} g^m \right]_{-1}^{1} = 0.$$

(ii) Let I be the integral we are computing. Letting $m = n$ in (i) we find

$$|I| / (K_n^2) = \left| \int_{-1}^{1} \frac{d^n}{dx^n} g^n \frac{d^n}{dx^n} g^n \, dx \right|.$$

Let L be this last integral. Integrating by parts we obtain

$$|L| = \left| \int_{-1}^{1} g^n \frac{d^{2n}}{dx^{2n}} g^n \, dx \right| = (2n!) \left| \int_{-1}^{1} g^n \, dx \right|.$$

This last integral can be evaluated by n successive integration by parts. Indeed,

$$\int_{-1}^{1} g^n \, dx = [xg^n]_{-1}^{1} - 2n \int_{-1}^{1} x^2 g^{n-1} \, dx = 2n \int_{-1}^{1} x^2 g^{n-1} \, dx,$$

so

$$\int_{-1}^{1} g^n \, dx = (-1)^k \frac{2^k n (n-1) \cdots (n-k-1)}{3 \cdot 5 \cdot 7 \cdots (2k-1)} \int_{-1}^{1} x^{2k} g^{n-k} \, dx.$$

Substituting $k = n$, we find that

$$\left| \int_{-1}^{1} g^n \, dx \right| = \frac{2^n n! 2^n n!}{2 \cdot 3 \cdot 4 \cdot (2n-1) \cdot 2n} = \frac{2}{2n+1},$$

so $(K_n)^2 L = 2/(2n+1)$.

Exercise V.2.3 *Show that*

$$\int_{-1}^{1} x^m P_n(x) \, dx = 0 \quad if \ m < n.$$

Evaluate

$$\int_{-1}^{1} x^n P_n(x) \, dx.$$

Solution. (i) Integrating by parts we get the following recurrence relation:

$$\int_{-1}^{1} x^m \frac{d^n}{dx^n} g^n dx = (-1)^k \frac{m!}{(m-k)!} \int_{-1}^{1} x^{m-k} \frac{d^{n-k}}{dx^{n-k}} g^n dx.$$

Now let $k = m$. Since

$$\int_{-1}^{1} \frac{d^{n-m}}{dx^{n-m}} g^n dx = 0,$$

the desired result follows at once.

(ii) Integrating by parts k times we also see that

$$\int_{-1}^{1} x^n P_n(x) dx = K_n (-1)^k \frac{n!}{(n-k)!} \int_{-1}^{1} x^{n-k} \frac{d^{n-k}}{dx^{n-k}} g^n dx.$$

So

$$\int_{-1}^{1} x^n P_n(x) dx = K_n (-1)^n n! \int_{-1}^{1} g^n dx = (-1)^{2n} \frac{n!}{2^n n!} \frac{2 \cdot 2^n n! 2^n n!}{(2n+1)!},$$

hence

$$\int_{-1}^{1} x^n P_n(x) dx = \frac{2^{n+1}(n!)^2}{(2n+1)!}.$$

Exercise V.2.4 *Let $a < b$. If f, g are continuous on $[a, b]$, let*

$$\langle f, g \rangle = \int_a^b f(x) g(x) dx.$$

Show that the symbol $\langle f, g \rangle$ satisfies the following properties.
(a) If f_1, f_2, g are continuous on $[a, b]$, then

$$\langle f_1 + f_2, g \rangle = \langle f_1, g \rangle + \langle f_2, g \rangle.$$

If c is a number, then $\langle cf, g \rangle = c \langle f, g \rangle$.
(b) We have $\langle f, g \rangle = \langle g, f \rangle$.
(c) We have $\langle f, f \rangle \geq 0$, and equality holds if and only if $f = 0$.

Solution. (a) We simply use the linearity of the integral, namely

$$\int_a^b (f_1(x) + f_2(x)) g(x) dx = \int_a^b f_1(x) g(x) dx + \int_a^b f_2(x) g(x) dx,$$

and

$$\int_a^b cf(x) g(x) dx = c \int_a^b f(x) g(x) dx.$$

(b) Obvious.
(c) Since $f(x) f(x) \geq 0$ we have $\langle f, f \rangle \geq 0$. If $f = 0$, then clearly we have $\langle f, f \rangle = 0$. The converse follows at once from Theorem 2.4.

Exercise V.2.5 *For any number $p \geq 1$ define*

$$\|f\|_p = \left[\int_a^b |f(x)|^p dx \right]^{1/p}.$$

Let q be a number such that $1/p + 1/q = 1$. Prove that

$$|\langle f, g \rangle| \leq \|f\|_p \|g\|_q.$$

[Hint: If $\|f\|_p$ and $\|g\|_q \neq 0$, let $u = |f|^p / \|f\|_p^p$ and $v = |g|^q / \|g\|_q^q$ and apply Exercise 10 of Chapter IV, §2.]

Solution. If $\|f\|_p = 0$ or $\|g\|_q = 0$ the inequality is trivially verified, so we assume that $\|f\|_p$ and $\|g\|_q$ are both $\neq 0$. Using the inequality

$$u^{1/p} v^{1/q} \leq \frac{u}{p} + \frac{v}{q}$$

of Exercise 10, Chapter IV, §2, we obtain

$$\frac{|f|}{\|f\|_p} \frac{|g|}{\|g\|_q} \leq \frac{|f|^p}{p\|f\|_p^p} + \frac{|g|^q}{q\|g\|_q^q}$$

so that

$$|f||g| \leq \frac{|f|^p}{p\|f\|_p^{p-1}} \|g\|_q + \frac{|g|^q}{q\|g\|_q^{q-1}} \|f\|_p.$$

Integrating both sides yields

$$\langle |f|, |g| \rangle \leq \frac{1}{p} \|f\|_p \|g\|_q + \frac{1}{q} \|f\|_p \|g\|_q = \|f\|_p \|g\|_q.$$

Corollary 2.2 implies that $|\langle f, g \rangle| \leq \langle |f|, |g| \rangle$, thus proving the desired inequality (called **Hölder's inequality**).

Exercise V.2.6 *Notation being as in the preceding exercise, prove that*

$$\|f + g\|_p \leq \|f\|_p + \|g\|_p.$$

[Hint: Let I denote the integral. Show that

$$\|f + g\|_p^p \leq I(|f + g|^{p-1}|f|) + I(|f + g|^{p-1}|g|)$$

and apply Exercise 5.]

Solution. Assume $|f + g| \neq 0$ otherwise the inequality is trivial. The triangle inequality for the absolute value implies,

$$I(|f + g|^p) = I(|f + g|^{p-1}|f + g|) \leq I(|f + g|^{p-1}|f|) + I(|f + g|^{p-1}|g|).$$

Since $1/p + (p-1)/p = 1$, Exercise 5 implies,

$$I(|f+g|^{p-1}|f|) \le \|(f+g)^{p-1}\|_{p/(p-1)}\|f\|_p,$$

so combined with a similar result for the second integral we get

$$\|f+g\|_p^p \le \|(f+g)^{p-1}\|_{p/(p-1)}(\|f\|_p + \|g\|_p).$$

But $\|f+g\|_p^p / \|(f+g)^{p-1}\|_{p/(p-1)} = \|f+g\|_p$, and the inequality drops out.

Exercise V.2.7 *Let $f\colon J \to \mathbf{C}$ be a complex valued function defined on an interval J. Write $f = f_1 + if_2$, where f_1, f_2 are real valued and continuous. Define the indefinite integral*

$$\int f(x)dx = \int f_1(x)dx + i \int f_2(x)dx,$$

and similarly for the definite integral. Show that the integral is linear, and prove similar properties for it with change of variables and integrating by parts.

Solution. The linearity of the integral and law of addition for the complex numbers imply

$$
\begin{aligned}
\int_a^b f(x) + g(x)dx &= \int_a^b f_1(x) + g_1(x)dx + i\int_a^b f_2(x) + g_2(x)dx \\
&= \int_a^b f_1 + \int_a^b g_1 + i\int_a^b f_2 + i\int_a^b g_2 \\
&= \int_a^b f(x)dx + \int_a^b g(x)dx.
\end{aligned}
$$

If $z = \sigma + it$ is a complex number, then

$$\int_a^b zf = \int_a^b (\sigma+it)(f_1+if_2) = \int_a^b (\sigma f_1 - tf_2) + i\int_a^b (\sigma f_2 + tf_1) = z\int_a^b f.$$

If f and g are complex valued functions, then a simple computation shows that $(fg)' = f'g + fg'$, therefore the integration by parts formula is the same as in the text.

The change of variable formula is the same because if f is a real valued function, we have

$$
\int_a^b g(f(x))f'(x)dx = \int_a^b g_1(f)f' + i\int_a^b g_2(f)f'
$$

$$
= G_1(f)|_a^b + iG_2(f)|_a^b = G_1|_{f(a)}^{f(b)} + iG_2|_{f(a)}^{f(b)} = \int_{f(a)}^{f(b)} g(x)dx.
$$

Exercise V.2.8 *Show that for real $a \neq 0$ we have*

$$\int e^{iax}\, dx = \frac{e^{iax}}{ia}.$$

Show that for every integer $n \neq 0$,

$$\int_0^{2\pi} e^{inx}\, dx = 0.$$

Solution. The first result is a consequence of

$$\frac{d}{dx}\left(\frac{e^{iax}}{ia}\right) = \frac{ia}{ia} e^{iax} = e^{iax}.$$

For the second result, note that

$$\int_0^{2\pi} e^{inx}\, dx = \left[\frac{e^{inx}}{in}\right]_0^{2\pi} = \frac{1}{in}(1-1) = 0.$$

V.3   Taylor's Formula

Exercise V.3.1 *Prove Theorem 3.4 by integrating*

$$\frac{1}{1+t} = 1 - t + t^2 - \cdots + (-1)^{n-1} t^{n-1} + (-1)^n \frac{t^n}{1+t}$$

from 0 to x with $-1 < x < 1$. Prove the estimates for the remainder to show that it tends to 0 as $n \to \infty$. If $0 < c < 1$, show that this estimate can be made independent of x in the interval $-c \leq x \leq c$, and that there is a constant K such that the remainder is bounded by $K|x|^{n+1}$.

Solution. Integrating the formula given by summing the geometric series we get the formula of Theorem 3.4, namely

$$\log(1+x) = x - \frac{x^2}{2} + \frac{x^3}{3} - \cdots + (-1)^{n-1}\frac{x^n}{n} + (-1)^n \int_0^x \frac{t^n}{1+t}\, dt.$$

Now we want to estimate the remainder, $R_{n+1}(x) = (-1)^n \int_0^x t^n/(1+t)dt$. In case 1 ($0 \leq x \leq 1$) we see that if t is positive, then $1 + t \geq 1$ so

$$|R_{n+1}(x)| \leq \int_0^x t^n\, dt = \frac{x^{n+1}}{n+1}.$$

In case 2 ($-1 < x \leq 0$), the inequality $|1+t| \geq 1 - |t| \geq 1 - |x| = 1+x$ implies

$$|R_{n+1}(x)| \leq \frac{1}{1+x} \int_0^{|x|} u^n\, du = \frac{|x|^{n+1}}{(n+1)(1+x)}.$$

If $|x| \leq c \leq 1$, then in case 1 we have $|R_{n+1}(x)| \leq c^{n+1}/(n+1)$ and in case 2 we have $|R_{n+1}(x)| \leq c^{n+1}/((n+1)(1-c))$. If $K = 1/(1-c)$, then for $|x| \leq c \leq 1$ the remainder is bounded by $K|x|^{n+1}$.

Exercise V.3.2 *Do the same type of things for the function $1/(1+t^2)$ to prove Theorem 3.5.*

Solution. Since $(d/dx)(\arctan x) = 1/(1+x^2)$ and

$$\frac{1}{1+t^2} = 1 - t^2 + t^4 - \cdots + (-t^2)^{n-1} + \frac{(-t^2)^n}{1+t^2},$$

integrating from 0 to x with $-1 < x < 1$ gives

$$\arctan x = x - \frac{x^3}{3} + \frac{x^5}{5} - \cdots + (-1)^{n-1}\frac{x^{2n-1}}{2n-1} + (-1)^n \int_0^x \frac{t^{2n}}{1+t^2}dt,$$

since $1 + t^2 \geq 1$, the remainder can be estimated as follows:

$$|R_{2n+1}(x)| \leq \int_0^{|x|} \frac{t^{2n}}{1+t^2}dt \leq \int_0^{|x|} t^{2n}dt = \frac{|x|^{2n+1}}{2n+1}.$$

Futhermore, if $0 < c < 1$ and $-c \leq x \leq c$, then the remainder is bounded by $c^{2n+1}/(2n+1)$.

Exercise V.3.3 *Let f, g be polynomials of degrees $\leq d$. Let $a > 0$. Assume that there exists $C > 0$ such that for all x with $|x| \leq a$ we have*

$$|f(x) - g(x)| \leq C|x|^{d+1}.$$

Show that $f = g$. (Show first that if h is a polynomial of degree $\leq d$ such that $|h(x)| \leq C|x|^{d+1}$, then $h = 0$.)

Solution. Suppose $h(x) = a_m x^m + a_{m-1}x^{m-1} + \cdots + a_0$ with $m \leq d$. Letting $x = 0$ shows that $a_0 = 0$. We now continue by induction. Suppose $a_0 = a_1 = \cdots = a_{k-1} = 0$ for some $1 \leq k \leq m$. Then we have for $x \neq 0$

$$|h(x)/x^k| \leq C|x|^{d+1-k}.$$

Letting $x \to 0$ proves that $a_k = 0$. Therefore $a_0 = \cdots = a_m = 0$ so $h = 0$. Letting $h = f - g$ proves the initial claim.

Exercise 3 shows that the polynomials obtained in Exercises 1 and 2 actually are the same as those obtained from the Taylor formula.

Exercise V.3.4 *Let $a > 1$. Prove that*

$$\lim_{n \to \infty} \frac{a^n}{n!} = 0.$$

Conclude that the remainder terms in the Taylor expansions for the sine, cosine, and exponential function tend to 0 as n tends to infinity.

Solution. Let $u_n = a^n/n!$, and select an integer $N > 2a - 1$. For all $n \geq N$ we have

$$\frac{u_{n+1}}{u_n} = \frac{a}{n+1} < \frac{1}{2},$$

thus for all $k \geq 0$ we have $u_{N+k} \leq (1/2)^k u_N$, hence $\lim_{n\to\infty} a^n/n! = 0$. Clearly, this limit is also true when $a \leq 1$. Form the expressions given after the formula for some cosine and the exponential we deduce at once that the remainder of these functions tends to 0 as $n \to \infty$.

Exercise V.3.5 *(a) Prove that* $\log 2 = \log(4/3) + \log(3/2)$, *or even better,*

$$\log 2 = 7 \log \frac{10}{9} - 2 \log \frac{25}{24} + 3 \log \frac{81}{80}.$$

(b) Find a rational number approximating $\log 2$ *to five decimals, and prove that it does so. The above trick is much more efficient than the slowly convergent expression of* $\log 2$ *as the alternating series.*

Solution. (a) The first equality follows from the fact that $2 = (4/3)(3/2)$. For the second equality, note that the right-hand side is equal to

$$\log \left(\frac{10}{9}\right)^7 \left(\frac{24}{25}\right)^2 \left(\frac{81}{80}\right)^3 = \log \frac{2^7 2^6 3^2 3^{12} 5^7}{2^{12} 3^{14} 5^7} = \log 2.$$

(b) We see that $10/9$, $25/24$, and $81/80$ are all close to 1. We use Theorem 3.4 to estimate each term. Let $P_n(x) = x - (x^2/2) + \cdots + (-1)^{n-1} x^n/n$. Then for the first term we have

$$7 \log \left(\frac{10}{9}\right) = 7 \log \left(1 + \frac{1}{9}\right) = 7 P_5(1/9) + 7 R_6(1/9)$$

with $|7 R_6(1/9)| \leq (7/2) \times 10^{-6}$. Similarly

$$2 \log \left(\frac{25}{24}\right) = 2 \log \left(1 + \frac{1}{24}\right) = 2 P_3(1/24) + 2 R_4(1/24)$$

with $|2 R_4(1/24)| \leq 2 \times 10^{-6}$ and

$$3 \log \left(\frac{81}{80}\right) = 3 \log \left(1 + \frac{1}{80}\right) = 3 P_2(1/80) + 3 R_3(1/80)$$

with $|3 R_6(1/9)| \leq 2 \times 10^{-6}$. Since

$$|7 R_6(1/9) - 2 R_4(1/24) + 3 R_6(1/9)| \leq \left(\frac{7}{2} + 2 + 2\right) \times 10^{-6} < 10^{-5}$$

it follows that $7 P_5(1/9) - 2 P_3(1/24) + 3 P_2(1/80) = 0.69314$ is a rational number, giving the desired approximation of $\log 2$.

Exercise V.3.6 *(a) Prove that*

$$\arctan u + \arctan v = \arctan \frac{u+v}{1-uv}.$$

(b) Prove that $\pi/4 = \arctan 1 = \arctan(1/2) + \arctan(1/3)$.
(c) Find a rational number approximating $\pi/4$ to 3 decimals.
(d) You will do so even faster if you prove that

$$\frac{\pi}{4} = 4\arctan(1/5) - \arctan(1/239).$$

Solution. (a) The addition formula for the tangent, which follows from the addition formula for the sine and cosine is

$$\tan(x+y) = \frac{\tan x + \tan y}{1 - \tan x \tan y}.$$

Let $u = \tan x$ and $v = \tan y$, then

$$\arctan \frac{u+v}{1-uv} = \arctan(\tan(x+y)) = x+y = \arctan u + \arctan v.$$

(b) Let $u = 1/2$ and $v = 1/3$ in the formula obtained in (a).
(c) We use the same method of approximation as in Exercise 5. If $u = 1/2$, then

$$\arctan u = u - \frac{u^3}{3} + \frac{u^5}{5} - \frac{u^7}{7} + R_9(u) = A + R_9(u),$$

with $|R_9(u)| \le 3 \times 10^{-4}$. If $v = 1/3$, then

$$\arctan v = v - \frac{v^3}{3} + \frac{v^5}{5} + R_7(u) = B + R_9(u),$$

with $|R_7(v)| \le 10^{-4}$. So $\pi/4 = A + B + R_9(u) + R_7(v)$ and since

$$|R_9(u) + R_7(v)| \le 10^{-3}$$

we conclude that the rational number $A + B = 0.785$ gives the desired approximation.
(d) Using the formula in (a) repeatedly we find

$$\begin{aligned}
\arctan(1/5) + \arctan(1/5) &= \arctan(5/12), \\
\arctan(1/5) + \arctan(5/12) &= \arctan(37/55), \\
\arctan(1/5) + \arctan(37/55) &= \arctan(120/119),
\end{aligned}$$

so $4\arctan(1/5) = \arctan(120/119)$ and therefore

$$4\arctan(1/5) - \arctan(1/239) = \arctan(120/119) + \arctan(-1/139).$$

One more application of the formula in (a) and the result drops out. Proceeding as in part (c) we see that the following number is rational and approximates $\pi/4$ to three decimals,

$$\left(\frac{1}{5} - \frac{(1/5)^3}{3} + \frac{(1/5)^5}{5}\right) - \frac{1}{239}.$$

Exercise V.3.7 *Let $A > 0$, and consider an interval $0 < \delta \le x \le 2A - \delta$. Show that there exists a constant C, and for each positive integer n, there exists a polynomial P_n such that for all x in the interval, one has*

$$|\log(x) - P_n(x)| \le C/n.$$

[Hint: Write $x = A + (x - A)$ so that

$$\log x = \log A + \log\left(1 + \frac{x - A}{A}\right).\Big]$$

Solution. Let $c = (A - \delta)/A$. Then $0 \le c < 1$, and $-c \le (x - A)/A \le c$ if and only if $\delta \le x \le 2A - \delta$. The estimates for the remainder in Theorem 3.4 show that if $-c \le X \le c$, then

$$|R_n(X)| \le \frac{1}{n}\frac{c}{1 - c},$$

so if $Q_{n-1}(X) = X - (X^2/2) + \cdots + (-1)^{n-2}X^{n-1}/(n-1)$, the polynomials

$$P_n(x) = \log A + Q_{n-1}\left(\frac{x - A}{A}\right)$$

satisfy the desired property with $C = c/(1-c)$. Note, that we can take more terms in the expansion of the logarithm so that we get a better uniform approximation. However, here we have shown that we can choose P_n to be of degree $n - 1$.

V.4   Asymptotic Estimates and Stirling's Formula

Exercise V.4.1 *Integrating by parts, prove the following formulas.*
(a) $\int \sin^n x\,dx = -\frac{1}{n}\sin^{n-1}x\cos x + \frac{n-1}{n}\int \sin^{n-2}x\,dx.$
(b) $\int \cos^n x\,dx = \frac{1}{n}\cos^{n-1}x\sin x + \frac{n-1}{n}\int \cos^{n-2}x\,dx.$

Solution. (a) Integrating by parts we have

$$\int \sin^{n-1}x\sin x\,dx = -\sin^{n-1}x\cos x + (n - 1)\int \cos^2 x\sin^{n-2}x\,dx,$$

but $\cos^2 x\sin^{n-2}x = (1 - \sin^2 x)\sin^{n-2}x = \sin^{n-2}x - \sin^n x$ which implies the desired result.

(b) Since

$$\int \cos^{n-1} x \cos x dx = \cos^{n-1} x \sin x + (n-1) \int \sin^2 x \cos^{n-2} x dx$$

and $\sin^2 x = 1 - \cos^2 x$, the equality drops out.

Exercise V.4.2 *Prove the formulas, where n is a positive integer:*
(a) $\int (\log x)^n dx = x(\log x)^n - n \int (\log x)^{n-1} dx.$
(b) $\int x^n e^x dx = x^n e^x - n \int x^{n-1} e^x dx.$

Solution. (a) Let $g'(x) = 1$ and $f(x) = (\log x)^n$ in the integration by parts formula.
(b) Let $g'(x) = e^x$ and $f(x) = x^n$ in the integration by parts formula.

Exercise V.4.3 *By induction, find the value $\int_0^\infty x^n e^{-x} dx = n!$. The integral to infinity is defined to be*

$$\int_0^\infty f(x) dx = \lim_{B \to \infty} \int_0^B f(x) dx.$$

Solution. Let $I_n = \int_0^\infty x^n e^{-x} dx$. Using the same argument as in Exercise 2(b) we find that

$$\int_0^B x^n e^{-x} dx = -B^n e^{-B} + n \int_0^B x^{n-1} e^{-x} dx.$$

Letting $B \to \infty$ shows that $I_n = nI_{n-1}$. So by induction $I_n = n!I_1$. But since

$$\int_0^B x e^{-x} dx = -Be^{-B} + \int_0^B e^{-x} dx = -Be^{-B} - e^{-B} + 1,$$

so we get $I_1 = 1$.

Exercise V.4.4 *Show that the relation of being asymptotic, i.e. $f(x) \sim g(x)$ for $x \to \infty$, is an equivalence relation.*

Solution. Clearly, $f(x) \sim f(x)$ because for all large x we have $f(x) \neq 0$ and $f(x)/f(x) = 1$. If $f(x) \sim g(x)$, then $g(x) \sim f(x)$ because if $\lim_{x \to \infty} f(x)/g(x) = 1$, then

$$\lim_{x \to \infty} g(x)/f(x) = \lim_{x \to \infty} \frac{1}{f(x)/g(x)} = 1.$$

Finally, $f(x) \sim h(x)$ whenever $f(x) \sim g(x)$ and $g(x) \sim h(x)$ because

$$\lim_{x \to \infty} f(x)/h(x) = \lim_{x \to \infty} \frac{f(x)}{g(x)} \frac{g(x)}{h(x)} = 1.$$

Exercise V.4.5 *Let r be a positive integer. Prove that*

$$\int_2^x \frac{1}{(\log x)^r} dx = O\left(\frac{x}{(\log x)^r}\right) \quad \text{for } x \to \infty.$$

[Hint: Integrate between 2 and $\sqrt{x}$, and then between $\sqrt{x}$ and x.]

Solution. Let $f(t) = 1/(\log t)^r$. The function f is decreasing and positive for $x \geq 2$, so

$$\int_2^{\sqrt{x}} f(t)dt \leq f(2)(\sqrt{x} - 2) \leq f(2)\sqrt{x},$$

and

$$\int_{\sqrt{x}}^x f(t)dt \leq \frac{x - \sqrt{x}}{(\log \sqrt{x})^r} \leq \frac{x}{(\log \sqrt{x})^r}.$$

But we have

$$\frac{x}{(\log \sqrt{x})^r} = 2^r \frac{x}{(\log x)^r} = O\left(\frac{x}{(\log x)^r}\right),$$

for $x \to \infty$. It is therefore sufficient to show that $\sqrt{x} = O\left(x/(\log x)^r\right)$ for $x \to \infty$. But

$$\frac{\sqrt{x}}{x/(\log x)^r} = \frac{(\log x)^r}{\sqrt{x}} \to 0$$

as $x \to \infty$ as was shown in Exercise 5, §2, of Chapter IV. Therefore, $\sqrt{x} = O\left(x/(\log x)^r\right)$ and we are done.

Exercise V.4.6 *(a) Define*

$$\text{Li}(x) = \int_2^x \frac{1}{\log x} dx.$$

Prove that

$$\text{Li}(x) = \frac{x}{\log x} + O\left(\frac{x}{(\log x)^2}\right) \quad \text{so} \quad \text{Li}(x) \sim \frac{x}{\log x}.$$

(b) Let r be a positive integer, and let

$$\text{Li}_r(x) = \int_2^x \frac{1}{(\log x)^r} dx.$$

Prove that $\text{Li}_r(x) \sim x/(\log x)^r$ for $x \to \infty$. Better, prove that

$$\text{Li}_r(x) = \frac{x}{(\log x)^r} + O\left(\frac{x}{(\log x)^{r+1}}\right).$$

(c) Give an asymptotic expansion of $\text{Li}(x)$ for $x \to \infty$.

Solution. (a) Let $r = 1$ in (b).

(b) Integrating by parts we find that

$$\operatorname{Li}_r(x) = \int_2^x \frac{1}{(\log t)^r} dt = \left[\frac{t}{(\log t)^r} \right]_2^x + r \int_2^x \frac{1}{(\log t)^{r+1}} dt$$

$$= \frac{x}{(\log x)^r} + C_r + r \int_2^x \frac{1}{(\log t)^{r+1}} dt,$$

where C_r is a number. By Exercise 5, §2, of Chapter IV we have

$$C_r = O \left(\frac{x}{(\log x)^{r+1}} \right)$$

for $x \to \infty$. Hence Exercise 5 implies the asymptotic estimate of $\operatorname{Li}_r(x)$.

(c) Let $f_r(x) = (r-1)! x / (\log x)^r$. Induction and the first equation obtained in (b) imply

$$\operatorname{Li}(x) = f_1(x) + \cdots + f_n(x) + n! \operatorname{Li}_{n+1}(x) + O(1).$$

Clearly, $f_{r+1}(x) = o(f_r(x))$ and the estimate of part (b) gives

$$n! \operatorname{Li}_{n+1}(x) = o(f_n(x)).$$

Since $f_r(x) \to \infty$ as $x \to \infty$, the term $O(1)$ causes no problem and therefore $\{f_r\}$ gives an asymptotic estimate of the function Li for $x \to \infty$.

Exercise V.4.7 *Let*

$$L(x) = \sum_{2 \leq k \leq x} \frac{1}{\log k}.$$

Show that $L(x) = \operatorname{Li}(x) + O(1)$ *for* $x \to \infty$, *so in particular,* $L(x) \sim \operatorname{Li}(x)$.

Solution. This is a particular case of Exercise 8 with $f(t) = 1/\log t$. Note that the assumptions of Exercise 8 are satisfied because $t \geq \log t$ implies

$$\int_2^x \frac{dt}{\log t} \geq \int_2^x \frac{dt}{t} = \log x - \log 2,$$

hence $\lim_{x \to \infty} \int_2^x (1/\log t) dt = \infty$.

Exercise V.4.8 *More generally, let* f *be a positive function defined, say, for all* $x \geq 2$. *Assume that* f *is decreasing, and let*

$$F(x) = \int_2^x f(t) dt.$$

Assume that $F(x)$ *is unbounded, i.e.* $\lim F(x) = \infty$ *for* $x \to \infty$. *Show that*

$$F(x) \sim \sum_{2 \leq k \leq x} f(k) \quad \text{for } x \to \infty.$$

In fact, if we denote the sum on the right by $S_f(x)$, *show that*

$$F(x) = S_f(x) + O(1).$$

Solution. Let N be the largest integer $\leq x$. The usual estimate with the upper and lower Riemann sums leads to the inequalities

$$\sum_{k=3}^{N} f(k) \leq \int_2^x f(t)dt \leq \sum_{k=2}^{N} f(k)$$

so that $-f(2) \leq F(x) - S_f(x) \leq 0$. Hence $F(x) = S_f(x) + O(1)$ and since $F(x) \to \infty$ as $x \to \infty$ we have the relation $F(x) \sim S_f(x)$.

For Exercises 9, 10, and 11, we let

$$F(x) = \int_2^x f(t)dt.$$

Exercise V.4.9 *Let f and h be two positive continuous functions on $\mathbf{R}$. Assume that $\lim_{x \to \infty} h(x) = 0$. Assume that $F(x) \to \infty$ as $x \to \infty$. Show that*

$$\int_2^x f(t)h(t)dt = o(F(x)) \quad for \ x \to \infty.$$

Solution. Given $\epsilon > 0$ select x_0 such that $x > x_0$ implies $h(x) < \epsilon$. Then for $x > x_0$

$$\int_2^x fh = \int_2^{x_0} fh + \int_{x_0}^x fh \leq M + \epsilon \int_{x_0}^x f,$$

where $M = \int_2^{x_0} fh$. Consequently,

$$\frac{\int_2^x fh}{\int_2^x f} \leq \frac{M}{\int_2^x f} + \epsilon.$$

Since $\int_2^x f \to \infty$ as $x \to \infty$ we can make the expression on the right $< 2\epsilon$ for all x sufficiently large, thereby proving that $\int_2^x f(t)dt = o(F(x))$.

Exercise V.4.10 *Assume that f, h are continuous positive, that $f(x) \to \infty$, and that $h(x) \to 0$ as $x \to \infty$. Show that*

$$\int_2^x f(t)h(t)dt = o(F(x)) \quad for \ x \to \infty.$$

Solution. If $\lim_{x \to \infty} f(x) = \infty$, then $\lim_{x \to \infty} F(x) = \infty$ so we are reduced to Exercise 9.

Exercise V.4.11 *Suppose that f is monotone positive (increasing or decreasing) and that*

$$f(x) = o(F(x)) \quad for \ x \to \infty.$$

Prove that

$$f(x)^{1/2} = o\left(\int_2^x f(t)^{1/2}dt\right) \quad for \ x \to \infty.$$

Solution. If f is decreasing, then $f(x)^{1/2}(x-2) \le \int_2^x f(t)^{1/2}dt$ so

$$0 \le \frac{f(x)^{1/2}}{\int_2^x f(t)^{1/2}dt} \le \frac{1}{x-2},$$

hence $f(x)^{1/2} = o\left(\int_2^x f(t)^{1/2}dt\right)$ for $x \to \infty$, and this result holds without the assumption that $f(x) = o(F(x))$ for $x \to \infty$. If f is increasing, then for all $2 \le t \le x$ we have

$$f(t)^{1/2}f(t)^{1/2} \le f(x)^{1/2}f(t)^{1/2}$$

hence $F(x) \le f(x)^{1/2}G(x)$ where $G(x) = \int_2^x f(t)^{1/2}dt$. We can write

$$f(x)/F(x) = \epsilon(x)$$

where $\lim_{x \to \infty} \epsilon(x) = 0$. Multiplying the last inequality by $f(x)^{1/2}$ and dividing by $F(x)$ we get

$$f(x)^{1/2} \le \frac{f(x)}{F(x)}G(x) = \epsilon(x)G(x)$$

and therefore

$$\frac{f(x)^{1/2}}{G(x)} \le \epsilon(x).$$

So in all cases we have the result $f(x)^{1/2} = o\left(\int_2^x f(t)^{1/2}dt\right)$ for $x \to \infty$.

VI
Normed Vector Spaces

VI.2 Normed Vector Spaces

Exercise VI.2.1 *Let S be a set. By a **distance function** on S one means a function $d(x, y)$ of pairs of elements of S, with values in the real numbers, satisfying the following conditions:*

$d(x, y) \geq 0$ *for all* $x, y \in S$, *and* $= 0$ *if and only if* $x = y$.

$d(x, y) = d(y, x)$ *for all* $x, y \in S$.

$d(x, y) \leq d(x, z) + d(z, y)$ *for all* $x, y, z \in S$.

Let E be a normed vector space. Define $d(x, y) = |x - y|$ for $x, y \in E$. Show that this is a distance function.

Solution. Condition **N 1** of the norm implies the first condition of the distance function. Setting $a = -1$ in **N 2** implies the second condition. Finally the triangle inequality for the norm implies

$$|x - y| = |x - z + z - y| \leq |x - z| + |z - y|$$

which is the third condition of the distance function.

Exercise VI.2.2 *(a) A set S with a distance function is called a **metric space**. We say that it is a **bounded** metric if there exists a number $C > 0$ such that $d(x, y) \leq C$ for all $x, y \in S$. Let S be a metric space with an arbitrary distance function. Let $x_0 \in S$. Let $r > 0$. Let S_r consist of all $x \in S$ such that $d(x, x_0) < r$. Show that the distance function of S defines a bounded metric on S_r.*

(b) Let S be a set with a distance function d. Define another function d' on S by $d'(x,y) = \min(1, d(x,y))$. Show that d' is a distance function, which is a bounded metric.

(c) Define

$$d''(x,y) = \frac{d(x,y)}{1 + d(x,y)}.$$

Show that d'' is a bounded metric.

Solution. (a) For all $x, y \in S_r$ we have

$$d(x,y) \le d(x, x_0) + d(x_0, y) < 2r.$$

(b) Clearly, $d'(x,y) \ge 0$ with equality if and only if $d(x,y) = 0$. The second condition is verified because $d(x,y) = d(y,x)$. To verify the triangle inequality consider separate cases: suppose $d(x,y) < 1$. Then since d is a distance, the result follows. If $d(x,y) \ge 1$, then $d'(x,y) = 1$ and $1 \le d(x,z) + d(z,y)$, so the result follows. This distance function is a bounded metric because we have $d'(x,y) \le 1$.

(c) We have $d''(x,y) = 0$ if and only if $d(x,y) = 0$ so the first property is verified. The second property $d''(x,y) = d''(y,x)$ holds because $d(x,y) = d(y,x)$. To prove the triangle inquality it suffices to show that

$$\frac{a}{1+a} \le \frac{b}{1+b} + \frac{c}{1+c}$$

whenever $a, b, c \ge 0$ and $a \le b + c$. But the above inequality is equivalent to

$$a \le b + c + abc + ab + bc$$

which is true. This implies that d'' is a distance function, and d'' is bounded by 1 because $d(x,y) \le 1 + d(x,y)$.

Exercise VI.2.3 *Let S be a metric space. For each $x \in S$, define the function $f_x : S \to \mathbf{R}$ by the formula*

$$f_x(y) = d(x, y).$$

(a) Given two points x, a in S show that $f_x - f_a$ is a bounded function on S.

(b) Show that $d(x,y) = \|f_x - f_y\|$.

(c) Fix an element a of S. Let $g_x = f_x - f_a$. Show that the map

$$x \mapsto g_x$$

*is a distance preserving embedding (i.e. injective map) of S into the normed vector space of bounded functions on S, with the **sup norm**. [If the metric on S originally was bounded, you can use f_x instead of g_x.] This exercise shows that the generality of metric spaces is illusory. In applications, metric spaces usually arise naturally as subsets of normed vector spaces.*

Solution. (a) The triangle inequality immediately implies

$$-d(a,x) \le f_x(y) - f_a(y) = d(x,y) - d(a,y) \le d(a,x).$$

(b) By (a) we know that $f_x - f_y$ is bounded and that

$$|f_x(z) - f_y(z)| \le d(y,x).$$

Letting $z = x$ or $z = y$ we see that $\|f_x - f_y\| = d(y,x)$.

(c) Let φ be the given map, that is $\varphi(x) = g_x$. The function φ is injective because if $\varphi(x_1) = \varphi(x_2)$, then $g_{x_1}(y) = g_{x_2}(y)$ for all y, and putting $y = x_1$ we get $0 = d(x_1, x_1) = d(x_1, x_2)$ hence $x_1 = x_2$. Furthermore, g_x is bounded and $\|\varphi(x) - \varphi(y)\| = \|g_x - g_y\| = d(x,y)$.

Exercise VI.2.4 *Let $|\cdot|$ be a norm on a vector space E. Let a be a number > 0. Show that the function $x \mapsto a|x|$ is also a norm on E.*

Solution. We have $a > 0$ so $a|x| \ge 0$ and $= 0$ if and only if $|x| = 0$. The second axiom holds because $a|bx| = a|b||x| = |b|a|x|$. Finally the triangle inequality follows from

$$a|x+y| \le a(|x| + |y|) \le a|x| + a|y|.$$

Exercise VI.2.5 *Let $|\cdot|_1$ and $|\cdot|_2$ be norms on E. Show that the functions $x \mapsto |x|_1 + |x|_2$ and $x \mapsto \max(|x|_1, |x|_2)$ are norms on E.*

Solution. (i) We have $|x|_1 + |x|_2 \ge 0$ and $= 0$ if and only if $|x|_1 = |x|_2 = 0$. Clearly, $|ax|_1 + |ax|_2 = a(|x|_1 + |x|_2)$. Finally, the triangle inequality

$$|x+y|_1 + |x+y|_2 \le |x|_1 + |y|_1 + |x|_2 + |y|_2.$$

(ii) For the second function only the triangle inequality requires a little work:

$$\begin{aligned}
\max(|x+y|_1, |x+y|_2) &\le \max(|x|_1 + |y|_1, |x|_2 + |y|_2) \\
&\le \max(|x|_1, |x|_2) + \max(|y|_1, |y|_2).
\end{aligned}$$

Exercise VI.2.6 *Let E be a vector space. By a **seminorm** on E one means a function $\sigma\colon E \to \mathbf{R}$ such that $\sigma(x) \ge 0$ for all $x \in E$, $\sigma(x+y) \le \sigma(x) + \sigma(y)$, and $\sigma(cx) = |c|\sigma(x)$ for all $c \in \mathbf{R}$, $x, y \in E$.*
(a) Let σ_1, σ_2 be seminorms. Show that $\sigma_1 + \sigma_2$ is a seminorm. If λ_1, λ_2 are numbers ≥ 0, show that $\lambda_1\sigma_1 + \lambda_2\sigma_2$ is a seminorm. By induction show that if $\sigma_1, \ldots, \sigma_n$ are seminorms and $\lambda_1, \ldots, \lambda_n$ are numbers ≥ 0, then $\lambda_1\sigma_1 + \cdots + \lambda_n\sigma_n$ is a seminorm.
(b) Let $\sigma = \max(\sigma_1, \sigma_2)$. Show that σ is a seminorm.

Solution. (a) The function $\sigma_1 + \sigma_2$ is a seminorm. Indeed, $\sigma_1(x) + \sigma_2(x) \ge 0$ and the other axioms follow from Exercise 5. The function $\lambda_1\sigma_1 + \lambda_2\sigma_2$ is a seminorm. Indeed, we have $\lambda_1 \sigma_1(x) \geq 0$ and $\lambda_2 \sigma_2(x) \geq 0$, thus by Exercise 4, we see that $\lambda_1 \sigma_1$ and $\lambda_2 \sigma_2$ are seminorms. By the first part of this exercise we conclude that $\lambda_1 \sigma_1 + \lambda_2 \sigma_2$ is seminorm. We prove the last statement by induction. If $\lambda_1 \sigma_1 + \cdots + \lambda_m \sigma_m$ is a seminorm with $0 \leq m < n$, then since $\lambda_{m+1} \sigma_{m+1}$ is a seminorm, we conclude that

$$\lambda_1 \sigma_1 + \cdots + \lambda_m \sigma_m + \lambda_{m+1} \sigma_{m+1}$$

is also a seminorm.
(b) The fact that $\max(\sigma_1, \sigma_2)$ is a seminorm follows from Exercise 5.

Exercise VI.2.7 *Let σ_1 be a norm and σ_2 a seminorm on a vector space. Show that $\sigma_1 + \sigma_2$ and $\max(\sigma_1, \sigma_2)$ are norms.*

Solution. From Exercise 6, we know that $\sigma_1 + \sigma_2$ and $\max(\sigma_1, \sigma_2)$ are seminorms. It is therefore sufficient to show that **N 1** holds in both cases. In the first case, $\sigma_1(x) + \sigma_2(x) = 0$ if and only if $\sigma_1(x) = \sigma_2(x) = 0$ and since σ_1 is a norm, we get the desired conclusion. In the second case, we have $\max(\sigma_1(x), \sigma_2(x)) = 0$ if and only if $\sigma_1(x) = \sigma_2(x) = 0$. Conclude.

Exercise VI.2.8 *Let σ be a seminorm on a vector space E. Show that the set of all $x \in E$ such that $\sigma(x) = 0$ is a subspace.*

Solution. Let $F = \{x \in E : \sigma(x) = 0\}$. Axiom 2 implies $\sigma(0 \cdot x) = 0 \cdot \sigma(x) = 0$ so $0 \in F$. If $x_1, x_2 \in F$, then $\sigma(x_1 + x_2) \leq \sigma(x_1) + \sigma(x_2) = 0$ so $x_1 + x_2 \in F$. Finally, if $x \in F$ we have $\sigma(ax) = |a|\sigma(x) = 0$ for all $x \in \mathbf{R}$, thereby proving that F is a subspace of E.

Exercise VI.2.9 (The C^p Seminorms) *Let p be an integer ≥ 0. Let $E = C^p([0,1])$ be the space of p-times continuously differentiable functions on $[0,1]$. Define σ_p and N_p by*

$$\sigma_p(f) = \sup_x |f^{(p)}(x)| \quad and \quad N_p(f) = \max_{0 \leq r \leq p} \sigma_r(f),$$

where the maximum is taken for $r = 0, \ldots, p$.
(a) Show that σ_p is a seminorm and N_p is a norm. Note that $\sigma_0 = N_0$ is just the ordinary sup norm. The norm N_p is called the C^p-norm.
(b) Describe the subspace of E consisting of those functions f such that $\sigma_p(f) = 0$ for $p = 0$ and also for $p > 0$. This is the subspace of Exercise 8.

Solution. (a) Clearly, we have $\sigma_p(f) \geq 0$ for all $f \in E$. We also have $\sigma_p(cf) = \sup |cf^{(p)}(x)| = |c|\sigma_p(f)$. Finally the triangle inequality holds because

$$\sigma_p(f_1 + f_2) \leq \sup(|f_1^{(p)}(x)| + |f_2^{(p)}(x)|) \leq \sup |f_1^{(p)}(x)| + \sup |f_2^{(p)}(x)|.$$

This proves that σ_p is a seminorm. Exercise 7 and an induction argument show that N_p is a seminorm. Suppose $N_p(f) = 0$, then $\sigma_0(f) = 0$ and since σ_0 is the sup norm we conclude that $f = 0$, hence N_p is a norm.

(b) If $p = 0$, then σ_0 is simply the sup norm, so E is reduced to the zero function. Suppose $p > 0$ and $\sigma_p(f) = 0$. Then $f^{(p)}(x) = 0$ for all $x \in [0,1]$. Integrating p-times we find that f is a polynomial of degree $\leq p - 1$. Conversely, all polynomials of degree $\leq p - 1$ have σ_p-seminorm equal to 0, hence E is the set of all polynomials of degree $\leq p - 1$.

Exercise VI.2.10 *Consider a scalar product on a vector space E which instead of satisfying* **SP 4** *(that is positive definiteness) satisfies the weaker condition that we only have $\langle v, v \rangle \geq 0$ for all $v \in E$. Let $w \in E$ be such that $\langle w, w \rangle = 0$. Show that $\langle w, v \rangle = 0$ for all $v \in E$. [Hint: Consider $\langle v + tw, v + tw \rangle \geq 0$ for large positive or negative values of t.]*

Solution. Suppose that for some $v_0 \in E$ we have $\langle w, v_0 \rangle \neq 0$. We then have
$$0 \leq \langle v_0 + tw, v_0 + tw \rangle = \langle v_0, v_0 \rangle + 2t\langle w, v_0 \rangle.$$

If $\langle w, v_0 \rangle > 0$ we see that for large negative values of t we get a contradiction. If $\langle w, v_0 \rangle < 0$ we get a contradiction for large positive values of t.

Exercise VI.2.11 *Notation as in the preceding exercise, show that the function*
$$w \mapsto \|w\| = \sqrt{\langle w, w \rangle}$$
is a seminorm, by proving the Schwarz inequality just as was done in the text.

Solution. We have $\|w\| \geq 0$ and $\|cw\| = \sqrt{c^2 \langle w, w \rangle} = |c|\|w\|$. To get the triangle inequality is suffices (by Theorem 2.2) to prove the Schwarz inequality. Suppose $\langle w, w \rangle \neq 0$, then we can divide by $\langle w, w \rangle$ and the desired inequality holds. If $\langle w, w \rangle = 0$, then $\langle v, w \rangle = 0$ (by the preceding exercise) and we see that the inequality still holds. This argument, together with the proof of the Schwarz inequality given in the text, concludes the exercise.

Exercise VI.2.12 *Let E be a vector space with a postive definite scalar product, and the corresponding norm $\|v\| = \sqrt{v \cdot v}$. Prove the* **parallelogram law** *for all $v, w \in E$:*
$$\|v + w\|^2 + \|v - w\|^2 = 2\|v\|^2 + 2\|w\|^2.$$

Draw a picture illustrating the law. For a follow up, see §4, Exercises 5, 6, and 7.

Solution. We prove the parallelogram law,
$$
\begin{aligned}
\|v + w\|^2 + \|v - w\|^2 &= \langle v + w, v + w \rangle + \langle v - w, v - w \rangle \\
&= \langle v, v \rangle + 2\langle v, w \rangle + \langle w, w \rangle + \langle v, v \rangle - 2\langle v, w \rangle + \langle w, w \rangle \\
&= 2\|v\|^2 + 2\|w\|^2.
\end{aligned}
$$

The parallelogram law is illustrated by the following picture:

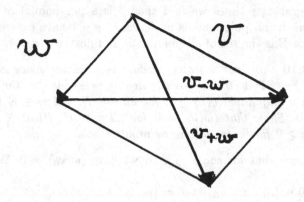

VI.3 n-Space and Function Spaces

Exercise VI.3.1 *Let E, F be normed vector spaces, with norms denoted by $|\cdot|$. Let $E \times F$ be the set of all pairs (x, y) with $x \in E$ and $y \in F$. Define addition componentwise:*

$$(x, y) + (x', y') = (x + x', y + y'),$$
$$c(x, y) = (cx, cy),$$

for $c \in \mathbf{R}$. Show that $E \times F$ is a vector space. Define

$$|(x, y)| = \max(|x|, |y|).$$

Show that this is a norm on $E \times F$. Generalize to n-factors, i.e. if $E_1, \ldots, E_n$ are normed vector spaces, define a similar norm on $E_1 \times \cdots \times E_n$ (the set of n-tuples $(x_1, \ldots, x_n)$ with $x_i \in E_i$).

Solution. One verifies without difficulty that all the axioms of a vector space hold for $E \times F$ with the operations defined above. The 0 element of $E \times F$ is given by $(0, 0)$.

We check that $|(x, y)|$ defines a norm on $E \times F$. If $|(x, y)| = 0$, then $|x| = 0$ and $|y| = 0$ so $x = y = 0$ and therefore $(x, y) = 0$. Conversely, $|(0, 0)| = 0$. Also we have

$$|c(x, y)| = |(cx, cy)| = \max(|cx|, |cy|) = |c| \max(|x|, |y|) = |c||(x, y)|.$$

Finally, the triangle inequality holds because

$$|(x_1, y_1) + (x_2, y_2)| = \max(|x_1 + x_2|, |y_1 + y_2|) \leq \max(|x_1| + |x_2|, |y_1| + |y_2|)$$
$$\leq \max(|x_1|, |y_1|) + \max(|x_2|, |y_2|) = |(x_1, y_1)| + |(x_2, y_2)|.$$

In the general case we simply set

$$|(x_1, \ldots, x_n)| = \max(|x_1|, \ldots, |x_n|).$$

Exercise VI.3.2 *Let $E = \mathbf{R}^n$, and for $A = (a_1, \ldots, a_n)$ define*

$$\|A\| = \sum_{i=1}^{n} |a_i|.$$

Show that this defines a norm. Prove directly that it is equivalent to the sup norm.

Solution. Obviously, $\|A\| \geq 0$ and if $\|A\| = 0$, then $|a_i| = 0$ for all $1 \leq i \leq n$ so that $A = 0$. Conversely, if $A = 0$, then we see at once that $\|A\| = 0$. For the second property we have

$$\|cA\| = \sum_{i=1}^{n} |ca_i| = \sum_{i=1}^{n} |c||a_i| = |c|\|A\|.$$

For the triangle inequality we have

$$\|A + B\| = \sum_{i=1}^{n} |a_i + b_i| \leq \sum_{i=1}^{n} |a_i| + \sum_{i=1}^{n} |b_i| = \|A\| + \|B\|.$$

The following inequalities show why this norm and the sup norm are equivalent

$$\max_{1 \leq i \leq n} |a_i| \leq \sum_{i=1}^{n} |a_i| \leq n \max_{1 \leq i \leq n} |a_i|.$$

Exercise VI.3.3 *Using properties of the integral, prove in detail that the symbol $\langle f, g \rangle$ defined by means of the integral is in fact a positive definite scalar product in the space of continuous functions on $[0, 1]$.*

Solution. See Exercise 4, §2, Chapter V.

Exercise VI.3.4 *Let E be the vector space of continuous functions on $[0, 1]$.*
(a) Show that the L^1-norm is indeed a norm on E.
(b) Show that the L^1-norm is not equivalent to the sup norm.
(c) Show that the L^1-norm is not equivalent to the L^2-norm. [Hint: Truncate the function $1/\sqrt{x}$ near 0.]
(d) Show that $\|f\|_1 \leq \|f\|_2$ for $f \in E$. [Hint: Use the Schwarz inequality.]

Solution. (a) Theorems 2.1 and 2.4 of Chapter V imply the first axiom. The second axiom holds because

$$\|cf\|_1 = \int_0^1 |cf(x)|dx = \int_0^1 |c||f(x)|dx = |c|\|f\|_1.$$

Finally, the triangle inequality holds because

$$\|f+g\|_1 = \int_0^1 |f(x)+g(x)|dx \le \int_0^1 |f(x)|dx + \int_0^1 |g(x)|dx = \|f\|_1 + \|g\|_1.$$

(b) Let $f_n(x) = x^n$. Then $\|f_n\| = 1$ (sup norm) but $\|f_n\|_1 = 1/(n+1)$, so we cannot find a constant C such that $\|f_n\| \le C\|f_n\|_1$ for all n.

(c) For each positive integer n, define a function g_n by $g_n(x) = \sqrt{n}$ if $x \in [0, 1/n]$ and $g_n(x) = 1/\sqrt{x}$ if $x \in [1/n, 1]$. Then we have

$$\|g_n\|_1 = \frac{\sqrt{n}}{n} + \int_{1/n}^1 \frac{1}{\sqrt{x}}dx = 2 - \frac{1}{\sqrt{n}},$$

and

$$\|g_n\|_2^2 = \frac{n}{n} + \int_{1/n}^1 \frac{1}{x}dx = 1 + \log n.$$

So we cannot find a constant C such that $\|g_n\|_2 \le C\|g_n\|_1$ for all n.

(d) Putting $v = |f|$ and $w = 1$ in the Schwarz inequality applied to the positive definite scalar product given in Exercise 3 we find

$$\int_0^1 |f(t)|dt \le \left(\int_0^1 |f(t)|^2 dt\right)^{1/2} \left(\int_0^1 1dt\right)^{1/2},$$

in other words, $\|f\|_1 \le \|f\|_2$.

Exercise VI.3.5 *Let E be a finite dimensional vector space. Show that the sup norms with respect to two different bases are equivalent.*

Solution. Let n be the dimension of E, let v be a vector in E, and $X_1(v) = (x_1, \ldots, x_n)$ and $X_2(v) = (y_1, \ldots, y_n)$ be the coordinate vectors of v in bases 1 and 2, respectively. Looking at the coordinates of the vectors of one basis in the other bases we see that there exists an invertible $n \times n$ matrix $M = (m_{ij})_{1 \le i, j \le n}$ such that $X_1(v) = MX_2(v)$ for all $v \in E$. Let $C_1 = \max_{1 \le i, j \le n} |m_{ij}| \ne 0$ and let $\| \cdot \|_k$ be the sup norm with respect to base k ($k = 1, 2$). Then for all $1 \le i \le n$ we have

$$|y_i| = \left|\sum_{j=1}^n m_{ij}x_j\right| \le \sum_{j=1}^n |m_{ij}||x_j| \le nC_1\|v\|_1.$$

Thus $\|v\|_2 \le nC_1\|v\|_1$. By symmetry, there exist $C_2 \ne 0$ such that $\|v\|_1 \le nC_2\|v\|_1$. This shows that $\| \cdot \|_1$ and $\| \cdot \|_2$ are equivalent. Of course one could simply apply Theorem 4.3 of the next section.

Exercise VI.3.6 *Give an example of a vector space with two norms, and a subset S of the vector space such that S is bounded for one norm but not for the other.*

Solution. Consider the vector space of C^1 functions on $[0,1]$. Let S be the set of all monomials, i.e. $S = \{1, x, x^2, \ldots, x^n, \ldots\}$. Then the sup norm of any element in S is 1, but the C^1-norm (see Exercise 9, §2) of $x \mapsto x^n$ is equal to n.

VI.4   Completeness

Exercise VI.4.1 *Give an example of a sequence in $C^0([0,1])$ which is L^2-Cauchy but not sup norm Cauchy. Is this sequence L^1-Cauchy? If it is, can you construct a sequence which is L^2-Cauchy but not L^1-Cauchy? Why?*

Solution. Let $f_n(x) = n^{1/4}x^n$. Then $|f_n(1) - f_m(1)| = |n^{1/4} - m^{1/4}|$ so $\{f_n\}$ is not sup norm Cauchy, but this sequence is L^2-Cauchy because

$$\|f_n\|_2^2 = \int_0^1 f_n^2 = \int_0^1 n^{1/2}x^{2n}dx = \frac{n^{1/2}}{2n+1} \le \frac{1}{n^{1/2}},$$

so

$$\|f_n - f_m\|_2 \le \|f_n\|_2 + \|f_m\|_2 \le \frac{1}{n^{1/4}} + \frac{1}{m^{1/4}},$$

and $1/n^{1/4} + 1/m^{1/4} \to 0$ as $n, m \to \infty$. The sequence $\{f_n\}$ is also L^1-Cauchy because of the inequality deduced in part (d) of Exercise 4, §3, namely, $\|f\|_1 \le \|f\|_2$. This inequality also shows that any L^2-Cauchy sequence in $C^0([0,1])$ is also L^1-Cauchy.

Exercise VI.4.2 *Let $f_n(x) = x^n$, and view $\{f_n\}$ as a sequence in $C^0([0,1])$. Show that $\{f_n\}$ approaches 0 in the L^1-norm and the L^2-norm, but not in the sup norm.*

Solution. We have

$$\|f_n\|_1 = \int_0^1 x^n dx = \frac{1}{n+1} \quad \text{and} \quad \|f_n\|_2^2 = \int_0^1 x^{2n}dx = \frac{1}{2n+1},$$

hence $\{f_n\}$ approaches 0 in the L^1-norm and in the L^2-norm. However, the sup norm of each f_n is 1.

Exercise VI.4.3 *Let $|\cdot|_1$ and $|\cdot|_2$ be two equivalent norms on a vector space E. Prove that limits, Cauchy sequences, convergent sequences are the same for both norms. For instance, a sequence $\{x_n\}$ has the limit v for one norm if and only if it has the limit v for the other norm.*

Solution. There exist positive numbers C_1 and C_2 such that for all $w \in E$ we have $C_1|w|_1 \leq |w|_2 \leq C_2|w|_1$. Suppose that $\{x_n\}$ converges to v in $|\cdot|_1$. Then given $\epsilon > 0$ there exists a positive integer N such that for all $n > N$ we have $|v-x_n|_1 < \epsilon$. Then for all $n > N$ we have $|v-x_n|_2 \leq C_2|v-x_n|_1 < C_2\epsilon$, which shows that $\{x_n\}$ converges to v in $|\cdot|_2$.

Suppose $\{x_n\}$ is a Cauchy sequence in the norm $|\cdot|_1$. Then given $\epsilon > 0$ there exists a positive integer N such that if $n, m > N$ we have $|x_n-x_m|_1 < \epsilon$. So for all $n, m > N$ we have $|x_n - x_m|_2 \leq C_2|x_n - x_m|_1 < C_2\epsilon$, which proves that $\{x_n\}$ is Cauchy for the norm $\|\cdot\|_1$.

Exercise VI.4.4 *Show that the space* $C^0([0,1])$ *is not complete for the* L^2*-norm.*

Solution. Consider the family of functions $\{f_n\}$ in $C^0([0,1])$ defined by

$$f_n(x) = \begin{cases} x^{-1/4} & \text{if } 1/n \leq x \leq 1 \\ n^{1/4} & \text{if } 0 \leq x \leq 1/n \end{cases}$$

The proof follows the one given in the text with $f(x) = x^{-1/4}$ for $0 < x \leq 1$. We have $\int_0^1 f^2 = 2$ and

$$\|f - f_n\|_2^2 = \int_0^{1/n} (x^{-1/4} - n^{1/4})^2 dx \leq \int_0^{1/n} x^{-1/2} + n^{1/2} dx = \frac{3}{\sqrt{n}},$$

hence

$$\|f_n - f_m\|_2 \leq \|f_n - f\|_2 + \|f - f_m\|_2 \leq \frac{\sqrt{3}}{n^{1/4}} + \frac{\sqrt{3}}{m^{1/4}},$$

which shows that $\{f_n\}$ is L^2-Cauchy. We now show that there is no continuous function g on $[0,1]$ such that $\{f_n\}$ is L^2-convergent to g. Suppose such a function g exists. Then

$$\|f - g\|_2 \leq \|f - f_n\| + \|f_n - g\|_2 \to 0$$

as $n \to \infty$. Hence $\|f-g\|_2 = 0$. Since both functions f and g are continuous on the open interval $(0, 1]$, if for some c we have $f(c) \neq g(c)$, then $\|f-g\|_2 > 0$. Since $f \notin C^0([0,1])$ we get a contradiction, which completes the proof that $\{f_n\}$ has no L^2-limit in $C^0([0,1])$.

Almost all the time it has been appropriate to deal with subsets of normed vector spaces. However, for the following exercises, it is clearer to formulate them in terms of metric spaces, as in Exercises 1 and 2 of §2. The notion of **Cauchy sequence** can be defined just as we did in the text, and a metric space X is said to be **complete** if every Cauchy sequence in X converges. We denote the distance between two points by $d(x_1, x_2)$.

Exercise VI.4.5 (The Semiparallelogram Law) *Let X be a complete metric space. We say that X satisfies the* **semiparallelogram law** *if for any two points x_1, x_2 in X, there is a point z such that for all $x \in X$ we have*

$$d(x_1, x_2)^2 + 4d(x, z)^2 \le 2d(x, x_1)^2 + 2d(x, x_2)^2.$$

(a) Prove that $d(z, x_1) = d(z, x_2) = d(x_1, x_2)/2$. [Hint: Substitute $x = x_1$ and $x = x_2$ in the law for one inequality. Use the triangle inequality for the other.] Draw a picture of the law when there is equality instead of an inequality.

(b) Prove that the point z is uniquely determined, i.e. if z' is another point satisfying the semiparallelogram law, then $z = z'$.

In light of (a) and (b), one calls z the **midpoint** *of x_1, x_2.*

Solution. Letting $x = x_1$ and then $x = x_2$ in the law, and taking square roots, we get

$$2d(x_1, z) \le d(x_1, x_2) \quad \text{and} \quad 2d(x_2, z) \le d(x_1, x_2). \qquad \text{(VI.1)}$$

The triangle inequality $d(x_1, x_2) \le d(x_1, z) + d(x_2, z)$ and the first inequality in (VI.1) implies $2d(x_1, z) \le d(x_1, z) + d(x_2, z)$ so $d(x_1, z) \le d(x_2, z)$ and by a similar argument we obtain $d(x_2, z) \le d(x_1, z)$, and therefore $d(x_1, z) = d(x_2, z)$. The triangle inequality now reads $d(x_1, x_2) \le 2d(x_1, z)$ so together with (VI.1) we get

$$2d(x_1, z) = 2d(x_2, z) = d(x_1, x_2),$$

as was to be shown.

With equality, the semiparallelogram law becomes the parallelogram law, and we have the following picture:

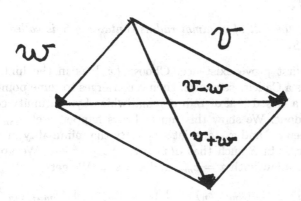

(b) Suppose z' also satisfies the semiparallelogram law. Then by (a) we must have

$$d(x_1, z') = d(x_2, z') = \frac{1}{2}d(x_1, x_2).$$

If we put $x = z'$ in the law and use the above inequalities we get

$$d(x_1, x_2)^2 + 4d(z, z')^2 \leq d(x_1, x_2)^2,$$

whence $d(z, z') = 0$, and therefore $z = z'$.

Exercise VI.4.6 (Bruhat-Tits-Serre) *Let X be a complete metric space, and let S be a bounded subset. Then S is contained in some closed ball $\overline{\mathbf{B}}_R(x)$ of some radius R and center $x \in X$. Define r (depending on S) to be the inf of all such radii R with all possible centers x. By definition, there exists a sequence $\{r_n\}$ of numbers $\geq r$ such that $\lim_{n \to \infty} r_n = r$, together with a sequence of balls $\overline{\mathbf{B}}_{r_n}(x_n)$ of centers x_n, such that $\overline{\mathbf{B}}_{r_n}(x_n)$ contains S. In general, it is not true that there exists a ball $\overline{\mathbf{B}}_r(x)$ with radius precisely r and some center x, containing S. If such a ball exists, it is called a **ball of minimal radius** containing S. Prove the following theorem:*

Let X be a complete metric space satisfying the semiparallelogram law. Let S be a bounded subset. Then there exists a unique closed ball $\overline{\mathbf{B}}_r(x_1)$ of minimal radius containing S.

[Hint: You have to prove two things: existence and uniqueness. Use the semiparallelogram law to prove each one. For existence, let $\{x_n\}$ be a sequence of points which are centers of balls of radius r_n approaching r, and $\overline{\mathbf{B}}_{r_n}(x_n)$ contains S. Prove that $\{x_n\}$ is a Cauchy sequence. Let c be its limit. Show that $\overline{\mathbf{B}}_r(c)$ contains S.

For uniqueness, again use the semiparallelogram law. Let $\overline{\mathbf{B}}_r(x_1)$ and $\overline{\mathbf{B}}_r(x_2)$ be balls of minimal radius centered at x_1, x_2. Let z be the midpoint, and use the fact that given ϵ, there exists an element $x \in S$ such that $d(x, z) \geq r - \epsilon.]$

*The center of the ball of minimal radius containing S is called the **circumcenter** of S.*

Solution. We first prove existence. Choose $\{x_n\}$ as in the hint. If the sequence $\{x_n\}$ is a Cauchy sequence, then it converges to some point which is the center of a closed ball or radius r and which by continuity contains S, and we are done. We show this must always happen. Let z_{mn} be the midpoint between x_n and x_m and let $\epsilon > 0$. By the minimality of r, there exists a point x_{nm} in S such that $d(x_{mn}, z_{mn})^2 \geq r^2 - \epsilon$. We apply the semiparallelogram law with $x = x_{mn}$ and $z = z_{mn}$. We get

$$\begin{aligned} d(x_m, x_n)^2 &\leq 2d(x_{mn}, x_n)^2 + 2d(x_{mn}, x_m)^2 - 4d(x_{mn}, z_{mn})^2 \\ &\leq 2r_n^2 + 2r_m^2 - 4r^2 + 4\epsilon, \end{aligned}$$

so for all large m and n we have $d(x_m, x_n)^2 \leq 5\epsilon$, thus proving that $\{x_n\}$ is a Cauchy sequence, and concluding the proof of existence.

We now prove uniqueness. Suppose there are two balls $\overline{\mathbf{B}}_r(x_1)$ and $\overline{\mathbf{B}}_r(x_2)$ of minimal radius containing S and $x_1 \neq x_2$. Let x be any point of S, so $d(x, x_1) \leq r$ and $d(x, x_2) \leq r$. Let z be the midpoint between x_1 and x_2. By the semiparallelogram law, we have

$$d(x_1, x_2)^2 \leq 4r^2 - 4d(x, z)^2.$$

Since by the definition of r, there are points x in S such that $d(x, z)$ is arbitrarily close to r, it follows that $d(x_1, x_2) = 0$ and therefore $x_1 = x_2$.

Let X be a metric space. By an **isometry** *of X we mean a bijection*

$$g \colon X \to X$$

such that g preserves distances. In other words, for all $x_1, x_2 \in X$ we have

$$d(g(x_1), g(x_2)) = d(x_1, x_2).$$

Note that if g_1, g_2 are isometries, so is the composite $g_1 \circ g_2$. Also if g is an isometry, then g has an inverse mapping (because g is a bijection), and the isometry condition immediately shows that $g^{-1} \colon X \to X$ is also an isometry. Note that the identity mapping $\mathrm{id} \colon X \to X$ is an isometry.

Let G be a set of isometries. We say that G is a **group** *of isometries if G contains the identity mapping, G is closed under composition (that is, if $g_1, g_2 \in G$, then $g_1 \circ g_2 \in G$), and is closed under inverse (that is, if $g \in G$ then $g^{-1} \in G$). One often writes $g_1 g_2$ instead of $g_1 \circ g_2$. Note that the set of all isometries is itself a group of isometries.*

Let $x' \in X$. The subset Gx' consisting of all elements $g(x')$ with $g \in G$ is called the **orbit** *of x' under G. Let S denote this orbit. Then for all $g \in G$ and all elements $x \in S$ it follows that $gx \in S$. Indeed, we can write $x = g_1 x'$ for some $g_1 \in G$, and then*

$$g(g_1 x') = g(g_1(x')) = (g \circ g_1)(x') \in S, \quad \text{and} \quad g \circ g_1 \in G \text{ by assumption.}$$

In fact, $G(S) = S$ because G contains the identity mapping.

After these preliminaries, prove the following major result.

Exercise VI.4.7 (Bruhat-Tits Fixed Point Theorem) *Let X be a complete metric space satisfying the semiparallelogram law. Let G be a group of isometries. Suppose that an orbit is bounded in X. Let x_1 be the circumcenter of this orbit. Then x_1 is a fixed point of G, that is, $g(x_1) = x_1$ for all $g \in G$.*

Solution. Let $\overline{\mathbf{B}}_r(x_1)$ be the unique closed ball of minimal radius containing the orbit Gx'. Then for any $g \in G$, the image $g\overline{\mathbf{B}}_r(x_1) = \overline{\mathbf{B}}_r(x_2)$ is a closed ball of the same radius containing the orbit Gx', and $x_2 = gx_1$, so by the uniqueness of the ball (Exercise 6) it follows that $x_2 = x_1$ and therefore x_1 is a fixed point for G.

VI.5   Open and Closed Sets

Exercise VI.5.1 *Let S be a subset of a normed vector space E, and let $\overline{S}$ denote the set of all points of E which are adherent to S.*
(a) Prove that $\overline{S}$ is closed. We call $\overline{S}$ the **closure** *of S.*
(b) If S, T are subsets of E, and $S \subset T$, show that $\overline{S} \subset \overline{T}$.
(c) If S, T are subsets of E, show that $\overline{S \cup T} = \overline{S} \cup \overline{T}$.
(d) Show that $\overline{\overline{S}} = \overline{S}$.
(e) If $S \subset T \subset \overline{S}$, prove that $\overline{T} = \overline{S}$.
(f) Let E, F be normed vector spaces, S a subset of E and T a subset of F. Take the sup norm on $E \times F$. Show that $(\overline{S \times T}) = \overline{S} \times \overline{T}$.

Solution. (a) Let $\{x_n\}_{n=1}^{\infty}$ be a sequence in $\overline{S}$ with limit x. Let $\epsilon > 0$ and consider the open ball $B_\epsilon(x)$ of radius ϵ centered at x. We must show that this ball contains an element of S. Select N such that $x_N \in B_{\epsilon/2}(x)$, and since x_N is adherent to S, there exists $y \in S$ such that $y \in B_{\epsilon/2}(x_N)$ which implies that $y \in B_\epsilon(x)$, as was to be shown.
(b) Let $x \in \overline{S}$. Then for all $\epsilon > 0$, the open ball $B_\epsilon(x)$ contains elements of S and therefore of T, so $x \in \overline{T}$.
(c) If $x \in \overline{S \cup T}$, then for all $\epsilon > 0$, the open ball $B_\epsilon(x)$ contains elements of S or of T, so $\overline{S \cup T} \subset \overline{S} \cup \overline{T}$. Conversely, if $x \in \overline{S} \cup \overline{T}$, then for all $\epsilon > 0$, $B_\epsilon(x)$ contains elements of $S \cup T$.
(d) The inclusion $\overline{S} \subset \overline{\overline{S}}$ is clear because the closure of a set always contains the set itself. Conversely, suppose $x \in \overline{\overline{S}}$. Then there exists a sequence $\{x_n\}$ in $\overline{S}$ converging to x. For each n there exists $y_n \in S$ such that $|x_n - y_n| < 1/n$. Then

$$|x - y_n| \leq |x - x_n| + |x_n - y_n| \leq |x - x_n| + \frac{1}{n}$$

and we see that $\{y_n\}$ converges to x. This shows that $x \in \overline{S}$.
(e) We apply (b) twice. First $S \subset T$ implies $\overline{S} \subset \overline{T}$ and applying (b) to $T \subset \overline{S}$ we obtain $\overline{T} \subset \overline{\overline{S}}$, so by (d) we get $\overline{T} \subset \overline{S}$, and we are done.
(f) Suppose $x = (x_1, x_2) \in \overline{S \times T}$. Then for all $\epsilon > 0$, the open ball $B_\epsilon(x)$ contains points of $S \times T$. Since the norm on the product is the sup norm, it follows that x_1 is adherent to S and x_2 is adherent to T. Conversely, suppose that $x = (x_1, x_2) \in \overline{S} \times \overline{T}$. Then given $\epsilon > 0$ we can find $y_1 \in S$ and $y_2 \in T$ such that $|x_1 - y_1|_S \leq \epsilon$ and $|x_2 - y_2|_T \leq \epsilon$. By the definition of the sup norm it follows that if $y = (y_1, y_2) \in S \times T$, then $|x - y|_{S \times T} < \epsilon$. So $x \in \overline{S \times T}$ as was to be shown.

Exercise VI.5.2 *A* **boundary point** *of S is a point $v \in E$ such that every open set U which contains v also contains an element of S and an element of E which is not in S. The set of boundary points is called the* **boundary** *of S, and is denoted by ∂S.*
(a) Show that ∂S is closed.

(b) Show that S is closed if and only if S contains all its boundary points.
(c) Show that the boundary of S is equal to the boundary of its complement.

Solution. One could simply note that $\partial S = \overline{S} \cap \overline{S^c} = \overline{S} \cap \overline{E - S}$. We can also argue straight from the definitions. Suppose $x \in \partial S$ and assume without loss of generality that $x \in S$. Then for some $\epsilon > 0$ the open ball $B_\epsilon(x)$ does not intersect the complement of S, so $B_\epsilon(x)$ is contained in the complement ∂S.

(b) Suppose S is closed. Let $x \in E - S$, then there exists $\epsilon > 0$ such that $B_\epsilon(x) \cap S$ is empty. So x is not a boundary point of S. Conversely, suppose S contains all of its boundary points. Then given $x \in E - S$ there exists $\epsilon > 0$ such that $B_\epsilon(x) \cap S$ is empty, otherwise x would be a boundary point of S. Thus S is closed.

(c) The fact that the boundary of S equals the boundary of its complement follows at once from the symmetry of the definition of a boundary point. Indeed, $x \in E$ is a boundary point of S if and only if x is a boundary point of the complement of S.

Exercise VI.5.3 *An element u of S is called an* **interior point** *of S if there exists an open ball B centered at u such that B is contained in S. The set of interior points of S is denoted by $\mathrm{Int}(S)$. It is obviously open. It is immediate that the intersection of $\mathrm{Int}(S)$ and ∂S is empty. Prove the formula*

$$\overline{S} = \mathrm{Int}(S) \cup \partial S.$$

In particular, a closed set is the union of its interior points and its boundary points. If S, T are subsets of normed vector spaces, then also show that

$$\mathrm{Int}(S \times T) = \mathrm{Int}(S) \times \mathrm{Int}(T).$$

Solution. Suppose $x \in \overline{S}$, so x is adherent to S. If for some $\epsilon > 0$, the open ball $B_\epsilon(x)$ is contained in S, then $x \in \mathrm{Int}(S)$. If not, then for all positive integers n, the open ball $B_{1/n}(x)$ intersects the complement of S, so that $x \in \partial S$. Conversely, it is immediate that any point in ∂S or $\mathrm{Int}(S)$ belongs to $\overline{S}$ so $\overline{S} = \mathrm{Int}(S) \cup \partial S$.

We now show that $\mathrm{Int}(S \times T) = \mathrm{Int}(S) \times \mathrm{Int}(T)$. Suppose $(x_1, y_1) \in \mathrm{Int}(S \times T)$, then for some $\epsilon > 0$ the open ball $B_\epsilon((x_1, y_1))$ is contained in $S \times T$. Suppose that $|x_1 - x_2| < \epsilon$, then the sup norm implies

$$|(x_1, y_1) - (x_2, y_1)|_{S \times T} = |x_1 - x_2| < \epsilon$$

so $x_2 \in S$ and therefore $x_1 \in \mathrm{Int}(S)$. A similar argument for the second factor proves that $(x_1, y_1) \in \mathrm{Int}(S) \times \mathrm{Int}(T)$. Conversely, if $(x_1, y_1) \in \mathrm{Int}(S) \times \mathrm{Int}(T)$, then for some $\epsilon > 0$, $|x_1 - x_2| < \epsilon$ implies $x_2 \in S$ and $|y_1 - y_2| < \epsilon$ implies $y_2 \in T$. So if $|(x_1, y_1) - (x_2, y_2)|_{S \times T} < \epsilon$, then it follows that $(x_2, y_2) \in S \times T$, so $(x_1, y_1) \in \mathrm{Int}(S \times T)$.

Exercise VI.5.4 *Let S, T be subsets of a normed vector space. Prove the following:*

(a) $\partial(S \cup T) \subset \partial S \cup \partial T$.

(b) $\partial(S \cap T) \subset \partial S \cup \partial T$.

(c) Let $S - T$ denote the set of elements $x \in S$ such that $x \notin T$. Then $\partial(S - T) \subset \partial S \cup \partial T$. [Note: You may save yourself some work if you use the fact that $\partial S^c = \partial S$ where S^c is the complement of S, and use properties like $S - T = S \cap T^c$, as well as $(S \cap T)^c = S^c \cup T^c$.]

(d) $\partial(S \times T) = (\partial S \times \overline{T}) \times (\overline{S} \times \partial T)$.

Solution. (a) Let $x \in \partial(S \cup T)$. For every $\epsilon > 0$, the open ball $B_\epsilon(x)$ contains points of S^c and T^c because $(S \cup T)^c = S^c \cap T^c$. If x is not a boundary point of S or T, then there exists $\epsilon_0 > 0$ such that $B_{\epsilon_0}(x) \cap (S \cup T)$ is empty. Contradiction.

(b) Every open ball centered at a point of $\partial(S \cap T)$ intersects S and T. Since $(S \cap T)^c = S^c \cup T^c$ we conclude that every open ball centered at a point of $\partial(S \cap T)$ intersects S^c or T^c. Conclude.

(c) We have

$$\partial(S - T) = \partial(S \cap T^c) \subset \partial S \cup \partial T^c$$

but $\partial T^c = \partial T$, so the result drops out.

(d) Suppose $(x, y) \in \partial(S \times T)$. Then since every open ball centered at (x, y) intersects $S \times T$ the definition of the sup norm implies that $x \in \overline{S}$ and $y \in \overline{T}$. If S and T are subsets of E we see that

$$(S \times T)^c = (S^c \times E) \cup (E \times T^c),$$

so we have $\partial(S \times T) \subset (\partial S \times \overline{T}) \times (\overline{S} \times \partial T)$. Conversely, if (x, y) belongs to $\partial S \times \overline{T}$ or $\overline{S} \times \partial T$, then by the definition of the sup norm we see that $(x, y) \in \partial(S \times T)$.

Exercise VI.5.5 *Let S be a subset of a normed vector space E. An element v of S is called* **isolated** *(in S) if there exists an open ball centered at v such that v is the only element of S in this open ball. An element x of E is called an* **accumulation point** *(or* **point of accumulation***) of S if x belongs to the closure of the set $S - \{x\}$.*

(a) Show that x is adherent to S if and only if x is either an accumulation point of S or an isolated point of S.

(b) Show that the closure of S is the union of S and its set of accumulation points.

Solution. (a) Suppose x is adherent to S. If there exists $\epsilon > 0$ such that the only point of $B_\epsilon(x)$ in S is x, then x is an isolated point. If no such ϵ exists, then every open ball centered at x contains points $\neq x$ that lie in S. Hence x is adherent to $S - \{x\}$ and is therefore a point of accumulation of S. Conversely, if x is isolated, then it is adherent to S by definition. If x is a point of accumulation of S, then every open ball centered at x contains points of $S - \{x\}$, thus x is adherent to S.

(b) If x belongs to $\overline{S}$, then either $x \in S$ or x is a point of accumulation of S because every open ball centered at x intersects S. Conversely, if x belongs to S, then $x \in \overline{S}$ and if x is a point of accumulation of S, then x is adherent to S.

Exercise VI.5.6 *Let U be an open subset of a normed vector space E, and let $v \in E$. Let U_v be the set of all elements $x + v$ where $x \in U$. Show that U_v is open. Prove a similar statement about closed sets.*

Solution. Let $w \in U_v$. Let B be the open ball centered at $w - v$ and contained in U. Then B_v is contained in U_v.

If U is closed, then U_v is closed because $(E - U)_v = E - U_v$.

Exercise VI.5.7 *Let U be open in E. Let t be a number > 0. Let tU be the set of all elements tx with $x \in U$. Show that tU is open. Prove a similar statement about closed sets.*

Solution. Let $w = tx$ where $x \in U$. There exists $r > 0$ such that $B_r(x)$ is contained in U. Then $B_{tr}(x)$ is contained in tU.

If U is closed, then tU is closed because $t(E - U) = E - tU$.

Exercise VI.5.8 *Show that the projection $\mathbf{R} \times \mathbf{R} \to \mathbf{R}$ given by $(x, y) \mapsto x$ is continuous. Find an example of a closed subset A of $\mathbf{R} \times \mathbf{R}$ such that the projection of A on the first factor is not closed. Find an example of an open set U in $\mathbf{R}^2$ whose projection is closed, and $U \neq \mathbf{R}^2$.*

Solution. The projection map is continuous because

$$|x - x_0| \leq \|(x - x_0, y - y_0)\|.$$

Let $A = \{(x, y) \in \mathbf{R}^2 : y = 1/x \text{ with } x \in (0, 1]\}$. Then A is closed in $\mathbf{R}^2$ and the projection of A is $(0, 1]$, which is not closed in $\mathbf{R}$.

Let $U = \{(x, y) \in \mathbf{R}^2 : y > 0\}$. Then U is the upper half plane which is open in $\mathbf{R}^2$, and the projection of U is $\mathbf{R}$ which is closed.

Exercise VI.5.9 *Prove the remark before Theorem 5.6.*

Solution. First suppose that S is open in E. Suppose X is open in S. Then there exists an open set U in E such that $X = S \cap U$. The intersection of open sets is open, so X is open in E. Conversely, if X is open in E, note that $X = S \cap X$ and therefore we can take $U = X$ in the definition, thereby proving that X is open in S.

Now suppose S is closed in E. Suppose that X is closed in E. Then there exists a closed set Z in E such that $X = S \cap Z$. The intersection of two closed sets is closed, so X is closed. Conversely, if X is closed in E, we note that $X = S \cap X$ so that we can take $Z = X$ in the definition, therefore X is closed in E.

Exercise VI.5.10 *Let U be open in a normed vector space E and let V be open in a normed vector space F. Let*

$$f : U \to V \quad and \quad g : V \to U$$

be continuous maps which are inverses of each other, that is

$$f \circ g = \mathrm{id}_V \quad and \quad g \circ f = \mathrm{id}_U,$$

where id means the identity map. Show that if U_1 is open in U, then $f(U_1)$ is open in V, and that the open subsets of U and V are in bijection under the association

$$U_1 \mapsto f(U_1) \quad and \quad V_1 \mapsto g(V_1).$$

Solution. Since f and g are inverses of each other, $f(U_1) = g^{-1}(U_1)$ so the continuity of g guarantees that $f(U_1)$ is open whenever U_1 is open.

The existence of the inverse map of f implies that $U_1 \mapsto f(U_1)$ is injective. Indeed, if $f(x) = f(y)$ we compose with g to conclude that $x = y$. Therefore, $U_1 \mapsto f(U_1)$ and $V_1 \mapsto f(V_1)$ are bijective and inverses of each other.

Exercise VI.5.11 *Let B be the closed ball of radius $r > 0$ centered at the origin in a normed vector space. Show that there exists an infinite sequence of open sets U_n whose intersection is B.*

Solution. Let U_n be the open ball of radius $r + 1/n$ centered at the origin. Then B is contained in $\bigcap_{n=1}^{\infty} U_n$. Conversely, if $x \in \bigcap_{n=1}^{\infty} U_n$ then $|x| < r + 1/n$ for all n, hence $|x| \le r$ which shows that $x \in B$. This proves that

$$B = \bigcap_{n=1}^{\infty} U_n.$$

Exercise VI.5.12 *Prove in detail that the following notions are the same for two equivalent norms on a vector space: (a) open set; (b) closed set; (c) point of accumulation of a sequence; (d) continuous function; (e) boundary of a set; and (f) closure of a set.*

Solution. Consider two norms $|\cdot|_1$ and $|\cdot|_2$ such that

$$C_1 |v|_1 \le |v|_2 \le C_2 |v|_1.$$

(a) Suppose U is open with respect to $|\cdot|_1$. Let x belong to U. Then for some $\epsilon > 0$, the open ball $B_\epsilon^1(x)$ with respect to $|\cdot|_1$ is contained in U. So the open ball $B_{C_1\epsilon}^2(x)$ with respect to $|\cdot|_2$ is contained in $B_\epsilon^1(x)$ and is therefore also contained in U.

(b) Suppose that C is closed with respect to $|\cdot|_1$. The complement of C is open with respect to $|\cdot|_1$, so by (a) the complement of C is also open with respect to $|\cdot|_2$, and therefore C is closed with respect to $|\cdot|_2$.

(c) Suppose that x is a point of accumulation of the sequence $\{x_n\}$ for $|\cdot|_1$. Let $\epsilon > 0$, then for some m we have $|x - x_m|_1 \leq \epsilon/C_2$ so $|x - x_m|_2 \leq \epsilon$, hence x is a point of accumulation of $\{x_n\}$ for $|\cdot|_2$.

(d) Suppose f is continuous at x_0 with respect to $|\cdot|_1$. Let $\epsilon > 0$, then there exists $\delta > 0$ such that $|f(x) - f(x_0)|_1 < \epsilon/C_2$ whenever $|x - x_0|_1 < \delta/C_1$. Then, if $|x - x_0|_2 < \delta$ we immediately conclude that $|f(x) - f(x_0)|_2 < \epsilon$, so f is continuous at x_0 with respect to $|\cdot|_2$.

(e) By Exercise 3, $\partial S = \overline{S} - \text{Int}(S)$ and by (a) and (b) we know that the closure of S is the same with respect to equivalent norms and similarly for the interior of S, so the boundary of S is also the same with respect to equivalent norms.

(f) By Exercise 3, if follows that the closure of a set is the same set with respect to equivalent norms.

Exercise VI.5.13 *Let $|\cdot|_1$ and $|\cdot|_2$ be two norms on a vector space E, and suppose that there exists a constant $C > 0$ such that for all $x \in E$ we have $|x|_1 \leq C|x_2|$. Let $f_1(x) = |x|_1$. Prove in detail: Given ϵ, there exists δ such that if $x, y \in E$, and $|x - y|_2 < \delta$, then $|f_1(x) - f_1(y)| < \epsilon$. [Remark. Since f_1 is real valued, the last occurrence of the signs $|\cdot|$ denotes the absolute value on $\mathbf{R}$.] In particular, f_1 is continuous for the norm $|\cdot|_2$.*

Solution. Given $\epsilon > 0$, let $\delta = \epsilon/C$. Then $|x - y|_2 < \delta$ implies $|x - y|_1 < \epsilon$. The triangle inequality for the norm implies $|x|_1 - |y|_1 \leq |x - y|_1$ and $|y|_1 - |x|_1 \leq |x - y|_1$ so $||x|_1 - |y|_1| \leq |x - y|_1$ and therefore

$$|f_1(x) - f_1(y)| \leq |x - y|_1 < \epsilon,$$

as was to be shown.

Exercise VI.5.14 *Let BS denote the space of all sequences of numbers*

$$X = (x_1, x_2, \ldots, x_n, \ldots)$$

which are bounded, i.e. there exists $C > 0$ (depending on X) such that $|x_n| \leq C$ for all $n \in \mathbf{Z}^+$. Then BS is a special case of Example 2 of §1, namely the space of bounded maps $\mathcal{B}(\mathbf{Z}^+, \mathbf{R})$. For $X \in BS$, the sup norm is

$$\|X\| = \sup_n |x_n|,$$

i.e. $\|X\|$ is the least upper bound of all absolute values of the components.

(a) Let E_0 be the set of all sequences X such that $x_n = 0$ for all but a finite number of n. Show that E_0 is a subspace of BS.

(b) Is E_0 dense in BS? Prove your assertion. [Note. In Theorem 3.1 of Chapter VII, it will be shown that BS is complete.]

Solution. (a) Addition and scalar multiplication is defined componentwise. Suppose X and Y belong to E_0, and that p and q are the number of nonzero elements in X and Y, respectively. Then $X + Y$ has at most $p + q$ nonzero elements, hence $X + Y$ belongs to E_0. Similarly, aX belongs to E_0 whenever $a \in \mathbf{R}$ and $X \in E_0$. Clearly, $(0, 0, \ldots)$ belongs to E_0, hence we conclude that E_0 is a subspace of BS.

(b) We contend that E_0 is not dense if BS. Let $X = (1, 1, 1, \ldots)$. Then we see that for all $Y \in E_0$ the vector $X - Y$ has at least one component equal to 1. In fact, $X - Y$ has only finitely many components $\neq 1$, therefore $\|X - Y\| \geq 1$ and this proves our contention.

VII
Limits

VII.1 Basic Properties

Exercise VII.1.1 *A subset S of a normed vector space E is said to be* **convex** *if given $x, y \in S$ the points*

$$(1 - t)x + ty, \quad 0 \le t \le 1,$$

are contained in S. Show that the closure of a convex set is convex.

Solution. Suppose x and y are adherent to S and let $\{x_n\}$ and $\{y_n\}$ be sequences of points in S converging to x and y, respectively. Then for each $0 \le t \le 1$, $(1 - t)x_n + ty_n$ belongs to S and converges to $(1 - t)x + ty$ as $n \to \infty$. This implies that for every $0 \le t \le 1$, the point $(1 - t)x + ty$ belongs to the closure of S.

Exercise VII.1.2 *Let S be a set of numbers containing arbitrarily large numbers (that is, given an integer $N > 0$, there exists $x \in S$ such that $x \ge N$). Let $f : S \to \mathbf{R}$ be a function. Prove that the following conditions are equivalent:*
(a) Given ϵ, there exists N such that whenever $x, y \in S$ and $x, y \ge N$, then

$$|f(x) - f(y)| < \epsilon.$$

(b) The limit

$$\lim_{x \to \infty} f(x)$$

exists.

(Your argument should be such that it applies as well to a map $f : S \to F$ of S into a complete normed vector space.)

Solution. Condition (b) implies (a) because if w is the limit in (b), and we choose N such that $x \geq N$ implies $|f(x) - w| < \epsilon/2$, then for all $x, y \geq N$ we have

$$|f(x) - f(y)| \leq |f(x) - f(w)| + |f(w) - f(y)| < \epsilon.$$

Conversely, assume (a) and choose a sequence of points $\{x_n\}$ such that $x_n \geq n$. Then $\{f(x_n)\}$ is a Cauchy sequence because if $n, m > N$, then $|f(x_n) - f(x_m)| < \epsilon$. The completeness of $\mathbf{R}$ (or F) implies that $\{f(x_n)\}$ converges to a limit point, say w. Then for all $x \geq N$ and x_n so large that $x_n > N$ and $|f(x_n) - w| \leq \epsilon$ we have

$$|f(x) - w| \leq |f(x) - f(x_n)| + |f(x_n) - w| < 2\epsilon.$$

Exercise VII.1.3 *Let F be a normed vector space. Let E be a vector space (not normed yet) and let $L : E \to F$ be a linear map, that is satisfying $L(x + y) = L(x) + L(y)$ and $L(cx) = cL(x)$ for all $c \in \mathbf{R}$, $x, y \in E$. Assume that L is injective. For each $x \in E$, define $|x| = |L(x)|$. Show that the function $x \mapsto |x|$ is a norm on E.*

Solution. Since L is injective, $L(x) = 0$ if and only if $x = 0$, so $|x| = 0$ if and only if $x = 0$. Clearly,

$$|ax| = |L(ax)| = |aL(x)| = |a||L(x)| = |a||x|.$$

Finally we check the triangle inequality

$$|x + y| = |L(x + y)| = |L(x) + L(y)| \leq |L(x)| + |L(y)| = |x| + |y|,$$

and this concludes the proof.

Exercise VII.1.4 *Let P_5 be the vector space of polynomial functions of degree ≤ 5 on the interval $[0, 1]$. Show that P_5 is closed in the space of all bounded functions on $[0, 1]$ with the sup norm. [Hint: If $f(x) = a_5 x^5 + \cdots + a_0$ is a polynomial, associate to it the point $(a_5, \ldots, a_0)$ in $\mathbf{R}^6$, and compare the sup norm on functions, with the norm on $\mathbf{R}^6$.]*

Solution. Consider the association $f(x) = a_5 x^5 + \cdots + a_0 \leftrightarrow (a_5, \ldots, a_0)$ and define a new norm on $\mathbf{R}^6$ to be the sup norm of the associated polynomial. This new norm is equivalent to the euclidean norm (Theorem 4.3). If $\{P_n\}$ converges, it is a Cauchy sequence for the sup norm, so the associated sequence of points in $\mathbf{R}^6$ is a Cauchy sequence for the euclidean norm and by completeness we conclude that $\{p_n\}$ converges to a polynomial of degree ≤ 5.

Exercise VII.1.5 *Let E be a complete normed vector space and let F be a subspace. Show that the closure of F in E is a subspace. Show that this closure is complete.*

Solution. The element 0 belongs to the closure. If x, y lie in the closure of F, choose sequences $\{x_n\}$ and $\{y_n\}$ of points in F that converge to x and y, respectively. Then, $x + y = \lim_{n \to \infty}(x_n + y_n)$, so $x + y$ belongs to the closure of F. Similarly, $cx = \lim_{n \to \infty} cx_n$ so cx belongs to the closure of F. This proves that the closure of F is a subspace.

A Cauchy sequence in the closure of F is a Cauchy sequence in E, so by completeness, this sequence converges in E. Since the closure of F is closed, the result follows at once.

Exercise VII.1.6 *Let E be a normed vector space and F a subspace. Assume that F is dense in E and that every Cauchy sequence in F has a limit in E. Prove that E is complete.*

Solution. Let $\{x_n\}$ be a Cauchy sequence in E. For each positive integer n choose $y_n \in F$ such that $|x_n - y_n| < 1/n$. Then using the triangle inequality we see that

$$
\begin{aligned}
|y_n - y_m| &\leq |y_n - x_n| + |x_n - x_m| + |x_m - y_m| \\
&\leq \frac{1}{n} + |x_n - x_m| + \frac{1}{m}
\end{aligned}
$$

so $\{y_n\}$ is a Cauchy sequence. Let y be its limit. Then $\{x_n\}$ converges to y because

$$
|x_n - y| \leq |x_n - y_n| + |y_n - y| \leq \frac{1}{n} + |y_n - y|.
$$

VII.2   Continuous Maps

Exercise VII.2.1 *(a) Prove Theorem 2.4(a).*
(b) Prove Theorem 2.4(b). [Hint: Given ϵ, and two points $x, y \in E$ with $|x - y| < \epsilon$, there exists $v \in S$ such that $|x - v| < d(x, S) + \epsilon$. Then

$$
d(y, S) \leq d(y, v) \leq d(x, S) + 2\epsilon.
$$

Take it from here.]
(c) Let S, T be two non-empty closed subsets of E, and assume that they are disjoint, i.e. have no points in common. Show that the function

$$
f(v) = \frac{d(S, v)}{d(S, v) + d(T, v)}
$$

is a continuous function, with values between 0 and 1, taking the value 0 on S and 1 on T.
(For a continuation of this exercise, cf. the next chapter, §2.)

Solution. (a) Suppose v lies in the closure, then there exists a sequence of points in S converging to v thus $d(S, v) = 0$. Conversely, if $d(S, v) = 0$, then some points of S come arbitrarily close to v, proving that v is in the closure of S.

(b) From the hint we find that $d(y, S) - d(x, S) < 2\epsilon$. There exists a vector w in S such that $|y - w| < d(y, S) + \epsilon$. Then we have

$$d(x, S) \le |x - w| \le |x - y| + |y - w| \le d(y, S)| + 2\epsilon$$

so that $d(x, S) - d(y, S) \le 2\epsilon$. Hence $|d(x, S) - d(y, S)| \le 2\epsilon$.

(c) Since both subsets are closed, their closures are disjoint, so the denominator of f is never 0 which together with (b) implies the continuity of f. Since $d(S, v) + d(T, v) \ge d(S, v)$ the function f takes values between 0 and 1. On S, the numerator is 0. On T, the numerator and denominator are both non-zero and they are equal.

Exercise VII.2.2 *(a) Show that a function f which is differentiable on an interval and has a bounded derivative is uniformly continuous on the interval.*

(b) Let $f(x) = x^2 \sin(1/x^2)$ for $0 < x \le 1$ and $f(0) = 0$. Is f uniformly continuous on $[0, 1]$? Is the derivative of f bounded on $(0, 1)$? Is f uniformly continuous on the open interval $(0, 1)$? Proofs?

Solution. (a) By the mean value theorem we have $|f(x) - f(y)| \le M|x - y|$, where M is a bound for the derivative.

(b) Since $|x^2 \sin(1/x^2)| \le |x^2|$ we see that f is continuous at 0. Hence f is continuous on $[0, 1]$ and therefore uniformly continuous on this interval because $[0, 1]$ is compact. The derivative is unbounded on $(0, 1)$ because

$$f'(x) = 2x \sin(1/x^2) - (2/x) \cos(1/x^2)$$

and if $x_n = 1/\sqrt{2\pi n}$, then $x_n \to 0$ and $|f'(x_n)| = 2\sqrt{2\pi n} \to \infty$ as $n \to \infty$. However, the function f is uniformly continuous on $(0, 1)$ because it is uniformly continuous on the larger interval $[0, 1]$.

Exercise VII.2.3 *(a) Show that for every $c > 0$, the function $f(x) = 1/x$ is uniformly continuous for $x \ge c$.*

(b) Show that the function $f(x) = e^{-x}$ is uniformly continuous for $x \ge 0$, but not on $\mathbf{R}$.

(c) Show that the function $\sin x$ is uniformly continuous on $\mathbf{R}$.

Solution. (a) For $x, y \ge c$ we have

$$|f(x) - f(y)| = \frac{|x - y|}{|xy|} \le \frac{1}{c^2}|x - y|.$$

(b) For $x \ge 0$, the function f has a bounded derivative and is therefore uniformly continuous. To see why f is not uniformly continuous on $\mathbf{R}$, suppose that given ϵ there exists $\delta > 0$ such that $|x - y| < \delta$ implies $|e^{-x} - e^{-y}| < \epsilon$. Let $x = y - \delta/2$. Then $|x - y| < \delta$ and

$$|f(x) - f(y)| = |e^{-y+\delta/2}e^{-y}| = e^{-y}|e^{\delta/2} - 1|.$$

This last expression tends to ∞ as y tends to $-\infty$ which gives the desired contradiction.

(c) The mean value theorem implies $|\sin x - \sin y| \leq |x - y|$ for all $x, y \in \mathbf{R}$.

Exercise VII.2.4 *Show that the function* $f(x) = \sin(1/x)$ *is not uniformly continuous on the interval* $0 < x \leq \pi$, *even though it is continuous.*

Solution. For $n \geq 0$, let $x_n = 1/(n\pi + \pi/2)$. Then $|x_{n+1} - x_n| \to 0$ as $n \to \infty$ and

$$|f(x_{n+1}) - f(x_n)| = 2,$$

so f is not uniformly continuous.

Exercise VII.2.5 *(a) Define for numbers* t, x:

$$f(t, x) = \frac{\sin tx}{t} \quad \text{if } t \neq 0, \quad f(0, x) = x.$$

Show that f *is continuous on* $\mathbf{R} \times \mathbf{R}$. *[Hint: The only problem is continuity at a point* $(0, b)$. *If you bound* x, *show precisely how* $\sin tx = tx + o(tx)$.]
(b) Let

$$f(x, y) = \begin{cases} \frac{(y^2-x)^2}{y^4+x^2} & \text{if } (x, y) \neq (0, 0), \\ 1 & \text{if } (x, y) = (0, 0), \end{cases}$$

Is f *continuous at* $(0, 0)$? *Explain.*

Solution. (a) Taylor's formula at the origin implies that $\sin(z) = z + R(z)$ with the esimate $|R(z)| \leq |z|^2/2!$. So for x bounded we have $|R(tx)|/t \to 0$ as $t \to 0$, therefore

$$\lim_{(t,x)\to(0,b)} \frac{\sin(tx)}{t} = \lim_{(t,x)\to(0,b)} x + \frac{R(tx)}{t} = b,$$

which proves the continuity of f at the point $(0, b)$.

(b) The function f is not continuous at $(0, 0)$ because $\lim_{a\to\infty} f(a^2, a) = 0$.

Exercise VII.2.6 *(a) Let* E *be a normed vector space. Let* $0 < r_1 < r_2$. *Let* $v \in E$. *Show that there exists a continuous function* f *on* E, *such that:*
    $f(x) = 1$ *if* x *is in the ball of radius* r_1 *centered at* v,
    $f(x) = 0$ *if* x *is outside the ball of radius* r_2 *centered at* v.
    We have $0 \leq f(x) \leq 1$ *for all* x.
[Hint: Solve first the problem on the real line, and then for the special case $v = 0$.]
(b) Let $v, w \in E$ *and* $v \neq w$. *Show that there exists a contiuous function* f *on* E *such that* $f(v) = 1$ *and* $f(w) = 0$, *and* $0 \leq f(x) \leq 1$ *for all* $x \in E$.

Solution. (a) The function defined by

$$f(x) = \begin{cases} 1 & \text{if } x \in B_{r_1}(v), \\ \frac{r_2 - |x - v|}{r_2 - r_1} & \text{if } r_1 \leq |x - v| \leq r_2, \\ 0 & \text{if } x \notin B_{r_2}(v), \end{cases}$$

satisfies all the conditions. Note that we can also apply Exercise 1(c) with T the closed ball of radius r_1 centered at v, and S the complement of the open ball of radius r_2 centered at v.

(b) Let $r_1 = |v - w|/3$ and $r_2 = 2|v - w|/3$. The function f defined in part (a) satisifes the desired conditions.

Exercise VII.2.7 *Let S be a subset of a normed vector space E, and let $f : S \to F$ be a continuous map of S into a normed vector space. Let S' consist of all points $v \in E$ such that v is adherent to S and*

$$\lim_{x \to v, \ x \in S} f(x) \ \text{exists.}$$

Define $\overline{f}(v)$ to be this limit. If $x \in S$, then $\overline{f}(v) = f(v)$ by definition, so $\overline{f}$ is an extension of f to S'. For simplicity one may therefore write $f(v)$ instead of $\overline{f}(v)$. Show that $\overline{f}$ is continuous on S'. [Hint: Select $v \in S'$. You have to consider separately the estimates

$$|f(x) - f(v)| \quad \text{and} \quad |f(v') - f(v)|$$

for $x \in S$ and $v' \in S'$. You thus run into a 2ϵ-proof.]

In Exercise 7, the set S' is contained in $\overline{S}$ essentially by definition, but is not necessarily equal to $\overline{S}$. The next exercise gives a condition under which $S' = \overline{S}$.

Solution. Let $v_0 \in S'$ and let $\epsilon > 0$. Select $\delta > 0$ such that if x belongs to S and $|x - v_0| < \delta$, then

$$|f(x) - f(v_0)| < \epsilon.$$

Suppose x' belongs to S' and $|x' - v_0| < \delta/2$. Select y in S near x' with $|y - v_0| < \delta/2$ and $|f(y) - f(x')| < \epsilon$. Then we have

$$|f(x') - f(v_0)| \leq |f(x') - f(y)| + |f(y) - f(v_0)| \leq 2\epsilon$$

thus the extension of f is continuous on S'.

Exercise VII.2.8 *Prove Theorem 2.6. [Hint: Show that the uniform continuity condition implies that for every $v \in \overline{S}$, the limit $\lim_{x \to v} f(x)$ exists. Define $\overline{f}(v)$ to be this limit.]*

Solution. Given $\epsilon > 0$, pick $\delta > 0$ such that $|f(x) - f(y)| < \epsilon$ whenever $|x - y| < 2\delta$ and $x, y \in S$. If $v_0 \in \overline{S}$, then for all $x, y \in S$ and $|x - v_0| < \delta$, $|y - v_0| < \delta$ we have

$$|f(x) - f(y)| < \epsilon.$$

Theorem 1.2 implies the existence of $\lim_{x \to v_0} f(x)$, $x \in S$.

Exercise VII.2.9 *Let S, T be closed subsets of a normed vector space, and let $A = S \cup T$. Let $f: A \to F$ be a map into some normed vector space. Show that f is continuous on A if and only if its restrictions on S and T are continuous.*

Solution. Suppose f is continuous on A. Let $x \in S$ and let $\{x_n\}$ be a sequence in S converging to x. Then since $\{x_n\}$ is a sequence in A converging to x it follows that $\lim f(x_n) = f(x)$, thereby proving that f is continuous on S. A similar argument proves that f is continuous on T.

Conversely, suppose that f is continuous on S and T. Let $x \in A$ and let $\epsilon > 0$. If $x \notin T$, then x is not adherent to T, so we can choose a ball of small radius centered at x which does not intersect T. This ball is contained in S so we see that f is continuous at x. Similarly if $x \notin S$. Now if $x \in S \cap T$, then choose δ_1 and δ_2 such that

$$|f(y_S) - f(x)| < \epsilon \quad \text{and} \quad |f(y_T) - f(x)| < \epsilon$$

whenever $|y_S - x| < \delta_1$, $y_S \in S$ and $|y_T - x| < \delta_2$, $y_T \in T$. Then put $\delta = \min(\delta_1, \delta_2)$ in the definition of continuity.

Continuous Linear Maps

Exercise VII.2.10 *Let E, F be normed vector spaces, and let $L : E \to F$ be a linear map.*
(a) Assume that there is a number $C > 0$ such that $|L(x)| \leq C|x|$ for all $x \in E$. Show that L is continuous.
(b) Conversely, assume that L is continuous at 0. Show that there exists such a number C. [Hint: See §1 of Chapter X.]

Solution. (a) Given $\epsilon > 0$, let $\delta = \epsilon/C$ because

$$|L(x) - L(y)| = |L(x - y)| \leq C|x - y|.$$

(b) Choose $\delta > 0$ such that $|x| \leq \delta$ implies $|L(x)| \leq 1$. Then for all $v \in E$, $v \neq 0$ we see that $|\delta v/|v|| = \delta$, so

$$|L(\delta v/|v|)| \leq 1$$

which implies $|L(v)| \leq C|v|$ where $C = 1/\delta$.

Exercise VII.2.11 *Let $L: \mathbf{R}^k \to F$ be a linear map of $\mathbf{R}^k$ into a normed vector space. Show that L is continuous.*

Solution. For $1 \le j \le k$, let $e_j = (0, 0, \ldots, 0, 1, 0, \ldots)$ be the vector with all entries 0 except 1 at the j-th entry. If $v = (a_1, \ldots, a_k)$, then

$$|L(v)| \le |a_1||L(e_1)| + \cdots + |a_n||L(e_n)|$$
$$\le |v|_{\mathbf{R}^k}(|L(e_1)| + \cdots + |L(e_n)|),$$

where $|\cdot|_{\mathbf{R}^k}$ is the sup norm on $\mathbf{R}^k$. Exercise 10 implies the continuity of L.

Exercise VII.2.12 *Show that a continuous linear map is uniformly continuous.*

Solution. Let $\epsilon > 0$. There exists $\delta > 0$ such that $|L(v)| < \epsilon$ whenever $|v| < \delta$. If $|x - y| < \delta$, then $|L(x) - L(y)| = |L(x - y)| < \epsilon$. This proves the uniform continuity of L.

Exercise VII.2.13 *Let $L: E \to F$ be a continuous linear map. Show that the values of L on the closed ball of radius 1 are bounded. If r is a number > 0, show that the values of L on any closed ball of radius r are bounded. (The closed balls are centered at the origin.) Show that the image under L of a bounded set is bounded.*

Because of Exercise 10, a continuous linear map L is also called bounded. If C is a number such that $|L(x)| \le C|x|$ for all $x \in E$, then we call C a **bound** *for L.*

Solution. Since L is continuous at 0, there exists a number $C > 0$ such that $|L(x)| \le C|x|$ for all $x \in E$. If $|x| \le r$, then $|L(x)| \le C|r|$ which proves that the values of L on the closed ball of radius r centered at the origin are bounded.

Given a bounded set S, there exists a closed ball such that S is contained in the closed ball. Therefore, the values of L on S are bounded.

Exercise VII.2.14 *Let L be a continuous linear map, and let $|L|$ denote the greatest lower bound of all numbers C such that $|L(x)| \le C|x|$ for all $x \in E$. Show that the continuous linear maps of E into F form a vector space, and that the function $L \mapsto |L|$ is a norm on this vector space.*

Solution. The sum of continuous linear maps is continuous and linear, and cL is continous and linear whenever L is continuous and linear and c is a scalar. It is easy to verify that all the properties of a vector space hold, so the space of continuous linear maps is a vector space.

We must show that $|\cdot|$ is a norm on the vector space of continuous linear maps. First note that $0 \le |L| < \infty$ because L is continuous. If $L = 0$, then $|L| = 0$. Conversely, suppose $|L| = 0$. Then for each x the inequality

$$|L(x)| \le \epsilon|x|$$

holds for every $\epsilon > 0$. Letting $\epsilon \to 0$ shows that $L = 0$. The second property of a norm holds because

$$|aL(x)| = |L(ax)| \leq C|ax| = C|a||x|$$

and therefore C is a bound for L if and only if $|a|C$ is a bound for aL. Finally, we prove the triangle inequality. Suppose $|L_i(x)| \leq C_i|x|$ for $i = 1, 2$, then

$$|L_1(x) + L_2(x)| \leq |L_1(x)| + |L_2(x)| \leq C_1|x| + C_2|x| = (C_1 + C_2)|x|$$

and therefore $|L_1 + L_2| \leq |L_1| + |L_2|$.

Exercise VII.2.15 *Let $a < b$ be numbers, and let $E = C^0([a,b])$ be the space of continuous functions on $[a,b]$. Let $I_a^b : E \to \mathbf{R}$ be the integral. Is I_a^b continuous: (a) for the L^1-norm; and (b) for the L^2-norm on E? Prove your assertion.*

Solution. (a) Let $f, g \in E$, and let $\epsilon > 0$. Suppose that $|f - g|_1 < \epsilon$, then

$$|I_a^b(f) - I_a^b(g)| \leq |I_a^b(f - g)| \leq \int_a^b |f(x) - g(x)|dx = |f - g|_1 < \epsilon,$$

whence I_a^b is continuous for the L_1-norm.

(b) The integral is also continuous for the L^2-norm because the Schwarz inequality gives

$$\begin{aligned}
|f - g|_1 &= \int_a^b |f(x) - g(x)|dx \\
&\leq \left(\int_a^b 1 dx \right)^{1/2} \left(\int_a^b |f(x) - g(x)|^2 dx \right)^{1/2} \\
&\leq C|f - g|_2.
\end{aligned}$$

Exercise VII.2.16 *Let X be a complete metric space satisfying the semi-parallelogram law. (Cf. Exercise 5 of Chapter VI, §4.) Let S be a closed subset, and assume that given $x_1, x_2 \in S$ the midpoint between x_1 and x_2 is also in S. Let $v \in X$. Prove that there exists an element $w \in S$ such that $d(v, w) = d(v, S)$.*

Solution. Let $d = d(v, S)$. We assume that $v \notin S$, otherwise choose $v = w$. Let $\{x_n\}$ be a sequence of points in S such that $d(v, x_n)$ converges to d. We now show that $\{x_n\}$ is a Cauchy sequence. Let z_{mn} be the midpoint of x_m and x_n. Applying the semiparallelogram law with $x = v$ we get

$$d(x_m, x_n)^2 \leq 2d(x_m, v)^2 + 2d(x_n, v)^2 - 4d(z_{mn}, v)^2.$$

By assumption, $z_{mn} \in S$ so $4d(z_{mn}, v)^2 \geq 4d^2$ and therefore

$$d(x_m, x_n)^2 \leq 2d(x_m, v)^2 + 2d(x_n, v)^2 - 4d^2,$$

whence $\{x_n\}$ is Cauchy. Since X is complete, and S is closed, the sequence $\{x_n\}$ converges to an element $w \in S$ which verifies $d(v, w) = d(v, S)$.

VII.3   Limits in Function Spaces

Exercise VII.3.1 *Let $f_n(x) = x^n/(1 + x^n)$ for $x \geq 0$.*
(a) Show that f_n is bounded.
(b) Show that the sequence $\{f_n\}$ converges uniformly on any interval $[0, c]$ for any number $0 < c < 1$.
(c) Show that this sequence converges uniformly on the interval $x \geq b$ if b is a number > 1, but not on the interval $x \geq 1$.

Solution. (a) For $x \geq 0$, we have $x^n \leq 1 + x^n$ so $0 \leq f_n(x) \leq 1$.
(b) The sequence converges uniformly to 0 because

$$\left| \frac{x^n}{1 + x^n} \right| \leq |x^n| < c^n$$

and $c^n \to 0$ as $n \to \infty$.
(c) If $x \geq b > 1$ we estimate

$$|f_n(x) - 1| = \frac{1}{|1 + x^n|} \leq \frac{1}{|x|^n - 1} \leq \frac{1}{b^n - 1}$$

so the sequence $\{f_n\}$ converges uniformly to 1.

For all $x > 1$ we have $\lim_{n \to \infty} f_n(x) = 1$ and $f_n(1) = 1/2$ for all n so the sequence $\{f_n\}$ does not converge uniformly for $x \geq 1$.

Exercise VII.3.2 *Let g be a function defined on a set S, and let a be a number > 0 such that $g(x) \geq a$ for all $x \in S$. Show that the sequence*

$$g_n = \frac{ng}{1 + ng}$$

converges uniformly to the constant function 1. Prove the same thing if the assumption is that $|g(x)| \geq a$ for all $x \in S$.

Solution. We prove the stronger result when $|g(x)| \geq a > 0$. The uniform convergence follows from the fact that for all large n we have

$$|g_n(x) - 1| = \frac{1}{|1 + ng(x)|} \leq \frac{1}{n|g(x)| - 1} \leq \frac{1}{na - 1}.$$

Exercise VII.3.3 *Let $f_n(x) = x/(1 + nx^2)$. Show that $\{f_n\}$ converges uniformly for $x \in \mathbf{R}$, and that each function f_n is bounded.*

Solution. For $x \geq 0$, $x \leq 1 + nx^2$ (separate cases $x \leq 1$ and $x > 1$), and each f_n is odd, so $|f_n(x)| \leq 1$ for all x and all n.

By symmetry we can reduce the proof of uniform convergence for $x \geq 0$. We then have

$$f_n'(x) = \frac{1 - nx^2}{(1 + nx^2)^2},$$

so f_n attains its maximum $1/(2\sqrt{n})$ at $x = 1/\sqrt{n}$, thus $|f_n(x)| \leq 1/(2\sqrt{n})$ proving that the sequence converges to 0 uniformly on $\mathbf{R}$.

Exercise VII.3.4 *Let S be the interval $0 \leq x < 1$. Let f be the function defined on S by $f(x) = 1/(1-x)$.*
(a) Determine whether f is uniformly continuous.
(b) Let $p_n(x) = 1 + x + \cdots + x^n$. Does the sequence $\{p_n\}$ converge uniformly to f on S?
(c) Let $0 < c < 1$. Show that f is uniformly continuous on the interval $[0, c]$, and that the sequence $\{p_n\}$ converges uniformly to f on this interval.

Solution. (a) The function f is not uniformly continuous because if it were, then there would exist $0 < \delta < 1/2$ such that $|f(x) - f(y)| < 1$ whenever $|x - y| < \delta$. Let $y = 1 - \delta$ and suppose $y < x < 1$. Then for values of x close to 1 we see that $|f(x) - f(y)| \geq 1$.
(b) We know that $p_n(x) = (1 - x^{n+1})/(1 - x)$ so

$$|p_n(x) - f(x)| = \frac{|x|^{n+1}}{|1 - x|}.$$

For fixed n we have $|x|^{n+1}/|1 - x| \to \infty$ as $x \to 1$ so the sequence $\{p_n\}$ does not converge uniformly to f on S.
(c) On $[0, c]$, the function f has a bounded derivative, so f is uniformly continuous. Furthermore, we have the estimate

$$|p_n(x) - f(x)| \leq \frac{c^{n+1}}{1 - c},$$

so $\{p_n\}$ converges uniformly to f on $[0, c]$.

Exercise VII.3.5 *Let $f_n(x) = x^2/(1 + nx^2)$ for all real x. Show that the sequence $\{f_n\}$ converges uniformly on $\mathbf{R}$.*

Solution. For each $n \geq 0$ the function f_n is even, positive, and differentiable on $\mathbf{R}$ with

$$f_n'(x) = \frac{2x}{(1 + nx^2)^2},$$

and $f_n(x)$ tends to $1/n$ as $x \to \pm\infty$. Therefore $|f_n(x)| \leq 1/n$ which proves that $\{f_n\}$ converges uniformly to 0 as $n \to \infty$.

Exercise VII.3.6 *Consider the function defined by*

$$f(x) = \lim_{m \to \infty} \lim_{n \to \infty} (\cos m! \pi x)^{2n}.$$

Find explicitly the values of f at rational and irrational numbers.

Solution. Suppose $x = p/q$, then for large m, $m \geq 2|q|$, hence $m!\pi x = 2\pi k$ for some $k \in \mathbf{Z}$ thus $f(x) = 1$.
If x is irrational, then $|\cos(m!\pi x)| < 1$ for all positive integers m so $f(x) = 0$.

Exercise VII.3.7 *As in Exercise 5 of* §2, *let*

$$f(x,y) = \begin{cases} \frac{(y^2-x)^2}{y^4+x^2} & \text{if } (x,y) \neq (0,0), \\ 1 & \text{if } (x,y) = (0,0). \end{cases}$$

Is f *continuous on* $\mathbf{R}^2$? *Explain, and determine all points where* f *is continuous. Determine the limits*

$$\lim_{x \to 0} \lim_{y \to 0} f(x,y) \quad \text{and} \quad \lim_{y \to 0} \lim_{x \to 0} f(x,y).$$

Solution. The function f is continuous on $\mathbf{R}^2 - \{0\}$. The only problem is at the origin and we have shown in Exercise 5, §2, that f is not continuous at the origin. However, $\lim_{y \to 0} f(x,y) = 1$ and $\lim_{x \to 0} f(x,y) = 1$ so

$$\lim_{x \to 0} \lim_{y \to 0} f(x,y) = \lim_{y \to 0} \lim_{x \to 0} f(x,y) = 1.$$

Exercise VII.3.8 *Let* S, T *be subsets of a normed vector spaces. Let* f: $S \to T$ *and* $g: T \to F$ *be mappings, with* F *a normed vector space. Assume that* g *is uniformly continuous. Prove: Given* ϵ, *there exists* δ *such that if* $f_1: S \to T$ *is a map such that* $\|f - f_1\| < \delta$, *and* $g_1: T \to F$ *is a map such that* $\|g - g_1\| < \epsilon$, *then* $\|g \circ f - g_1 \circ f_1\| < 2\epsilon$.

In other words, if f_1 *approximates* f *uniformly and* g_1 *approximates* g *uniformly, then* $g_1 \circ f_1$ *approximates* $g \circ f$ *uniformly. One can apply this result to polynomial approximations obtained from Taylor's formula, to reduce computations to polynomial computations, within a given degree of approximation.*

Solution. For $x \in S$ the triangle inequality gives

$$|g \circ f(x) - g_1 \circ f_1(x)| \leq |g \circ f(x) - g \circ f_1(x)| + |g \circ f_1(x) - g_1 \circ f_1(x)|.$$

Pick δ such that if $y, y_1 \in T$ and $|y - y_1| < \delta$, then $|g(y) - g(y_1)| < \epsilon$. Apply this with $y = f(x)$ and $y_1 = f_1(x)$. Then the first term on the right is $< \epsilon$ and the second term is also $< \epsilon$ by the hypothesis that $\|g - g_1\| < \epsilon$.

Exercise VII.3.9 *Give a Taylor formula type proof that the absolute value can be approximated uniformly by polynomials on a finite closed interval* $[-c, c]$. *First, reduce it to the interval* $[-1, 1]$ *by multiplying the variable by* c *or* c^{-1} *as the case may be. Then write* $|t| = \sqrt{t^2}$. *Select* δ *small,* $0 < \delta < 1$. *If we can approximate* $(t^2 + \delta)^{1/2}$, *then we can approximate* $\sqrt{t^2}$. *Now to get* $(t^2 + \delta)^{1/2}$ *either use the Taylor series approximation for the square root function, or if you don't like the binomial expansion, first approximate*

$$\log(t^2 + \delta)^{1/2} = \frac{1}{2} \log(t^2 + \delta)$$

by a polynomial P. *This works because the Taylor formula for the log converges uniformly for* $\delta \leq u \leq 2A - \delta$. *Then take a sufficiently large number* *of terms from the Taylor formula for the exponential function, say a poly-nomial Q, and use $Q \circ P$ to solve your problems. Cf. Exercise 7 of Chapter V, §3.*

Solution. Given ϵ, uniform continuity implies that there exists δ which we select small such that for all $t \in [0, 1]$ we have

$$|\sqrt{t^2} - \sqrt{t^2 + \delta}| < \epsilon.$$

By the estimates given in Chapter 5, we see that the Taylor polynomial of $\sqrt{1 + X}$ converges uniformly on $-1 + \delta \leq X \leq \delta$. We then substitute $X = t^2 + \delta - 1$.

We can use Exercise 8, with $f = \frac{1}{2} \log$, $g = \exp$, and f_1, g_1 their respective Taylor polynomials. The detailed proof runs as follows: by uniform conti-nuity on a closed and bounded interval $[-C, C]$ where C is large positive, we see that there exists ϵ_1 such that if $|a - b| < \epsilon_1$, then $|e^a - e^b| < \epsilon$. By the uniform convergence of the Taylor polynomial for the log on $[\delta, \delta + 1]$ we see that given ϵ_1 there exists a polynomial P such that $|\frac{1}{2} \log X - \frac{1}{2} P(X)| < \epsilon_1$ (we have in mind $X = t^2 + \delta$). Thus

$$\left| e^{\frac{1}{2} \log X} - e^{\frac{1}{2} P(X)} \right| < \epsilon. \tag{VII.1}$$

Similarly, the uniform convergence of the Taylor polynomial for the expo-nential on an arbitrarily large closed interval containing the origin, guar-antees the existence of a polynomial Q such that $|e^Y - Q(Y)| < \epsilon$. So

$$\left| e^{\frac{1}{2} P(X)} - Q(\frac{1}{2} P(X)) \right| < \epsilon. \tag{VII.2}$$

Combining equations (VII.1) and (VII.2) and subsituting $X = t^2 + \delta$ we get

$$\left| \sqrt{t^2 + \delta} - Q(\frac{1}{2} P(t^2 + \delta)) \right| < 2\epsilon.$$

Exercise VII.3.10 *Give another proof for the preceding fact, by using the sequence of polynomials $\{P_n\}$, starting with $P_0(t) = 0$ and letting*

$$P_{n+1}(t) = P_n(t) + \frac{1}{2}(t - P_n(t)^2).$$

Show that $\{P_n\}$ tends to $\sqrt{t}$ uniformly on $[0, 1]$, showing by induction that

$$0 \leq \sqrt{t} - P_n(t) \leq \frac{2\sqrt{t}}{2 + n\sqrt{t}},$$

whence $0 \leq \sqrt{t} - P_n(t) \leq 2/n$.

Solution. We prove by induction the inequalitites and the fact that $P_n(t) \geq 0$. Since $0 \leq \sqrt{t} - 0 \leq \sqrt{t}$ and $P_0(t) = 0$ the assertion is true when $n = 0$. Assume the assertion is true for an integer n. Since

$$\sqrt{t} - P_{n+1}(t) = (\sqrt{t} - P_n(t))\left(1 - \frac{1}{2}(\sqrt{t} + P_n(t))\right)$$

and $P_n(t) \leq \sqrt{t}$ we have

$$\frac{1}{2}(\sqrt{t} + P_n(t)) \leq \sqrt{t} \leq 1,$$

so $0 \leq \sqrt{t} - P_{n+1}(t)$. Furthermore, if $P_n(t) \geq 0$, then $0 \leq P_n^2(t) \leq t$ which implies $P_{n+1}(t) \geq 0$. Finally, for the second inequality note that

$$1 - \frac{1}{2}\left(\sqrt{t} + P_n(t)\right) \leq 1 - \frac{\sqrt{t}}{2},$$

so

$$0 \leq \sqrt{t} - P_{n+1}(t) \leq \frac{2\sqrt{t}}{2 + n\sqrt{t}}\left(1 - \frac{\sqrt{t}}{2}\right).$$

But

$$\frac{2\sqrt{t}}{2 + n\sqrt{t}}\left(1 - \frac{\sqrt{t}}{2}\right) = \frac{2\sqrt{t}}{2 + (n+1)\sqrt{t}}\frac{2(2 + n\sqrt{t}) - (n+1)t}{2(2 + n\sqrt{t})},$$

so the inequality drops out.

VIII
Compactness

VIII.1 Basic Properties of Compact Sets

Exercise VIII.1.1 *Let S be a compact set. Show that every Cauchy sequence of elements of S has a limit in S.*

Solution. Let $\{x_n\}$ be a Cauchy sequence. This sequence has a subsequence $\{x_{n_k}\}$ which converges to some x in S. Given $\epsilon > 0$, there exists positive integers N and M such that for all $n, m > N$ we have $|x_n - x_m| < \epsilon$ and such that for all $k > M$ we have $|x_{n_k} - x| < \epsilon$. Select k such that $k > M$ and $n_k > N$. Then for all $n > N$ we have

$$|x_n - x| \leq |x_n - x_{n_k}| - |x_{n_k} - x| < 2\epsilon.$$

Exercise VIII.1.2 *(a) Let $S_1, \ldots, S_m$ be a finite number of compact sets in E. Show that the union $S_1 \cup \cdots \cup S_m$ is compact.*
(b) Let $\{S_i\}_{i \in I}$ be a family of compact sets. Show that the intersection $\bigcap_{i \in I} S_i$ is compact. Of course, it may be empty.

Solution. (a) We prove the result for two sets. Let S be an infinite subset of $S_1 \cup S_2$. Assume that $S \cap S_1$ is infinite (if not, then $S \cap S_2$ must be infinite and the argument is the same) so that $S \cap S_1$ has a point of accumulation in S_1. Therefore S has a point of accumulation in $S_1 \cup S_2$. By induction we conclude that $S_1 \cup \cdots \cup S_m$ is compact.
(b) Let S be any member of the family. Then $\bigcap_{i \in I} S_i$ is a closed subset of S and is therefore compact by Theorem 1.2.

Exercise VIII.1.3 *Show that a denumerable union of compact sets need not be compact.*

Solution. Each interval $I_n = [-n, n]$ is compact but the union $\bigcup_n I_n$ is unbounded and therefore not compact.

Exercise VIII.1.4 *Let $\{x_n\}$ be a sequence in a normed vector space E such that $\{x_n\}$ converges to v. Let S be the set consisting of all x_n and v. Show that S is compact.*

Solution. Let $\{v_k\}$ be a sequence in S. If there are infinititely many k's such that v_k is equal to the same element of S, then $\{v_k\}$ has a converging subsequence in S. If one element of S is not repeated infnitely many times in $\{v_k\}$, then we can find a subsequence of $\{v_k\}$ which is a subsequence of $\{x_n\}$ and which therefore converges to v in S.

VIII.2   Continuous Maps on Compact Sets

Exercise VIII.2.1 *Let $S \subset T$ be subsets of a normed vector space E. Let $f : T \to F$ be a mapping into some normed vector space. We say that f is **relatively uniformly continuous** on S if given ϵ there exists δ such that whenever $x \in S$, $y \in T$, and $|x - y| < \delta$, then $|f(x) - f(y)| < \epsilon$. Assume that S is compact and f is continuous at every point of S. Verify that the proof of Theorem 2.3 yields that f is relatively uniformly continuous on S.*

Solution. Suppose that f is not relatively uniformly continuous on S. Then there exists an $\epsilon > 0$ and for each positive integer n there exists $x_n \in S$ and $y_n \in T$ such that $|x_n - y_n| < 1/n$ but $|f(x_n) - f(y_n)| > \epsilon$. There exists an infinite subset J_1 of $\mathbf{Z}^+$ and a $v \in S$ such that $x_n \to v$ as $n \to \infty$, $n \in J_1$.
  The inequality
$$|v - y_n| \leq |v - x_n| + |x_n - y_n|$$
implies that $y_n \to v$ as $n \to \infty$, $n \in J_1$. Then the continuity of f at v implies that $|f(x_n) - f(y_n)|$ approaches 0 as $n \to \infty$, $n \in J_1$, which gives us the desired contradiction.

Exercise VIII.2.2 *Let S be a subset of a normed vector space. Let $f : S \to F$ be a map of S into a normed vector space. Show that f is continuous on S if and only if the restriction of f to every compact subset of S is continuous. [Hint: Given $v \in S$, consider sequences of elements of S converging to v.]*

Solution. Suppose the restriction of f to every compact subset is continuous. Let $v \in S$ and consider a sequence $\{x_n\}$ which converges to v. Then the subset $T = \{v, x_1, \ldots, x_n, \ldots\}$ of S is compact and therefore $\lim_{n \to \infty} f(x_n) = f(v)$. This proves that f is continuous on S.
  The converse is clear.

Exercise VIII.2.3 *Prove that two norms on $\mathbf{R}^n$ are equivalent by the following method. Use the fact that a continuous function on a compact set has a minimum. Take a norm to be the function, and let the compact set be the unit sphere for the sup norm.*

Solution. Suppose that we want to show that the norm $|\cdot|_1$ is equivalent to the sup norm $|\cdot|_\infty$. Equip $\mathbf{R}^n$ with the sup norm. The function $f(v) = |v|_1$ is continuous because

$$||x|_1 - |y|_1| \leq |x - y|_1 \leq C|x - y|_\infty,$$

where $C = \sum_{i=1}^n |e_i|_1$. The unit sphere is compact for the sup norm, therefore f attains its minimum and maximum, say at x and y, respectively. Then

$$|x|_1 \leq |v|_1 \leq |y|_1.$$

Given any $v \in \mathbf{R}^n$, $v \neq 0$ we have $|v/|v|_\infty|_\infty = 1$, so if we let $C_1 = |x|_1$ and $C_2 = |y|_1$ we obtain

$$C_1 \leq \left| \frac{v}{|v|_\infty} \right|_1 \leq C_2$$

which implies

$$C_1|v|_\infty \leq |v|_1 \leq C_2|v|_\infty.$$

Exercise VIII.2.4 *(Continuation of Exercise 1, Chapter VII, §2). Let $E = \mathbf{R}^k$ and let S be a closed subset of $\mathbf{R}^k$. Let $v \in \mathbf{R}^k$. Show that there exists a point $w \in S$ such that*

$$d(S, v) = |w - v|.$$

[Hint: Let B be a closed ball of some suitable radius, centered at v, and consider the function $x \mapsto |x - v|$ for $x \in B \cap S$.]

Solution. Consider a closed ball centered at v of radius large enough so that $B \cap S$ is not empty. Since $E = \mathbf{R}^k$, and $B \cap S$ is closed and bounded, we conclude that $B \cap S$ is compact. The function defined by $f_v(x) = |x - v|$ is continuous, and therefore f_v has a minimum on $B \cap S$.

Exercise VIII.2.5 *Let K be a compact set in $\mathbf{R}^k$ and let S be a closed subset of $\mathbf{R}^k$. Define*

$$d(K, S) = \text{glb}_{x \in K, y \in S}|x - y|.$$

Show that there exist elements $x_0 \in K$ and $y_0 \in S$ such that

$$d(K, S) = |x_0 - y_0|.$$

[Hint: Consider the continuous map $x \mapsto d(S, x)$ for $x \in K$.]

Solution. The function defined by $f(x) = d(x, S)$ is continuous (Exercise 1, §2, of Chapter 7) and Theorem 2.2 implies that for some $x_0 \in K$, the function f attains its minimum. Exercise 4 implies that there exists $y_0 \in S$ such that $f(x_0) = |x_0 - y_0|$. Then $d(K, S) = |x_0 - y_0|$. Indeed, $d(x, S) \leq |x - y|$ for all $x \in K$ and $y \in S$, thus $f(x_0) \leq d(K, S)$ and the reverse inequality, $d(K, S) \leq f(x_0)$ is obvious.

Exercise VIII.2.6 *Let K be a compact set, and let $f : K \to K$ be a continuous map. Suppose that f is expanding, in the sense that*

$$|f(x) - f(y)| \geq |x - y|$$

for all $x, y \in K$.
(a) Show that f is injective and that the inverse map $f^{-1} : f(K) \to K$ is continuous.
(b) Show that $f(K) = K$. [Hint: Given $x_0 \in K$, consider the sequence $\{f^n(x_0)\}$, where f^n is the n-th iterate of f. You might use Corollary 2.3.]

Solution. (a) If $f(x) = f(y)$, then $0 \geq |x - y|$ so $x = y$. The inverse function is continuous because

$$|x - y| = |f(f^{-1}(x)) - f(f^{-1}(y))| \geq |f^{-1}(x) - f^{-1}(y)|.$$

(b) Suppose there exists $x \in K$ such that $x \notin f(K)$. Since $f(K)$ is compact, Exercise 5 implies that $d(f(K), x) = d > 0$. Let $x_0 = x$ and define the sequence $\{x_n\}$ by $x_{n+1} = f(x_n)$. Since K is compact, there exists a subsequence $\{x_{n_k}\}$ which converges in K. So there exists $m \in \mathbf{N}$ such that $|x_{n_{m+1}} - x_{n_m}| < d/2$. We can write $x_{n_{m+1}} = x_p$ and $x_{n_m} = x_q$ for some positive integers p, q with $p > q$. Then

$$|x_{n_{m+1}} - x_{n_m}| = |f(x_{p-1}) - f(x_{q-1})| \geq |x_{p-1} - x_{q-1}|,$$

and therefore repeating the process we obtain

$$|x_{n_{m+1}} - x_{n_m}| \geq |x_{p-1} - x_{q-1}| \geq \cdots \geq |x_{p-q} - x_0| \geq d,$$

and the last inequality holds because x_{p-q} belongs to $f(K)$. We get a contradiction which proves that $f(K) = K$.

Exercise VIII.2.7 *Let U be an open subset of $\mathbf{R}^n$. Show that there exists a sequence of compact subsets K_j of U such that $K_j \subset \mathrm{Int}(K_{j+1})$ for all j, and such that the union of all K_j is U. [Hint: Let $\overline{B}_j$ be the closed ball of radius j, and let K_j be the set of points $x \in \overline{U} \cap \overline{B}_j$ such that $d(x, \partial U) \geq 1/j.]$*

Solution. Define K_j as in the hint. Let $f(x) = d(x, \partial U)$ so that f is a continuous function on $\mathbf{R}^n$. Since $[1/j, \infty)$ is closed, it follows that $f^{-1}([1/j, \infty))$ is closed, namely, the set of all $x \in \mathbf{R}^n$ such that $d(x, \partial U) \geq 1/j$ is closed.

Hence its intersection with the closed and bounded set $\overline{U} \cap \overline{B}_j$ is compact, so K_j is compact. As for various inclusions, we have

$$K_j \subset \{x \in U \cap B_{j+1} \text{ such that } d(x, \partial U) > 1/(j+1)\} \quad \text{(VIII.1)}$$
$$\subset \{x \in \overline{U} \cap \overline{B}_{j+1} \text{ such that } d(x, \partial U) \geq 1/(j+1)\} \quad \text{(VIII.2)}$$
$$= K_{j+1}. \quad \text{(VIII.3)}$$

Again, since f is continuous, the set on the right of the first inclusion is open because it is the intersection with $U \cap B_{j+1}$ with the inverse image under f of the open interval $(1/(j+1), \infty)$. Hence (VIII.1) shows that $K_j \subset \text{Int}(U)$ and (VIII.2) shows that $K_j \subset \text{Int}(K_{j+1})$. Finally we show that the union of all the K_j is equal to U. Observe that every point of U is at distance > 0 from the boundary of U because the boundary is closed. Pick j such that $1/j < d(x, \partial U)$ and $|x| < j$, so by definition we have $x \in K_j$, which concludes the proof.

VIII.4   Relation with Open Coverings

Exercise VIII.4.1 *Let $\{U_1, \ldots, U_m\}$ be an open covering of a compact subset S of a normed vector space. Prove that there exists a number $r > 0$ such that if $x, y \in S$ and $|x - y| < r$, then x and y are contained in U_i for some i.*

Solution. Suppose that such a number r does not exists. Then for each n there exists a point $x_n \in S$ such that $B_{1/n}(x_n)$ is not contained in U_i for any i. There exists a subsequence $\{x_{n_k}\}$ of $\{x_n\}$ which converges to x. Pick m' such that $x \in U_{m'}$. Since $U_{m'}$ is open, there exists a positive integer N such that $B_{1/N}(x) \subset U_{m'}$. We can find p such that for all $k > p$ we have $|x_{n_k} - x| < 1/2N$. Then for $k > \max(p, 2N)$ we have $B_{1/n_k}(x_{n_k}) \subset U'_m$, a contradiction.

Exercise VIII.4.2 *Let $\{S_i\}_{i \in I}$ be a family of compact subsets of a normed vector space E. Suppose the intersection $\bigcap_{i \in I} S_i$ is empty. Prove that there is a finite number of indices $i_1, \ldots, i_n$ such that*

$$S_{i_1} \cap \cdots \cap S_{i_n} \text{ is empty.}$$

This is the "dual" property of the finite covering property.

Solution. Let S be one member of the family. Let $U_i = S_i^c$, so that U_i is open. Then by hypothesis, the family $\{U_i\}$ covers S, whence a finite number $U_{i_1}, \ldots, U_{i_n}$ covers S by Theorem 4.2, that is $S \subset U_{i_1} \cup \cdots \cup U_{i_n}$. Taking complements shows that $S_{i_1} \cap \cdots \cap S_{i_n} \subset S^c$, so the intersection $S \cap S_{i_1} \cap \cdots \cap S_{i_n}$ is empty, as was to be shown.

Exercise VIII.4.3 *Let S be a compact set and let R be the set of continuous real valued functions on S. Let I be a subset of R containing 0, and having the following properties:*

(i) If $f, g \in I$, then $f + g \in I$.

(ii) If $f \in I$ and $h \in R$, then $hf \in I$.

Such a subset is called an **ideal** *of R. Let Z be the set of points $x \in S$ such that $f(x) = 0$ for all $f \in I$. We call Z the set of* **zeros** *of I.*

(a) Prove that Z is closed, expressing Z as an intersection of closed sets.

(b) Let $f \in R$ be a function which vanishes on Z, i.e. $f(x) = 0$ for all $x \in Z$. Show that f can be uniformly approximated by elements of I. [Hint: Given ϵ, let C be the closed set of elements $x \in S$ such that $|f(x)| \geq \epsilon$. For each $x \in C$, there exists $g \in I$ such that $g(x) \neq 0$ in a neighborhood of C. Cover C with a finite number of them, corresponding to functions $g_1, \ldots, g_r$. Let $g = g_1^2 + \cdots + g_r^2$. Then $g \in I$. Furthermore, g has a minimum on C, and for n large, the function

$$f \frac{ng}{1 + ng}$$

is close to f on C, and its absolute value is $< \epsilon$ on the complement of C in S. Justify all the details of this proof.]

Solution. (a) Given $f \in I$, let $X_{f,n} = \{x \in S : |f(x)| \leq 1/n\}$. Since f is continuous $X_{f,n}$ is closed. So $Z(f) = \bigcap_{n=1}^{\infty} X_{f,n}$ is closed and $Z = \bigcap_{f \in I} Z(f)$ is also closed.

(b) If $x \in C$, then $x \notin Z$, so we can find a function $g \in I$ such that $g(x) \neq 0$. By continuity, g is non-zero in a neighborhood V_x of x. Then $\bigcup_{x \in C} V_x$ is an open covering of C from which we can select a finite subcovering of C because C is compact. Let $g_1, \ldots, g_r$ be the functions corresponding to the finite subcover, and let $g = g_1^2 + \cdots + g_r^2$. The function g belongs to I because of properties (i) and (ii). Furthermore, g is continuous on C, it is ≥ 0 and nowhere 0, otherwise $g_1(y) = \cdots = g_r(y) = 0$ for an element $y \in C$ where y belongs to one of the sets of the subcover. This is a contradiction. The function g attains a minimum $a > 0$ on C. Consider the function

$$\frac{ng}{1 + ng}.$$

This function belongs to I and Exercise 2, §3, of Chapter 7 implies that $ng/(1 + ng)$ tends uniformly to 1 on C. Thus

$$f \frac{ng}{1 + ng}$$

which belongs to I tends uniformly to f on C, so for all large n we have

$$\left| f - f \frac{ng}{1 + ng} \right| < \epsilon.$$

On the complement we have the estimate

$$\left| f - f\frac{ng}{1+ng} \right| \le |f| + \left| f\frac{ng}{1+ng} \right| < 2\epsilon.$$

IX

Series

IX.2 Series of Positive Numbers

Exercise IX.2.1 *(a) Prove the convergence of the series $\sum 1/n(\log n)^{1+\epsilon}$ for every $\epsilon > 0$.*
(b) Does the series $\sum 1/n \log n$ converge? Proof?
(c) Does the series $\sum 1/n(\log n)(\log \log n)$ converge? Proof? What if you stick an exponent of $1 + \epsilon$ to the $(\log \log n)$?

Solution. (a) We have $\int_2^\infty 1/x(\log x)^{1+\epsilon} dx < \infty$ because

$$\int_2^B \frac{1}{x(\log x)^{1+\epsilon}} dx = \left[\frac{-1}{\epsilon}(\log x)^{-\epsilon}\right]_2^B,$$

so the series $\sum 1/n(\log n)^{1+\epsilon}$ converges.
(b) The integral test shows that the series diverges. Indeed,

$$\int_2^B \frac{1}{x \log x} dx = [\log \log x]_2^B = \log \log B - \log \log 2 \to \infty$$

as $B \to \infty$.
(c) Again, the integral test shows that the series diverges,

$$\int_3^B \frac{1}{x(\log x)(\log \log x)} dx = [\log \log \log x]_3^B = \log \log \log B - \log \log \log 3$$

and this last expression $\to \infty$ as $B \to \infty$.

For any $\epsilon > 0$, the series $\sum 1/n(\log n)(\log \log n)^{1+\epsilon}$ converges because

$$\int_3^B \frac{1}{x(\log x)(\log \log x)^{1+\epsilon}} dx = \left[-\frac{1}{\epsilon}(\log \log x)^{-\epsilon} \right]_3^B.$$

Exercise IX.2.2 *Let $\sum a_n$ be a series of terms ≥ 0. Assume that there exist infinitely many integers n such that $a_n > 1/n$. Assume that the sequence $\{a_n\}$ is decreasing. Show that $\sum a_n$ diverges.*

Solution. Let $E = \{n \in \mathbf{N} : a_n > 1/n\}$. Then E is unbounded. Let $n_0 = 1$ and let n_{j+1} be the smallest integer in E such that $n_{j+1} > 2n_j$. Then using the fact that $\{a_n\}$ is decreasing we conclude that for every positive integer m we have

$$\sum_{k=1}^{n_m} a_k \geq \sum_{j=1}^{m}(n_j - n_{j-1})\frac{1}{n_j} > \sum_{j=1}^{m} 1 - \frac{1}{2}.$$

But $n_j \to \infty$ as $j \to \infty$ whence the partial sums are unbounded, and the series diverges.

Exercise IX.2.3 *Let $\sum a_n$ be a convergent series of numbers ≥ 0, and let $\{b_1, b_2, b_3, \dots\}$ be a bounded sequence of numbers. Show that $\sum a_n b_n$ converges.*

Solution. Let B be a bound for the sequence $\{b_n\}$. Given $\epsilon > 0$, there exists a positive integer N such that for all $n > m > N$ we have $0 \leq \sum_{k=m+1}^n a_k \leq \epsilon/B$. Furthermore, we have $|a_n b_n| \leq B|a_n|$, so for all $n > m > N$ we have

$$\left| \sum_{k=m+1}^n a_k b_k \right| \leq \sum_{k=m+1}^n |a_k b_k| < \epsilon.$$

Thus the partial sums of $\sum a_n b_n$ form a Cauchy sequence, as was to be shown.

Exercise IX.2.4 *Show that $\sum(\log n)/n^2$ converges. If $s > 1$, does $\sum(\log n)^3/n^s$ converge? Given a positive integer d, does $\sum(\log n)^d/n^s$ converge?*

Solution. Let $s = 2$ in the text. Now we prove the general result. Let d be a positive integer, and write $s = 1 + 2\epsilon$ with $\epsilon > 0$. Then

$$\frac{(\log n)^d}{n^{1+2\epsilon}} = \frac{1}{n^{1+\epsilon}}\frac{(\log n)^d}{n^\epsilon},$$

and for all large n we have $(\log n)^d \leq n^\epsilon$ whence we conclude that the series $\sum(\log n)^d/n^s$ converges.

Exercise IX.2.5 *(a) Let $n! = n(n-1)(n-2)\cdots 1$ be the product of the first n integers. Using the ratio test, show that $\sum 1/n!$ converges.*
(b) Show that $\sum 1/n^n$ converges. For any number x, show that $\sum x^n/n!$ converges and so does $\sum x^n/n^n$.

Solution. (a) Let $a_n = 1/n!$. For all $n \geq 1$, $a_{n+1} \leq \frac{1}{2}a_n$ because

$$\frac{a_{n+1}}{a_n} = \frac{1/(n+1)!}{1/n!} = \frac{1}{n+1}.$$

(b) Since $n^n \geq n!$ the comparison test implies that $\sum 1/n^n$ converges.

Suppose $x > 0$ and let $a_n = x^n/n!$. Then $a_{n+1}/a_n = x/(n+1)$ which tends to 0 as $n \to \infty$, so the ratio test implies the convergence of the series $\sum x^n/n!$.

If $x < 0$, let $a_n = |x|^n/n!$ and $b_n = (-1)^n$. Exercise 3 implies the convergence of the series $\sum x_n/n!$.

Finally, if $x = 0$, then $\sum x_n/n! = 0$.

Note that for $x > 0$, the series $\sum x^n/n^n$ converges, and this result also holds for $x \leq 0$.

Exercise IX.2.6 *Let k be an integer ≥ 2. Show that*

$$\sum_{n=k}^{\infty} 1/n^2 < 1/(k-1).$$

Solution. Using an upper sum estimate, as in the integral test, we find that

$$\sum_{n=k}^{\infty} \frac{1}{n^2} < \int_{k-1}^{\infty} \frac{dx}{x^2} = \frac{1}{k-1}.$$

We could also argue as follows. For each $n \geq 2$ we have $n^2 > n(n-1)$, so for all $m > k$ we obtain

$$\sum_{n=k}^{m} \frac{1}{n^2} < \sum_{n=k}^{m} \frac{1}{n(n-1)} = \sum_{n=k}^{m} \frac{1}{n-1} - \frac{1}{n} = \frac{1}{k-1} - \frac{1}{m}.$$

Letting $m \to \infty$ we see that

$$\sum_{n=k}^{\infty} \frac{1}{n^2} \leq \frac{1}{k-1}.$$

Suppose that for some k, this last inequality is an equality. Then we have

$$\sum_{n=k}^{\infty} \frac{1}{n^2} = \sum_{n=k}^{\infty} \frac{1}{n(n-1)},$$

which implies

$$\sum_{n=k}^{\infty} \left(\frac{1}{n(n-1)} - \frac{1}{n^2} \right) = 0.$$

But $1/n(n-1) - 1/n^2 > 0$ for all n, so we get a contradiction. We conclude that the inequality is always strict, that is

$$\sum_{n=k}^{\infty} \frac{1}{n^2} < 1/(k-1),$$

for all $k \geq 2$.

Exercise IX.2.7 *Let $\sum a_n^2$ and $\sum b_n^2$ converge, assuming $a_n \geq 0$ and $b_n \geq 0$ for all n. Show that $\sum a_n b_n$ converges. [Hint: Use the Schwarz inequality but be careful: The Schwarz inequality has so far been proved only for finite sequences.]*

Solution. For each integer $m > 0$, consider the m-tuples $(a_1, \ldots, a_m)$ and $(b_1, \ldots, b_m)$. The standard inner product in $\mathbf{R}^m$ and the Schwarz inequality imply

$$\sum_{n=1}^{m} |a_n b_n| \leq \left(\sum_{n=1}^{m} a_n^2\right)^{1/2} \left(\sum_{n=1}^{m} b_n^2\right)^{1/2} \leq \left(\sum a_n^2\right)^{1/2} \left(\sum b_n^2\right)^{1/2}$$

so the partial sums are bounded.

Exercise IX.2.8 *Let $\{a_n\}$ be a sequence of numbers ≥ 0, and assume that the series $\sum a_n/n^s$ converges for some number $s = s_0$. Show that the series converges for $s \geq s_0$.*

Solution. If $s \geq s_0$, then $1/n^s \leq 1/n^{s_0}$ so $\sum a_n/n^s$ converges by comparision with $\sum a_n/n^{s_0}$.

Exercise IX.2.9 *Let $\{a_n\}$ be a sequence of numbers ≥ 0 such that $\sum a_n$ diverges. Show that:*
(a) $\sum \frac{a_n}{1+a_n}$ diverges.
(b) $\sum \frac{a_n}{1+n^2 a_n}$ converges.
(c) $\sum \frac{a_n}{1+na_n}$ sometimes converges and sometimes diverges.
(d) $\sum \frac{a_n}{1+a_n^2}$ sometimes converges and sometimes diverges.

Solution. (a) Suppose the series $\sum \frac{a_n}{1+a_n}$ converges. There is some n_0 such that for all $n \geq n_0$ we have

$$0 < \frac{a_n}{1+a_n} = b_n < \frac{1}{2}.$$

Since $a_n = b_n/(1-b_n)$ we conclude that for all $n \geq n_0$, $0 \leq a_n \leq 2b_n$ which implies the convergence of $\sum a_n$. This gives us the desired contradiction.
(b) The series $\sum \frac{a_n}{1+n^2 a_n}$ converges because

$$0 \leq \sum \frac{a_n}{1+n^2 a_n} \leq \frac{1}{n^2}.$$

(c) If we put $a_n = 1$, then $\sum \frac{a_n}{1+na_n}$ diverges. Suppose $a_n = 1$ whenever n is a perfect square, i.e. $\sqrt{n} \in \mathbf{Z}^+$ and $a_n = 0$ otherwise. Then $\sum \frac{a_n}{1+na_n}$ converges.

(d) If we put $a_n = 1$, then $\sum \frac{a_n}{1+a_n^2}$ diverges. Choose a number $c > 1$ and let $a_n = c^n$. Then

$$\frac{a_n}{1 + a_n^2} \leq \frac{1}{a_n} = \frac{1}{c^n},$$

so $\sum \frac{a_n}{1+a_n^2}$ converges by comparision to the geometric series.

Exercise IX.2.10 *Let $\{a_n\}$ be a sequence of real numbers ≥ 0 and assume that $\lim a_n = 0$. Let*

$$\prod_{k=1}^{n}(1 + a_k) = (1 + a_1)(1 + a_2) \cdots (1 + a_n).$$

We say that the product **converges** *as $n \to \infty$ if the limit of the preceding product exists, in which case it is denoted by*

$$\prod_{k=1}^{\infty}(1 + a_k).$$

Assume that $\sum a_n$ converges. Show that the product converges. [Hint: Take the log of the finite product, and compare $\log(1 + a_k)$ with a_k. Then take exp.]

Solution. Taking the log of the partial product we obtain $\sum_{k=1}^{n} \log(1+a_k)$. But $\log(1 + a_k) \leq a_k$, so $\sum \log(1 + a_k)$ converges. The exponential map is continuous and

$$\exp\left(\sum_{k=1}^{n} \log(1 + a_k)\right) = \prod_{k=1}^{n}(1 + a_k),$$

so the infinite product converges.

Exercise IX.2.11 (Decimal Expansions) *(a) Let α be a real number with $0 \leq \alpha \leq 1$. Show that there exist integers a_n with $0 \leq a_n \leq 9$ such that*

$$\alpha = \sum_{n=1}^{\infty} \frac{a_n}{10^n}.$$

The sequence $(a_1, a_2, \dots)$ or the series $\sum a_n/10^n$ is called a **decimal expansion** *of α. [Hint: Cut $[0,1]$ into 10 pieces, then into 10^2, etc.]*
(b) Let $\alpha = \sum_{k=m}^{\infty} a_k/10^k$ with numbers a_k such that $|a_k| \leq 9$. Show that $|\alpha| \leq 1/10^{m-1}$.
(c) Conversely, let $\{a_k\}$ be integers with $|a_k| \leq 9$, $a_k \neq \pm 1$ for all k. Let

$$\alpha = \sum_{k=1}^{\infty} \frac{a_k}{10^k}.$$

*Suppose that $|\alpha| \leq 1/10^N$ for some positive integer N. Show that $a_k = 0$
for $k = 1, \ldots, N - 1$.*
*(d) Let $\alpha = \sum_{k=1}^{\infty} a_k/10^k = \sum_{k=1}^{\infty} b_k/10^k$ with integers a_k, b_k such that
$0 \leq a_k \leq 9$ and $0 \leq b_k \leq 9$. Assume that there exist arbitrarily large k such
that $a_k \neq 9$ and similarly $b_k \neq 9$. Show that $a_k = b_k$ for all k.*

Solution. (a) Cut up the interval $[0, 1]$ in 10 equal segments of length $1/10$,
namely, let

$$I_0^1 = \left[0, \frac{1}{10}\right], \ldots, I_j^1 = \left[\frac{j}{10}, \frac{j+1}{10}\right], \ldots, I_9^1 = \left[\frac{9}{10}, 1\right].$$

Pick j such that $\alpha \in I_j^1$, and let $a_1 = j$. Then $|\alpha - a_1| < 1/10$. Then cut
up the interval $I_{a_1}^1$ in 10 , namely consider

$$I_j^2 = \left[\frac{a_1}{10} + \frac{j}{10^2}, \frac{a_1}{10} + \frac{j+1}{10^2}\right]$$

for $j = 0, \ldots, 9$. Pick j such that $\alpha \in I_j^2$, and let $a_2 = j$. Then

$$\left|\alpha - \left(\frac{a_1}{10} + \frac{a_2}{10^2}\right)\right| \leq \frac{1}{10^2}.$$

Proceeding by induction we get a sequence $(a_1, a_2, \ldots)$ such that $0 \leq a_n \leq 9$
and such that

$$\left|\alpha - \sum_{n=1}^{m} \frac{a_n}{10^n}\right| \leq \frac{1}{10^m}.$$

Hence $\alpha = \sum a_n/10^n$.
(b) The inequality follows from

$$|\alpha| \leq \sum_{k=m}^{\infty} \frac{|a_k|}{10^k} \leq \frac{9}{10^m} \sum_{k=0}^{\infty} \frac{1}{10^k}$$

and the fact that $\sum_{k=0}^{\infty} 1/10^k = 10/9$.
(c) Write $\alpha = \sum_{k=m}^{\infty} a_k/10^k$ with $a_m \neq 0$ and $m \leq N - 1$. Then

$$|a_m| \leq |10^m \alpha - a_m| + |10^m \alpha|.$$

On the one hand, we have

$$|10^m \alpha - a_m| = \left|\sum_{k=1}^{\infty} \frac{a_{m+k}}{10^k}\right| \leq \sum_{k=1}^{\infty} \frac{|a_{m+k}|}{10^k} \leq \frac{9}{10} \frac{1}{1 - 1/10} = 1,$$

and on the other hand, we have

$$|10^m \alpha| \leq 10^{N-1}|\alpha| \leq \frac{1}{10}.$$

Hence $|a_m| \leq 1 + 1/10$ and since $a_m \neq \pm 1$ by assumption, we conclude that $a_k = 0$ for $k = 1, \ldots, N-1$.

(d) We prove the result by induction. The assumption implies that for arbitrarily large k we have $|a_k - b_k| < 9$ so

$$\left| \frac{a_1 - b_1}{10} \right| = \left| \sum_{k=2}^{\infty} \frac{a_k - b_k}{10^k} \right| < \sum_{k=2}^{\infty} \frac{9}{10^k} = \frac{1}{10}$$

and therefore, $|a_1 - b_1| < 1$ which implies $a_1 = b_1$. Suppose $a_k = 0$ for all $0 \leq k \leq n$, then

$$\left| \frac{a_{n+1} - b_{n+1}}{10^{n+1}} \right| = \left| \sum_{k=n+2}^{\infty} \frac{a_k - b_k}{10^k} \right| < \sum_{k=n+2}^{\infty} \frac{9}{10^k} = \frac{1}{10^{n+1}}$$

and therefore $a_{n+1} = b_{n+1}$.

Exercise IX.2.12 *Let S be a subset of $\mathbf{R}$. We say that S has* **measure zero** *if given ϵ there exists a sequence of intervals $\{J_n\}$ such that*

$$\sum_{n=1}^{\infty} \text{length}(J_n) < \epsilon,$$

and such that S is contained in the union of these intervals.

(a) If S and T are sets of measure 0, show that their union has measure 0.

(b) If $S_1, S_2, \ldots$ is a sequence of sets of measure 0, show that the union of all

$$S_i (i = 1, 2, \ldots)$$

has measure 0.

Solution. (a) Special case of (b).

(b) For each i, we can find a sequence of intervals $\{J_{i,n}\}_{n=1}^{\infty}$ which cover S_i and such that $\sum_{n=1}^{\infty} |J_{i,n}| < \epsilon/2^i$. Then $\bigcup_{i=1}^{\infty} \bigcup_{n=1}^{\infty} J_{i,n}$ covers $\bigcup_{i=1}^{\infty} S_i$. Moreover,

$$\sum_{i=1}^{\infty} \sum_{n=1}^{\infty} |J_{i,n}| \leq \sum_{i=1}^{\infty} \frac{\epsilon}{2^i} = \epsilon.$$

Conclude.

Exercise IX.2.13 (The Space ℓ^2) *Let ℓ^2 be the set of sequences of numbers*

$$X = (x_1, x_2, \ldots, x_n, \ldots)$$

such that

$$\sum_{n=1}^{\infty} x_n^2$$

converges.

(a) Show that ℓ^2 is a vector space.
(b) Using Exercise 7, show that one can define a product between two elements X and $Y = (y_1, y_2, \dots)$ by

$$\langle X, Y \rangle = \sum_{n=1}^{\infty} x_n y_n.$$

Show that this product satisfies all the conditions of a positive definite scalar product, whose associated norm is given by

$$\|X\|_2^2 = \sum x_n^2.$$

(c) Let E_0 be the space of all sequences of numbers such that all but a finite number of components are equal to 0, i.e. sequences

$$X = (x_1, x_2, \dots, x_n, 0, 0, 0, \dots).$$

Then E_0 is a subspace of ℓ^2. Show that E_0 is dense in ℓ^2.
(d) Let $\{X_i\}$ be an ℓ^2-Cauchy sequence in E_0. Show that $\{X_i\}$ is ℓ^2-convergent to some element in ℓ^2.
(e) Prove that ℓ^2 is complete. [Cf. Exercise 6 of Chapter VII, §1.]

Solution. (a) In ℓ^2 define addition and scalar multiplication component-wise. If $X, Y \in E$, then $X + Y \in E$ because the Schwartz inequality implies

$$0 \le \sum_{n=1}^{m}(x_n + y_n)^2 = \sum_{n=1}^{m} x_n^2 + y_n^2 + 2x_n y_n$$

$$\le \sum x_n^2 + \sum y_n^2 + 2\left(\sum x_n^2\right)^{1/2}\left(\sum y_n^2\right)^{1/2} < \infty,$$

and clearly, $cX \in E$. The zero element is simply $0 = (0, 0, \dots)$ and belongs to E and $X + (-X) = 0$, so we conclude that E is a vector space.
(b) The product $\langle \cdot, \cdot \rangle : E \times E \to \mathbf{R}$ is well defined because by Exercise 7 we know that if $\sum x_n^2$ and $\sum y_n^2$ converge, then so does $\sum |x_n||y_n|$. We contend that this product is a positive definite scalar product. The linearity properties follows from standard properties of the real numbers and because of the theorems on limits. If $X = 0$, then clearly, $\langle X, X \rangle = 0$, and conversely, if $\langle X, X \rangle = 0$, then $\sum x_n^2 = 0$ which implies $x_n = 0$ for all n hence $X = 0$. So the scalar product is postitive definite. The associated norm is given by $\|X\|_2^2 = \langle X, X \rangle = \sum x_n^2$.
(c) It is clear that E_0 is a subspace of ℓ^2. We contend that this subspace is dense in ℓ^2. Let $X = (x_1, x_2, \dots) \in \ell^2$ and let $\epsilon > 0$. The series $\sum x_n^2$ converges, so for some N we have

$$\sum_{n=N+1}^{\infty} x_n^2 < \epsilon.$$

If we let $Y = (x_1, x_2, \ldots, x_N, 0, 0, \ldots)$, then $Y \in E_0$ and $\|X - Y\|_2^2 < \epsilon$. This proves our contention.

(d) Let $X_n = \{x_{n,j}\}_{j=1}^{\infty}$. Then given $\epsilon > 0$, there is a positive integer N such that for all $n, m > N$ we have $\|X_n - X_m\|_2 < \epsilon$. This implies that for each j, the sequence of real numbers $\{x_{n,j}\}_{n=1}^{\infty}$ is Cauchy and therefore, it converges to a real number which we denote by y_j. Let $Y = \{y_j\}_{j=1}^{\infty}$. We contend that $Y \in \ell^2$ and that X_n converges to Y in the ℓ^2-norm. For each positive integer M and all $n, m > N$ we have

$$\sum_{j=1}^{M} |x_{n,j} - x_{m,j}|^2 < \epsilon^2,$$

so letting $m \to \infty$ we get

$$\sum_{j=1}^{M} |x_{n,j} - y_j|^2 \leq \epsilon^2.$$

Therefore, for all $n > N$ we get that $\|X_n - Y\|_2 < \epsilon$. Moreover,

$$\sum_{j=1}^{k} |y_j|^2 = \sum_{j=1}^{k} |y_j - x_{n,j} + x_{n,j}|^2$$

$$\leq \|X_n - Y\|_2 + \|X_n\|_2 + 2\sum_{j=1}^{k} |y_j - x_{n,j}||x_{n,j}|$$

so Exercise 7 implies that $Y \in \ell^2$.

(e) Immediate consequence from (d) and Exercise 6, §1, of Chapter 7 (although you should note that (d) applies directly because we never assumed that $\{X_n\}$ was in E_0.)

Exercise IX.2.14 *Let S be the set of elements e_n in the space ℓ^2 of Exercise 13 such that e_n has component 1 in the n-th coordinate and 0 for all other coordinates. Show that S is a bounded set in E but is not compact.*

Solution. For all n we have $\|e_n\|_2 = 1$, so the set S is bounded. However, if $n \neq m$, then $\|e_n - e_m\|_2^2 = 2$ so the sequence $\{e_n\}$ has no converging subsequence. Thus S is not compact.

Exercise IX.2.15 (The Space ℓ^1) *Let ℓ^1 be the set of all sequences of numbers $X = \{x_n\}$ such that the series*

$$\sum_{n=1}^{\infty} |x_n|$$

converges. Define $\|X\|_1$ to be the value of this series.

(a) Show that ℓ^1 is a vector space.

(b) Show that $X \mapsto \|X\|_1$ defines a norm on this space.

(c) Let E_0 be the same space as in Exercise 13(c). Show that E_0 is dense in ℓ^1.

(d) Let $\{X_i\}$ be an ℓ^1-Cauchy sequence in E_0. Show that $\{X_i\}$ is ℓ^1-convergent to an element of ℓ^1.

(e) Prove that ℓ^1 is complete.

Solution. (a) Define addition and scalar multiplication componentwise. The triangle inequality and the theorems on limits imply that if $X, Y \in \ell^1$, then $X + Y \in \ell^1$ and if c is a scalar, then $cX \in \ell^1$. Hence E is a vector space.

(b) The function $\| \cdot \|_1$ is a norm. Indeed, $\|0\|_1 = 0$ and if $\|X\|_1 - 0$, then $\sum |x_n| = 0$ which implies $x_n = 0$ for all n, hence $X = 0$. Clearly, $\|cX\|_1 = |c|\|X\|_1$. For the triangle inequality we simply have on the partial sums

$$\sum_{n=1}^{N} |x_n + y_n| \le \sum_{n=1}^{N} |x_n| + \sum_{n=1}^{N} |y_n| \le \|X\|_1 + \|Y\|_1.$$

Letting $N \to \infty$ gives the desired inequality, and we conclude that $\| \cdot \|_1$ is a norm on ℓ^1.

(c) Given $\epsilon > 0$ and $X \in \ell^1$, there exists N such that $\sum_{n=N+1}^{\infty} |x_n| < \epsilon$, so if we let $Y = (x_1, x_2, \ldots, x_N, 0, 0, \ldots)$ we get $\|X - Y\|_1 < \epsilon$. Thus E_0 is dense in ℓ^1.

(d) Suppose $X_n = \{x_{n,j}\}_{j=1}^{\infty}$ is a Cauchy sequence. Then given $\epsilon > 0$, there is a positive integer N such that for all $n, m > N$ we have $\|X_n - X_m\|_1 < \epsilon$. This implies that for each j, the sequence of real numbers $\{x_{n,j}\}_{n=1}^{\infty}$ is Cauchy and therefore converges to a real number which we denote by y_j. Let $Y = \{y_j\}_{j=1}^{\infty}$. We contend that $Y \in \ell^1$ and that X_n converges to Y in the ℓ^1-norm. For each positive integer M and all $n, m > N$ we have

$$\sum_{j=1}^{M} |x_{n,j} - x_{m,j}| < \epsilon,$$

so letting $m \to \infty$ we get

$$\sum_{j=1}^{M} |x_{n,j} - y_j| \le \epsilon.$$

Therefore, for all $n > N$, we get that $\|X_n - Y\|_1 < \epsilon$. Moreover $Y \in \ell^1$ because

$$\sum_{j=1}^{M} |y_j| \le \sum_{j=1}^{M} |y_j - x_{n,j}| + \sum_{j=1}^{M} |x_{n,j}| \le \|X_n - Y\|_1 + \|X_n\|_1,$$

for all $M \ge 1$.

(e) Immediate from (d) and Exercise 6, §1, of Chapter 7.

Exercise IX.2.16 *Let $\{a_n\}$ be a sequence of positive numbers such that $\sum a_n$ converges. Let $\{\sigma_n\}$ be a sequence of seminorms on a vector space E. Assume that for each $x \in E$ there exists $C(x) > 0$ such that $\sigma_n(x) \leq C(x)$ for all n. Show that $\sum a_n \sigma_n$ defines a seminorm on E.*

Solution. Let $\sigma(x) = \sum a_n \sigma_n(x)$. For each $x \in E$ we have $\sigma(x) = C(x) \sum a_n$ so $\sigma(x)$ is well defined. Since $\sigma_n(x) \geq 0$ for all x and all n we see that $\sigma(x) \geq 0$ for all $x \in E$. Furthermore

$$\sum_{n=1}^{M}(a_n \sigma_n(x+y)) \leq \sum_{n=1}^{M}(a_n \sigma_n(x) + a_n \sigma_n(y)) \leq \sigma(x) + \sigma(y)$$

so letting $M \to \infty$ we obtain the triangle inequality. Finally, the property $\sigma(cx) = |c|\sigma(x)$ follows from

$$\sum_{n=1}^{M} a_n \sigma_n(cx) = \sum_{n=1}^{M} a_n |c| \sigma_n(x) = |c| \sum_{n=1}^{M} a_n \sigma_n(x)$$

and letting $M \to \infty$.

Exercise IX.2.17 (Khintchine) *Let f be a positive function, and assume that*

$$\sum_{q=1}^{\infty} f(q)$$

converges. Let S be the set of numbers x such that $0 \leq x \leq 1$, and such that there exist infinitely many integers $q, p > 0$ such that

$$\left| x - \frac{p}{q} \right| < \frac{f(q)}{q}.$$

Show that S has measure 0. [Hint: Given ϵ, let q_0 be such that

$$\sum_{q \geq q_0} f(q) < \epsilon.$$

Around each fraction $0/q, 1/q, \ldots, q/q$ consider the interval of length $f(q)/q$. For $q \geq q_0$, the set S is contained in the union of such intervals....]

Solution. Given ϵ, let q_0 be such that $\sum_{q \geq q_0} f(q) < \epsilon$. For each $q \geq q_0$ consider the intervals of length $2f(q)/q$ centered at $0/q, 1/q, \ldots, q/q$. Let U be the union of all such intervals. Then, given any $x \in S$, there exist $q \geq q_0$ and an integer p with $0 \leq p \leq q$, such that

$$\left| x - \frac{p}{q} \right| < \frac{f(q)}{q},$$

so S is contained in U, and the sum of the length of the intervals is bounded by

$$\sum_{q \geq q_0} 2(q+1)\frac{f(q)}{q} \leq \sum_{q \geq q_0} 4q\frac{f(q)}{q} < 4\epsilon.$$

Exercise IX.2.18 *Let α be a real number. Assume that there is a number $C > 0$ such that for all integers $q > 0$ and integers p we have*

$$\left|\alpha - \frac{p}{q}\right| > \frac{C}{q}.$$

Let ψ be a positive decreasing function such that the sum $\sum_{n=1}^{\infty} \psi(n)$ converges. Show that the inequality

$$\left|\alpha - \frac{p}{q}\right| < \psi(q)$$

has only a finite number of solutions. [Hint: Otherwise, $\psi(q) > C/q$ for infinitely many q. Cf. Exercise 2.]

Solution. Suppose that the inequality $|\alpha - p/q| < \psi(q)$ has infinitely many solutions. Then $C/q < \left|\alpha - \frac{p}{q}\right| < \psi(q)$ for infinitely many q. Exercise 2 implies that the series $\sum \psi(q)/C$ diverges, a contradiction.

Exercise IX.2.19 (Schanuel) *Prove the converse of Exercise 18. That is, let α be a real number. Assume that for every positive decreasing function ψ with convergent sum $\sum \psi(n)$, the inequality $|\alpha - p/q| < \psi(q)$ has only a finite number of solutions. Show that there is a number $C > 0$ such that $|\alpha - p/q| > C/q$ for all intergers p, q with $q > 0$. [Hint: If not, there exists a sequence $1 < q_1 < q_2 < \cdots$ such that*

$$|\alpha - p_i/q_i| < 1/(2^i q_i).$$

Let

$$\psi(t) = \sum_{i=1}^{\infty} \frac{e}{2^i q_i} e^{-t/q_i}.]$$

Solution. Suppose the conclusion of the exercise is false. Then for each integer $i \geq 1$, there exists a pair of integers (p_i, q_i) with $q_i > 0$ such that

$$|\alpha - p_i/q_i| < 1/(2^i q_i).$$

We now show that we can choose the q_i's such that $1 < q_1 < q_2 < \cdots$. To do so, it suffices to prove the following lemma:

Lemma. For each i, there is infinitely many choices of $q > 0$ such that

$$\left| \alpha - \frac{p}{q} \right| < \frac{1}{2^i q},$$

for some p (which depends on q).

Proof. Suppose that for some i, there is only finitely many choices. Then, since α is irrational, there exists $\delta > 0$ such that if $|\alpha - p/q| < \delta$, then

$$\left| \alpha - \frac{p}{q} \right| \geq \frac{1}{2^i q}.$$

Select k such that $k > i$ and $1/2^k < \delta$. By assumption, there exists p', q' with $q' > 0$ such that

$$\left| \alpha - \frac{p'}{q'} \right| < \frac{1}{2^k q'} < \delta,$$

so $|\alpha - p'/q'| \geq 1/(2^i q')$. We choose $k > i$ so

$$\frac{1}{2^i q'} \leq \left| \alpha - \frac{p'}{q'} \right| < \frac{1}{2^k q'} < \frac{1}{2^i q'},$$

and this contradiction proves the lemma, and shows that for each i we can find a pair (p_i, q_i) with $q_i > 0$ such that

$$|\alpha - p_i/q_i| < 1/(2^i q_i) \quad \text{and} \quad 1 < q_1 < q_2 < \cdots.$$

Now let

$$\psi(t) = \sum_{i=1}^{\infty} \frac{e}{2^i q_i} e^{-t/q_i}.$$

Then $\psi(t) \leq e \sum (1/2^i) < \infty$, ψ is positive and decreasing because $t \mapsto e^{-t}$ is decreasing. Furthermore,

$$\psi(q_j) > \frac{e}{2^j q_j} e^{-q_j/q_j} = \frac{1}{2^j q_j}$$

for all j, and the series $\sum \psi(n)$ converges because

$$\sum_{n=0}^{\infty} \frac{1}{2^i q_i} e^{-n/q_i} = \frac{1}{2^i q_i} \frac{1}{1 - e^{-1/q_i}},$$

and for all small x we have the inequality, $e^{-x} \leq 1 - \frac{1}{2} x$ so for all large q_i we get

$$\frac{1}{2^i q_i} \frac{1}{1 - e^{-1/q_i}} \leq \frac{2q_i}{2^i q_i},$$

and therefore, $\sum \psi(n) < \infty$. This contradiction concludes the exercise.

IX.3 Non-Absolute Convergence

Exercise IX.3.1 *Let $a_n \geq 0$ for all n. Assume that $\sum a_n$ converges. Show that $\sum \sqrt{a_n}/n$ converges.*

Solution. Since $\sum 1/n^2$ and $\sum a_n$ converge, Exercise 7, §2, implies the convergence of $\sum \sqrt{a_n}/n$.

Exercise IX.3.2 *Show that for x real, $0 < x < 2\pi$, $\sum e^{inx}/n$ converges. Conclude that*

$$\sum \frac{\sin nx}{n} \quad and \quad \sum \frac{\cos nx}{n}$$

converge in the same interval.

Solution. If $0 < x < 2\pi$, then $e^{ix} \neq 1$. We estimate the sum $G(k) = e^{ix} + \cdots + e^{kix}$, using the standard formula for a geometric series, namely

$$|G(k)| = \left| \frac{e^{ix} - e^{(k+1)ix}}{1 - e^{ix}} \right| \leq \frac{2}{|1 - e^{ix}|}.$$

Since $x \mapsto 1/x$ decreases to 0 as $x \to \infty$, summation by parts implies the convergence of the series $\sum e^{inx}/n$.

A sequence of complex numbers $\{z_n\}$ converges to z, if and only if the real sequences $\{\operatorname{Re}(z_n)\}$ and $\{\operatorname{Im}(z_n)\}$ converge to $\operatorname{Re}(z)$ and $\operatorname{Im}(z)$, respectively. But $e^{inx} = \cos(nx) + i\sin(nx)$ so the series

$$\sum \frac{\sin nx}{n} \quad and \quad \sum \frac{\cos nx}{n}$$

converge.

Exercise IX.3.3 *A series of numbers $\sum a_n$ is said to **converge absolutely** if $\sum |a_n|$ converges. Determine which of the following series converge absolutely, and which just converge.*
(a) $\sum \frac{(-1)^n}{n^{1+1/n}}$.   (b) $\sum (-1)^n \frac{\sin n}{n}$
[Hint for (b): Show that among three consecutive positive integers, for at least one of them, say n, one has $|\sin n| \geq 1/2$.]
(c) $\sum (-1)^n \frac{\sqrt{n+1} - \sqrt{n}}{n}$.   (d) $\sum \frac{n}{2^n}$
(e) $\sum \frac{\sin n}{2n^2 - n}$.   (f) $\sum (-1)^n \frac{n^2 - 4n}{2n^3 + n - 5}$
(g) $\sum \frac{2^n + 1}{3^n - 4}$.   (h) $\sum \frac{n \cos n}{n^5 - n^3 + 1}$
(i) $\sum (-1)^n \frac{1}{\log n}$.   (j) $\sum (-1)^n \frac{1}{n(\log n)^2}$

Solution. (a) For large n, $n^{1/n} \leq 2$ so $n^{1+1/n} \leq 2n$ hence the series is not absolutely convergent.
(b) Given any three consecutive positive integers at least one must verify $|\sin n| \geq 1/2$ because

$$\{x > 0 : |\sin x| \le 1/2\} \subset \bigcup_{k \in \mathbf{Z}^+} \left[-\frac{\pi}{6} + k\pi, \frac{\pi}{6} + k\pi \right].$$

So for each $j \in \mathbf{Z}^+$,

$$a_{3j+1} + a_{3j+2} + a_{3j+3} \ge \frac{1}{2(3j+3)}.$$

We conclude that $\sum |a_n|$ diverges.
(c) Since

$$\frac{\sqrt{n+1} - \sqrt{n}}{n} = \frac{1}{n(\sqrt{n+1} + \sqrt{n})} \le \frac{1}{n\sqrt{n}},$$

the sum $\sum |a_n|$ converges.
(d) Since $a_{n+1}/a_n \to 1/2$ the ratio test implies the convergence and absolute convergence of $\sum a_n$.
(e) The series converges absolutely because for all n, $2n^2 - n \ge n^2$, thus

$$\sum |a_n| \le \sum 1/n^2.$$

(f) For all large n, $2n^3 \le 2n^3 + n - 5 \le 3n^3$, so

$$\frac{1}{3n} = \frac{n^2}{3n^3} \le \frac{n^2}{2n^3 + n - 5} \quad \text{and} \quad \frac{4n}{2n^3 + n - 5} \le \frac{2}{n^2},$$

thus $\sum |a_n|$ diverges.
(g) For all large n we have $2^n + 1 \le 2^{n+1}$ and $3^n - 4 \ge 3^n/2$ so

$$\frac{2^n + 1}{3^n - 4} \le 4 \frac{2^n}{3^n}.$$

Since $\sum (2/3)^2$ converges we conclude that $\sum a_n$ converges absolutely.
(h) For all large n we have $n^5 - n^3 + 1 \ge n^5/2$ so that

$$\left| \frac{n \cos n}{n^5 - n^3 + 1} \right| \le \frac{2}{n^4}.$$

Thus $\sum |a_n|$ converges.
(i) Since $\log n \le n$ we conclude at once that $\sum |a_n|$ diverges.
(j) Put $\epsilon = 1$ in Exercise 1, §2, to prove that $\sum |a_n|$ converges.

Exercise IX.3.4 *For which values of x does the following series converge?*

$$\sum \frac{x^n}{x^{2n} - 1}.$$

Solution. If $-1 < x < 1$ the series converges because $\sum x^n$ converges and $x^{2^n} \to 0$ as $n \to \infty$.

Assume that $x > 1$. Then for all large n we have $x^{2^n} - 1 \geq x^{2^n}/2$ and $2^n \geq 2n$. This implies that for all large n we have

$$\frac{x^n}{x^{2^n} - 1} \leq 2\frac{x^n}{x^{2n}} \leq \frac{2}{x^n},$$

so if $x > 1$ the series converges. This convergence is also true for $x < -1$ as one sees from putting absolute values.

We conclude that the series converges for $x \in \mathbf{R} - \{1, -1\}$.

Exercise IX.3.5 *Let $\{a_n\}$ be a sequence of real numbers such that $\sum a_n$ converges. Let $\{b_n\}$ be a sequence of real numbers which converges monotonically to infinity. (This means that $\{b_n\}$ is an unbounded sequence such that $b_{n+1} \geq b_n$ for all n.) Show that*

$$\lim_{N \to \infty} \frac{1}{b_N} \sum_{n=1}^{N} a_n b_n = 0.$$

Does this conclusion still hold if we only assume that the partial sums of $\sum a_n$ are bounded?

Solution. Given $\epsilon > 0$, select a positive integer n_0 such that for all $m > n_0$ we have $|\sum_{n=n_0+1}^{m} a_n| < \epsilon$ and $a_{n_0} \geq 0$. Then for $N > n_0$ splitting the sum we obtain

$$\left| \frac{1}{b_N} \sum_{n=1}^{N} a_n b_n \right| \leq \left| \frac{1}{b_N} \sum_{n=1}^{n_0} a_n b_n \right| + \left| \frac{1}{b_N} \sum_{n=n_0+1}^{N} a_n b_n \right|.$$

The first sum will be $\leq \epsilon$ for all large N. For the second sum we use summation by parts to obtain, after some elementary computations,

$$\sum_{n=n_0+1}^{N} a_n b_n = b_N(A_N - A_{n_0}) - \sum_{k=n_0+1}^{N-1} (A_k - A_{n_0})(b_{k+1} - b_k),$$

where $A_n = \sum_{k=1}^{n} a_k$ are the partial sums. Therefore by the triangle inequality, the fact that $|A_k - A_{n_0}| < \epsilon$ for all $k \geq n_0$ and that $\{b_k\}$ increases we get

$$\left| \sum_{n=n_0+1}^{N} a_n b_n \right| \leq |b_N|\epsilon + \epsilon(b_N - b_{n_0}),$$

hence for all large N we have

$$\left| \frac{1}{b_N} \sum_{n=n_0+1}^{N} a_n b_n \right| \leq 3\epsilon$$

which concludes the proof.

If we only assume that the partial sums of $\sum a_n$ are bounded we cannot conclude that the limit is 0. Indeed, let $a_n = (-1)^n$ and $b_n = 2^n$. Then

$$\frac{1}{b_N} \sum_{n=1}^{N} a_n b_n = \frac{1}{2^N} \frac{-2 - (-2)^{N+1}}{1+2} = \frac{-2}{3 \times 2^N} + \frac{2 \times (-1)^N}{3}$$

and the above expression does not have a limit as $N \to \infty$.

Exercise IX.3.6 (The Cantor Set) *Let K be the subset of $[0,1]$ consisting of all numbers having a trecimal expansion*

$$\sum_{n=1}^{\infty} \frac{a_n}{3^n},$$

where $a_n = 0$ or $a_n = 2$. This set is called the Cantor set.
(a) Show that the numbers a_n in the trecimal expansion of a given number in K are uniquely determined.
(b) Show that K is compact.

Solution. The two lemmas will show that $\sum c_n/3^n$ with $c_n = 0, 2$ or -2 is close to 0 if and only if $c_n = 0$ for $n = 1, \dots, N$ with N large.

Lemma 1 *Let $\gamma = \sum_{k=m}^{\infty} c_k/3^k$ with $c_k = 0, 2$ or -2. Then $|\gamma| \leq 1/3^{m-1}$.*

Lemma 2 *Suppose that $\gamma = \sum c_n/3^n$ with $c_n = 0, 2$ or -2. If $|\gamma| \leq 1/3^{N+1}$, then $c_k = 0$ for $1 \leq k \leq N - 1$.*

In Exercise 11, §2, we proved the analogue results for decimal expansions. The proofs of the above lemmas are an ϵ-variation of the proofs given in Exercise 11, §2.

(a) Let $\alpha = \sum_{n=1}^{\infty} a_n/3^n$ and $\beta = \sum_{n=1}^{\infty} b_n/3^n$ and suppose that $\alpha = \beta$. Let $\gamma = \alpha - \beta = \sum_{n=1}^{\infty} c_n/3^n$. Then $\gamma = 0$ and $c_n = 0, 2$ or -2 so by Lemma 2 it follows that $c_n = 0$ for all n.

(b) We first show that K is closed. Let $\{\alpha_j\}$ be a Cauchy sequence in the Cantor set, with $\alpha_i = \sum a_{in}/3^n$, where $a_{in} = 0$ or 2. By Lemma 2, if $|\alpha_i - \alpha_j| \leq 1/3^{N+1}$, then $a_{in} = a_{jn}$ for $n = 1, \dots, N - 1$. Hence for each n, $\lim_{i \to \infty} a_{in}$ exists, say a_n. Then $a_n = 0$ or 2. Let $\alpha = \sum a_n/3^n$. Then α is in the Cantor set and $\alpha_i \to \alpha$ as $i \to \infty$, because given $\epsilon > 0$, we can find an integer n such that $\sum_{n=N+1}^{\infty} \frac{2}{3^n} < \epsilon$, and then

$$|\alpha - \alpha_i| \leq \sum_{n=1}^{\infty} \frac{|a_n - a_{in}|}{3^n} \leq \sum_{n=1}^{N} \frac{|a_n - a_{in}|}{3^n} + \epsilon$$

and the right-hand side can be made $< \epsilon$ for all large i. Hence K is closed. The Cantor set is contained in the compact interval $[0,1]$ and is therefore compact.

Exercise IX.3.7 (Peano Curve) *Let K be the Cantor set. Let $S = [0,1] \times [0,1]$ be the unit square. Let $f \colon K \to S$ be the map which to each element $\sum a_n/3^n$ of the Cantor set assigns the pair of numbers*

$$\left(\sum \frac{b_{2n+1}}{2^n}, \sum \frac{b_{2n}}{2^n} \right),$$

where $b_m = a_m/2$. Show that f is continuous, and is surjective.

Solution. We first prove surjectivity. Let $(x,y) \in [0,1] \times [0,1]$ and write

$$x = \sum \frac{x_n}{2^n} \quad \text{and} \quad y = \sum \frac{y_n}{2^n},$$

where x_n and y_n take on values of 0 or 1. Let

$$c = 2 \left(\sum_{n=1}^{\infty} \frac{x_n}{3^{2n+1}} + \sum_{n=1}^{\infty} \frac{y_n}{3^{2n}} \right).$$

Then $c \in K$ and $f(c) = (x,y)$.

For continuity, suppose $\alpha = \sum a_n/3^n$ and $\alpha' = \sum a'_n/3^n$ are elements of K which are close together. Let $\gamma = \alpha - \alpha'$. Then Lemma 2 of the previous exercise shows that $a_n = a'_n$ for $n = 1, \ldots, N-1$ with large N, so if

$$f(\alpha) = \left(\sum \frac{b_{2k+1}}{2^k}, \sum \frac{b_{2k}}{2^k} \right) \quad \text{and} \quad f(\alpha') = \left(\sum \frac{b'_{2k+1}}{2^k}, \sum \frac{b'_{2k}}{2^k} \right)$$

then $b_m = b'_m$. By the analogue of Lemma 1 of the previous exercise for binary expansions it follows that $|f(\alpha) - f(\alpha')|$ is small, and we even directly proved the uniform continuity of f.

IX.5 Absolute and Uniform Convergence

Exercise IX.5.1 *Show that the following series converge uniformly and absolutely in the stated interval for x.*
(a) $\sum \frac{1}{n^2+x^2}$ for $0 \le x$.   (b) $\sum \frac{\sin nx}{n^{3/2}}$ for all x.
(c) $\sum x^n e^{-nx}$ on every bounded interval $0 \le x \le C$.

Solution. (a) Since $1/(n^2 + x^2) \le 1/n^2$, the uniform convergence of the desired series follows at once.
(b) $|\sin nx|/n^{3/2} \le 1/n^{3/2}$ and $\sum 1/n^{3/2} < \infty$.
(c) For all $x \ge 0$, $x/e^x \le 1/e$, so $x^n e^{-nx} \le (1/e)^n$.

Exercise IX.5.2 *Show that the series*

$$\sum \frac{x^n}{1 + x^n}$$

converges uniformly and absolutely for $0 \le |x| \le C$, where C is any number with $0 < C < 1$. Show that the convergence is not uniform in $0 \le x < 1$.

Solution. The inequalities $0 \le |x|^n \le C^n \le C < 1$ imply

$$\left| \frac{x^n}{1+x^n} \right| \le \frac{|x|^n}{1-|x|^n} \le \frac{|x|^n}{1-C} \le \frac{C^n}{1-C},$$

so the convergence of the series is absolute and uniform on $[-C, C]$. The convergence is not uniform on $[0, 1)$ because

$$\sum_{n=k}^{\infty} \frac{x^n}{1+x^n} \ge \frac{1}{2} \sum_{n=k}^{\infty} x^n = \frac{1}{2} \frac{x^k}{1-x},$$

and $x^k/(1-x) \to \infty$ as $x \to 1$ and $x < 1$.

Exercise IX.5.3 *Let*

$$f(x) = \sum_{n=1}^{\infty} \frac{1}{1+n^2 x}.$$

Show that the series converges uniformly for $x \ge C > 0$. Determine all points x where f is defined, and also where f is continuous.

Solution. The inequality $1 + n^2 x \ge 1 + n^2 C$ whenever $x \ge C > 0$ implies the uniform convergence of the series for $x \ge C > 0$.

Let $S = \{y \in \mathbf{R} : y = -1/n^2 \text{ for some } n \in \mathbf{N}^*\} \cup \{0\}$. The function f is defined for all $x \in \mathbf{R} - S$ because the series $\sum 1/n^2$ converges. The convergence is uniform for $x \le C < -1$ and on every closed interval not intersecting S because

$$\frac{1}{|1+n^2 x|} = \frac{1}{n^2} \frac{1}{|1/n^2 + x|} \le \frac{1}{n^2} \frac{1}{||x| - 1/n^2|},$$

whence the function f is continuous on $\mathbf{R} - S$.

Exercise IX.5.4 *Show that the series*

$$\sum \frac{1}{n^2 - x^2}$$

converges absolutely and uniformly on any closed interval which does not contain an integer.

Solution. The distance between a closed interval not containing any integer and the set of integers is strictly positive. Therefore, if x belongs to a closed interval not intersecting the set of integers, there exists a number $C > 0$ such that for all integers n and all x in the interval, we have $|1 - x^2/n^2| \ge C$. Then

$$\frac{1}{|n^2 - x^2|} \le \frac{1}{n^2 C}.$$

Exercise IX.5.5 *(a) Show that*

$$\sum \frac{nx^2}{n^3 + x^3}$$

converges uniformly on any interval $[0, C]$ with $C > 0$.
(b) Show that the series

$$\sum_{n=1}^{\infty} (-1)^n \frac{x^2 + n}{n^2}$$

converges uniformly in every bounded interval, but does not converge absolutely for any value of x.

Solution. (a) For all x in the interval $[0, C]$, the following inequality holds

$$\frac{nx^2}{n^3 + x^3} \le \frac{C^2}{n^2},$$

which proves the uniform and absolute convergence of the series on $[0, C]$ with $C > 0$.
(b) Let C be a number and assume that $|x| \le C$. Then for all $n > C^2$ we have

$$\frac{x^2 + n}{n^2} \le \frac{C^2}{n^2} + \frac{1}{n} \le \frac{2}{n}.$$

By the tail-end estimate given in Theorem 3.3, we conclude that the series converges uniformly in every bounded interval. However the inequality

$$\frac{1}{n} \le \frac{x^2 + n}{n^2}$$

shows that the series does not converge absolutely for any value of x.

Exercise IX.5.6 *Show that the series $\sum e^{inx}/n$ is uniformly convergent in every interval $[\delta, 2\pi - \delta]$ for every δ such that*

$$0 < \delta < \pi.$$

Conclude the same for $\sum (\sin nx)/n$ and $\sum (\cos nx)/n$.

Solution. There exists a number $C > 0$ such that if $x \in [\delta, 2\pi - \delta]$, then $|1 - e^{ix}| \ge C$. We then have the following estimate

$$|e^{inx} + e^{i(n+1)x} + \cdots + e^{imx}| = \left| \frac{e^{inx} - e^{i(m+1)x}}{1 - e^{ix}} \right| \le \frac{2}{C}$$

which combined with summation by parts implies

$$\left| \sum_{k=n}^{m} \frac{e^{ikx}}{k} \right| \le \frac{2}{C} \left(\frac{1}{n} - \frac{1}{m} \right) \le \frac{2}{Cn}.$$

This tail-end estimate of the series $\sum e^{inx}/n$ proves its uniform convergence in every interval $[\delta, 2\pi - \delta]$ for every δ such that $0 < \delta < \pi$.

Since

$$\left| \sum_{k=n}^{m} \frac{e^{ikx}}{k} \right| \geq \left| \sum_{k=n}^{m} \frac{\cos kx}{k} \right|$$

and

$$\left| \sum_{k=n}^{m} \frac{e^{ikx}}{k} \right| \geq \left| \sum_{k=n}^{m} \frac{\sin kx}{k} \right|,$$

we conclude the same for the series $\sum (\sin nx)/n$ and $\sum (\cos nx)/n$.

Exercise IX.5.7 *Let ℓ^1 be the set of sequences $\alpha = \{a_n\}$, $a_n \in \mathbf{R}$, such that*

$$\sum_{n=1}^{\infty} |a_n|$$

converges. This is the space of §2, Exercise 15, with the ℓ^1-norm $\|\alpha\|_1 = \sum |a_n|$.
(a) Show that the closed ball of radius 1 in ℓ^1 is not compact.
(b) Let $\alpha = \{a_n\}$ be an element of ℓ^1, and let A be the set of all sequences $\beta = \{b_n\}$ in ℓ^1 such that $|b_n| \leq |a_n|$ for all n. Show that every sequence of elements of A has a point of accumulation in A, and hence A is compact.

Solution. (a) Consider the sequence $\{e_n\}$ in ℓ^1 where $e_n = (0, 0, \ldots, 0, 1, 0, \ldots)$ has components all 0 except for 1 in the n-th coordinate. Then $\|e_n - e_m\|_2^2 = 2$ whenever $n \neq m$ so the sequence $\{e_n\}$ has no converging subsequence. Thus the closed unit ball in ℓ^1 is not compact.

(b) Consider a sequence $\{\beta_n\}$ in A where $\beta_n = \{b_{n,j}\}_{j=1}^{\infty}$. Then for all n and all j we have $|b_{n,j}| \leq |a_j|$. From $\{b_{n,1}\}$ we can extract a subsequence $s_1 = \{b_{n_i,1}\}_{i=1}^{\infty}$ which converges to a limit b_1 (Bolzano-Weierstrass). Call m_1 the first index of s_1. Then from the truncated sequence of second terms $\{b_{n_i,2}\}_{i=1}^{\infty}$ extract a subsequence s_2 which converges to a limit b_2. Call m_2 its second index. Continue this process, find a converging subsequence s_{j+1} of the subsequence $\{b_{n,j+1}\}_n$ indexed by s_j and call b_{j+1} its limit, and m_{j+1} its $j+1$ index. Then $\beta = \{b_1, b_2, \ldots\}$ belongs to A and we contend that this sequence is a point of accumulation of $\{\beta_n\}$. Indeed, given $\epsilon > 0$ choose k such that $\sum_{n=k+1}^{\infty} |a_n| < \epsilon$. Then choose N such that $j \geq N$ implies

$$|b_{m_j,1} - b_1| < \epsilon/k, |b_{m_j,2} - b_2| < \epsilon/k, \ldots, |b_{m_j,k} - b_k| < \epsilon/k.$$

Then for all $j \geq N$ we have $|\beta_{m_j} - \beta| < 3\epsilon$. This concludes the exercise.

Exercise IX.5.8 *Let F be the complete normed vector space of continuous functions on $[0, 2\pi]$ with the sup norm. For $\alpha = \{a_n\}$ in ℓ^1, let*

$$L(\alpha) = \sum_{n=1}^{\infty} a_n \cos nx.$$

Show that L is a continuous linear map of ℓ^1 into F, and that $\|L(\alpha)\| \leq \|\alpha\|_1$ for all $\alpha \in \ell^1$.

Solution. Since $|\cos nx| \leq 1$ we have

$$\|L(\alpha)\| \leq \sum |a_n| = \|\alpha\|_1.$$

If $\beta = \{b_n\} \in \ell^1$ and c is a number we get

$$
\begin{aligned}
L(\alpha + \beta) &= \sum_{n=1}^{\infty} a_n \cos nx + b_n \cos nx \\
&= \sum_{n=1}^{\infty} a_n \cos nx + \sum_{n=1}^{\infty} b_n \cos nx \\
&= L(\alpha) + L(\beta)
\end{aligned}
$$

and

$$L(c\alpha) = \sum_{n=1}^{\infty} c a_n \cos nx = c \sum_{n=1}^{\infty} a_n \cos nx = cL(\alpha).$$

So L is linear. Finally, L is continuous because

$$\|L(\alpha) - L(\beta)\| \leq \|\alpha - \beta\|_1.$$

Exercise IX.5.9 *For $z \in \mathbf{C}$ (complex numbers) and $|z| \neq 1$, show that the following limit exists and give the values:*

$$f(z) = \lim_{n \to \infty} \left(\frac{z^n - 1}{z^n + 1} \right).$$

Is it possible to define $f(z)$ when $|z| = 1$ in such a way to make f continuous?

Solution. Suppose $|z| < 1$, then

$$\left| \frac{z^n - 1}{z^n + 1} - (-1) \right| = \left| \frac{2z^n}{z^n + 1} \right| \to 0$$

as $n \to \infty$, so $f(z) = -1$. If $|z| > 1$, then

$$\left| \frac{z^n - 1}{z^n + 1} - 1 \right| = \frac{2}{|z^n + 1|} \to 0$$

as $n \to \infty$, so $f(z) = 1$. From these results we see that we cannot define $f(z)$ when $|z| = 1$ so as to make f continuous.

Exercise IX.5.10 *For z complex, let*

$$f(z) = \lim_{n \to \infty} \frac{z^n}{1 + z^n}.$$

(a) What is the domain of definition of f, that is for which complex numbers z does the limit exist?

(b) Give explicitly the values of f(z) for the various z in the domain of f.

Solution. If $|z| < 1$, then $z^n \to 0$ as $n \to \infty$ so $f(z) = 0$. If $|z| > 1$, then $f(z) = 1$ because

$$\left| \frac{z^n}{z^n + 1} - 1 \right| = \frac{1}{|z^n + 1|} \to 0$$

as $n \to \infty$.

We now investigate what happens on the unit circle. Let $z = e^{i\theta}$ with $0 \le \theta < 2\pi$. Then $1 + z^n = 1 + e^{ni\theta}$, so if $\theta = 0$ we immediately get $f(1) = 1/2$. If $\theta \ne 0$, then

$$f(z) = \frac{e^{ni\theta}}{1 + e^{ni\theta}} = \frac{1}{1 + e^{-ni\theta}},$$

and since $e^{-ni\theta}$ goes around the circle, we cannot define f at the points $z = e^{i\theta}$ with $\theta \ne 0$. So the domain of definition of f is the set $z \in \mathbf{C}$ such that $|z| \ne 1$ or $z = 1$.

Exercise IX.5.11 *(a) For z complex, show that the series*

$$\sum_{n=1}^{\infty} \frac{z^{n-1}}{(1 - z^n)(1 - z^{n+1})}$$

converges to $1/(1-z)^2$ for $|z| < 1$ and to $1/z(1-z)^2$ for $|z| > 1$. [Hint: This is mostly a question of algebra. Formally, factor out $1/z$, then at first add 1 and subtract 1 in the numerator, and use a partial fraction decomposition, pushing the thing through algebraically, before you worry about convergence. Use partial sums.]

(b) Prove that the convergence is uniform for $|z| \le c < 1$ in the first case, and $|z| \ge b > 1$ in the second.

Solution. (a) Let $U_n = z^n/(1 - z^n)(1 - z^{n+1})$ and let $D(z) = (1 - z^n)(1 - z^{n+1})$. Then

$$U_n = \frac{z^n(1-z)}{D(z)(1-z)} = \frac{1}{1-z} \left[\frac{z^n - z^{n+1}}{D(z)} \right]$$

$$= \frac{1}{1-z} \left[\frac{(z_n - 1) + (1 - z^{n+1})}{(1 - z^n)(1 - z^{n+1})} \right]$$

$$= \frac{1}{1-z} \left[-\frac{1}{1 - z^{n+1}} + \frac{1}{1 - z^n} \right].$$

We get a telescopic sum, whence

$$\sum_{k=1}^{n} U_k = \frac{1}{1-z} \left[\frac{1}{1-z} - \frac{1}{1-z^{n+1}} \right]$$

and therefore

$$S_n(z) = \sum_{k=1}^{n} \frac{z^{k-1}}{(1-z^k)(1-z^{k+1})} = \frac{1}{z(1-z)} \left[\frac{1}{1-z} - \frac{1}{1-z^{n+1}} \right].$$

If $|z| < 1$, then $1/(1 - z^{n+1}) \to 1$ as $n \to \infty$ and therefore $S_n(z) \to 1/(1 - z^2)$ as $n \to \infty$. If $|z| > 1$, then $1/(1 - z^{n+1}) \to 0$ as $n \to \infty$ so $S_n(z) \to 1/z(1 - z^2)$.

(b) Suppose $|z| \leq c < 1$. A little algebra and part (a) imply that

$$\left| S_n(z) - \frac{1}{(1-z)^2} \right| = \left| \frac{1}{z(1-z)} \right| \left| \frac{z^{n+1}}{1-z^{n+1}} \right|.$$

But $|1 - z^{n+1}| \geq 1 - |z|^{n+1} \geq 1 - c^{n+1}$ so we get the estimate

$$\left| S_n(z) - \frac{1}{(1-z)^2} \right| \leq \frac{1}{1-c} \frac{c^n}{1-c^{n+1}}$$

for all z in the region $|z| \leq c < 1$. Now $c^n \to 0$, hence the convergence is uniform in the region $|z| \leq c < 1$.

If $|z| \geq b > 1$, then

$$\left| S_n(z) - \frac{1}{z(1-z)^2} \right| = \left| \frac{1}{z(1-z)} \right| \left| \frac{1}{1-z^{n+1}} \right| \leq \frac{1}{b(b-1)} \frac{1}{b^{n+1}-1}$$

and $b^{n+1} \to \infty$, so the convergence is uniform in the region $|z| \geq b > 1$.

IX.6   Power Series

Exercise IX.6.1 *Determine the radii of convergence of the following power series.*

(a) $\sum n x^n$.   *(b)* $\sum n^2 x^n$.   *(c)* $\sum \frac{x^n}{n}$.

(d) $\sum \frac{x^n}{n^n}$.   *(e)* $\sum 2^n x^n$.   *(f)* $\frac{x^n}{2^n}$.

(g) $\sum \frac{x^n}{(n^2+2n)}$.   *(h)* $\sum (\sin n\pi) x^n$.

Solution. (a) The radius of convergence is 1 because $n^{1/n} \to 1$ as $n \to \infty$.

(b) The radius of convergence is 1 because $n^{2/n} \to 1$ as $n \to \infty$.

(c) The radius of convergence is 1 because $1/n^{1/n} \to 1$ as $n \to \infty$.

(d) The radius of convergence is ∞ because $1/n \to 0$ as $n \to \infty$.

(e) The radius of convergence is $1/2$.

(f) The radius of convergence is 2.
(g) The radius of convergence is 1 because $n^{2/n}(1+2/n)^{1/n} \to 1$ as $n \to \infty$.
(b) The radius of convergence is ∞ because $\sin(n\pi) = 0$ for all n hence the series is identically zero.

Exercise IX.6.2 *Determine the radii of convergence of the following series.*
(a) $\sum (\log n) x^n$.   (b) $\sum \frac{\log n}{n} x^n$.
(c) $\sum \frac{1}{n^{\log n}} x^n$.   (d) $\sum \frac{1}{n(\log n)^2} x^n$.
(e) $\sum \frac{x^n}{(4n-1)!}$.   (f) $\sum \frac{2^n}{(2n+7)!} x^n$.

Solution. (a) The radius of convergence is 1 because $1 \leq (\log n)^{1/n} \leq n^{1/n}$ for $n \geq 3$.
(b) The radius of convergence is 1 because $(\log n)^{1/n} \to 1$ and $n^{1/n} \to 1$ as $n \to \infty$.
(c) The radius of convergence is 1 because

$$\frac{1}{n} \log \left(\frac{1}{n^{\log n}} \right) = -\left(\frac{\log n}{\sqrt{n}} \right)^2 \to 0$$

as $n \to \infty$.
(d) The radius of convergence is 1 because $(\log n)^{1/n} \to 1$ and $n^{1/n} \to 1$.
(e) The radius of convergence is ∞ because of Example 1 in the text and the inequality $1/(4n-1)! \leq 1/n!$.
(f) The radius of convergence is ∞ because for any $r > 0$,

$$\frac{2^{n+1} r^{n+1}}{(2(n+1)+7)!} \frac{(2n+7)!}{2^n r^n} \to 0$$

as $n \to \infty$.

Exercise IX.6.3 *Suppose that $\sum a_n z^n$ has a radius of convergence $r > 0$. Show that given $A > 1/r$ there exists $C > 0$ such that*

$$|a_n| \leq CA^n \quad \text{for all } n.$$

Solution. Since $A > 1/r = \limsup |a_n|^{1/n}$ the inequality $|a_n|^{1/n} \leq A$ holds for all but finitely many n. Hence $|a_n| \leq A^n$ for all but finitely many n so we can choose C so large that $|a_n| \leq CA^n$ for all n.

Exercise IX.6.4 *Let $\{a_n\}$ be a sequence of positive numbers, and assume that $\lim a_{n+1}/a_n = A \geq 0$. Show that $\lim a_n^{1/n} = A$.*

Solution. Suppose first $A > 0$ for simplicity. Given $\epsilon > 0$, let n_0 be such that $A - \epsilon \leq a_{n+1}/a_n \leq A + \epsilon$ if $n \geq n_0$. Without loss of generality, we can assume $\epsilon < A$ so $A - \epsilon > 0$. Write

$$a_n = a_1 \prod_{k=1}^{n_0-1} \frac{a_{k+1}}{a_k} \prod_{k=n_0}^{n} \frac{a_{k+1}}{a_k}.$$

then by induction, there exists constants $C_1(\epsilon)$ and $C_2(\epsilon)$ such that

$$C_1(\epsilon)(A - \epsilon)^{n-n_0} \le a_n \le C_2(\epsilon)(A + \epsilon)^{n-n_0}.$$

Put $C_1'(\epsilon) = C_1(\epsilon)(A - \epsilon)^{-n_0}$ and $C_2'(\epsilon) = C_2(\epsilon)(A + \epsilon)^{-n_0}$. Then

$$C_1'(\epsilon)^{1/n}(A - \epsilon) \le a_n^{1/n} \le C_2'(\epsilon)^{1/n}(A + \epsilon).$$

There exists $N \ge n_0$ such that for $n \ge N$ we have

$$C_1'(\epsilon)^{1/n} = 1 + \delta_1(n) \quad \text{where } |\delta_1(n)| \le \epsilon/(A - \epsilon)$$

and similarly

$$C_2'(\epsilon)^{1/n} = 1 + \delta_2(n) \quad \text{where } |\delta_2(n)| \le \epsilon/(A + \epsilon)$$

Then

$$A - \epsilon + \delta_1(n)(A - \epsilon) \le a^{1/n} \le A + \epsilon + \delta_2(n)(A + \epsilon).$$

This shows that $|a^{1/n} - A| \le 2\epsilon$ and concludes the proof when $A > 0$. If $A = 0$ one can simply write the terms on the left of the inequalities as 0 throughout.

Exercise IX.6.5 *Determine the radius of convergence of the following series.*

(a) $\sum \frac{n!}{n^n} z^n$. *(b)* $\sum \frac{(n!)^3}{(3n)!} z^n$.

Solution. (a) By Chapter IV, we know that $n! \equiv n^n e^n$ for $n \to \infty$, where $a_n \equiv b_n$ means that there exists a sequence $\{u_n\}$ such that $b_n = u_n a_n$ and $\lim u_n^{1/n} = 1$. See Exercise 18, §2, of Chapter IV. So

$$\lim_{n\to\infty} \left(\frac{n!}{n^n}\right)^{1/n} = \lim_{n\to\infty} (e^{-n})^{1/n} = \frac{1}{e}$$

and the radius of convergence of the series is e. Note that we can also use Exercise 4 because

$$\frac{(n+1)!}{(n+1)^{n+1}} \frac{n^n}{n!} = \frac{1}{(1 + \frac{1}{n})^n} \to \frac{1}{e}.$$

(b) In Exercise 18, §2, of Chapter IV, we proved that $(3n)! \equiv (3n)^{3n} e^{-3n}$ so that

$$\lim_{n\to\infty} \left(\frac{(n!)^3}{(3n)!}\right)^{1/n} = \lim_{n\to\infty} \left(\frac{n^{3n} e^{-3n}}{(3n)^{3n} e^{-3n}}\right)^{1/n} = \frac{1}{27}.$$

Hence the power series has a radius of convergence equal to 27.

Exercise IX.6.6 *Let* $\{a_n\}$ *be a decreasing sequence of positive numbers appoaching 0. Prove that the power series* $\sum a_n z^n$ *is uniformly convergent on the domain of complex z such that*

$$|z| \leq 1 \quad \text{and} \quad |z - 1| \geq \delta,$$

where $\delta > 0$. *Remember summation by parts.*

Solution. Let $T_n(z) = \sum_{k=0}^{n} a_k z^k$ and $S_n(z) = \sum_{k=0}^{n} z^k$. The summation by parts formula gives

$$T_n(z) = a_n S_n(z) - \sum_{k=0}^{n-1} S_k(z)(a_{k+1} - a_k),$$

hence if $n > m$ some straightforward computations show that the difference $T_n(z) - T_m(z)$ is equal to

$$a_n(S_n(z) - S_m(z)) + \sum_{k=m+1}^{n-1} (S_k(z) - S_m(z))(a_k - a_{k+1}). \qquad \text{(IX.1)}$$

Summing a geometric series we find $S_n(z) = (z^{n+1} - 1)/(z - 1)$, so using the assumption that $|z| \leq 1$ and $|z - 1| \geq \delta$ we get the uniform bound $|S_n(z)| \leq 2/\delta$ for all n. Therefore $|S_n(z) - S_m(z)| \leq 4/\delta$ for all m and n. Putting absolute values in (IX.1), using the triangle inequality and the fact that $\{a_n\}$ is positive, and decreasing we get

$$
\begin{aligned}
|T_n(z) - T_m(z)| &\leq a_n \frac{4}{\delta} + \frac{4}{\delta} \sum_{k=m+1}^{n-1} (a_k - a_{k+1}) \\
&= a_n \frac{4}{\delta} + \frac{4}{\delta}(a_{m+1} - a_n) \\
&= \frac{4}{\delta} a_{m+1}.
\end{aligned}
$$

Since $a_m \to 0$ as $m \to \infty$ we conclude that the series $\sum a_n z^n$ is uniformly convergent in the domain $|z| \leq 1$ and $|z - 1| \geq \delta$ of the complex plane.

Exercise IX.6.7 (Abel's Theorem) *Let* $\sum_{n=1}^{\infty} a_n z^n$ *be a power series with radius of convergence* ≥ 1. *Assume that the series* $\sum_{n=1}^{\infty} a_n$ *converges. Let* $0 \leq x < 1$. *Prove that*

$$\lim_{x \to 1} \sum_{n=1}^{\infty} a_n x^n = \sum_{n=1}^{\infty} a_n.$$

Solution. Let $f(x) = \sum_{k=1}^{\infty} a_k x^k$, $A = \sum_{k=1}^{\infty} a_k$ and $A_n = \sum_{k=1}^{n} a_k$. Consider the partial sums

$$s_n(x) = \sum_{k=1}^{n} a_k x^k.$$

We first prove that the sequence of partial sums $\{s_n(x)\}$ converges uniformly for $0 \leq x \leq 1$. For $m < n$, applying the summation by parts formula, we get

$$s_n(x) - s_m(x) = \sum_{k=m+1}^{n} x^k a_k$$

$$= x^n(A_n - A_{m+1}) + \sum_{k=m+1}^{n-1} (A_k - A_{m+1})(x^k - x^{k+1}).$$

There exists N such that for $k, m \geq N$ we have $|A_k - A_{m+1}| \leq \epsilon$. Hence for $0 \leq x \leq 1$ and $n, m \geq N$ we have

$$|s_n(x) - s_m(x)| \leq \epsilon + \epsilon \sum_{k=m+1}^{n-1} (x^k - x^{k+1})$$

$$= \epsilon + \epsilon(x^{m+1} - x^n)$$

$$\leq 3\epsilon.$$

This proves the uniform convergence of $\{s_n(x)\}$.

Now given ϵ, pick N as above. Choose δ (depending on N) such that if $|x - 1| < \delta$, then

$$|S_N(x) - A_N| < \epsilon.$$

By combining the above results we find that

$$|f(x) - A| \leq |f(x) - s_n(x)| + |s_n(x) - s_N(x)| + |s_N(x) - A_N| + |A_N - A|$$

$$\leq |f(x) - s_n(x)| + 5\epsilon$$

for all $n \geq N$ and $|x - 1| < \delta$. For a given x, pick n so large (depending on x!) so that the first term is also $< \epsilon$, to conclude the proof.

IX.7   Differentiation and Integration of Series

Exercise IX.7.1 *Show that if $f(x) = \sum 1/(n^2 + x^2)$, then $f'(x)$ can be obtained by differentiating this series term by term.*

Solution. Let $f_n(x) = 1/(n^2 + x^2)$. Then for all $|x| \leq C$ we have

$$|f_n'(x)| = \left| \frac{-2x}{(n^2 + x^2)^2} \right| \leq 2\frac{C}{n^4},$$

so $\sum f_n'(x)$ converges uniformly on every compact interval. Clearly, $\sum f_n(x)$ converges absolutely for each x, thus $f'(x) = \sum f_n'(x)$.

Exercise IX.7.2 *Same problem if* $f(x) = \sum 1/(n^2 - x^2)$, *defined when* x *is not equal to an integer.*

Solution. In Exercise 4, §5, we prove that the series converges absolutely and uniformly on any closed interval not containing an integer. The derived series is simply $\sum -2x/(n^2-x^2)^2$ which converges absolutely and uniformly on every compact interval not containing an integer. The argument is the same as the one given in Exercise 4, §5. We conclude that $f'(x) = \sum f_n(x)$ whenever x is not an integer.

Exercise IX.7.3 *Let F be the vector space of continuous functions on $[0, 2\pi]$ with the sup norm. On F define the scalar product*

$$\langle f, g \rangle = \int_0^{2\pi} f(x)g(x)dx.$$

Two functions f, g are called **orthogonal** *if $\langle f, g \rangle = 0$. Let*

$$\varphi_n(x) = \cos nx \quad and \quad \psi_n(x) = \sin nx.$$

(Take $n \geq 1$ except for $\varphi_0(x) = 1$.) Show that the functions $\varphi_0, \varphi_n, \psi_m$ are pairwise orthogonal. [Hint: Use the formula

$$\sin nx \cos mx = \frac{1}{2}[\sin(n + m)x + \sin(n - m)x]$$

and similar ones.] Find the norms of $\varphi_n, \varphi_0, \psi_m$.

Solution. For all $n \geq 1$ we have

$$\langle \varphi_0, \varphi_n \rangle = \int_0^{2\pi} \cos nx dx = \frac{1}{n} [\sin nx]_0^{2\pi} = 0$$

and

$$\langle \varphi_0, \psi_n \rangle = \int_0^{2\pi} \sin nx dx \frac{-1}{n} [\cos nx]_0^{2\pi} = 0.$$

Since $\sin 0 = 0$ we also find that $\langle \varphi_0, \psi_0 \rangle = 0$. Furthermore,

$$\langle \varphi_n, \psi_m \rangle = \int_0^{2\pi} \cos nx \sin mx dx = \frac{1}{2} \int_0^{2\pi} \sin(n + m)x + \sin(n - m)x dx = 0$$

so the functions $\varphi_0, \varphi_n, \psi_m$ are pairwise orthogonal.

For the norms, we have $\|\varphi_0\|^2 = \langle \varphi_0, \varphi_0 \rangle = 2\pi$ and for all $n \geq 1$,

$$\|\varphi_n\|^2 = \langle \varphi_n, \varphi_n \rangle = \int_0^{2\pi} \cos^2 nx dx = \frac{1}{2} \int_0^{2\pi} 1 + \cos 2nx dx = \pi,$$

and

$$\|\psi_n\|^2 = \langle \psi_n, \psi_n \rangle = \int_0^{2\pi} \sin^2 nx dx = \frac{1}{2} \int_0^{2\pi} 1 - \cos 2nx dx = \pi.$$

Exercise IX.7.4 *Let $\{a_n\}$ be a sequence of numbers such that $\sum a_n$ converges absolutely. Prove that the series*

$$f(x) = \sum a_n \cos nx$$

converges uniformly. Show that

$$\langle f, \varphi_0 \rangle = 0, \quad \langle f, \psi_m \rangle = 0 \; for \; all \; m, \quad \langle f, \varphi_k \rangle = \pi a_k.$$

Solution. The series converges uniformly because $|a_n \cos nx| \leq |a_n|$ and $\sum |a_n| < \infty$. Theorem 7.1 implies

$$\langle f, \varphi_0 \rangle = \sum \int_0^{2\pi} a_n \cos nx \, dx = 0.$$

Since the series $\sum a_n \cos nx \sin nx$ converges uniformly we have

$$\langle f, \psi_m \rangle = \sum \int_0^{2\pi} a_n \cos nx \sin nx \, dx = \sum a_n \langle \varphi_n, \psi_m \rangle = 0.$$

Similarly, we have

$$\langle f, \varphi_k \rangle = \sum \int_0^{2\pi} a_n \cos nx \cos kx \, dx = \sum a_n \langle \varphi_n, \varphi_k \rangle = a_k \|\varphi_k\|^2 = a_k \pi.$$

Exercise IX.7.5 *Let $\{a_n\}$ be a sequence of numbers. Show that there exists an infinitely differentiable function g defined on some open interval containing 0 such that*

$$g^{(n)}(0) = a_n.$$

[Hint: (Tate) Given $n \geq 0$ and ϵ, there exists a function $f = f_{n,\epsilon}$ which is C^∞ on $-1 < x < 1$ such that:

(1) $f(0) = f'(0) = \cdots = f^{(n-1)}(0) = 0$ and $f^{(n)}(0) = 1$.
(2) $|f^{(k)}(x)| \leq \epsilon$ for $k = 0, \ldots, n-1$ and $|x| \leq 1$.

Indeed, let φ be a C^∞ function on $(-1, 1)$ such that:

$\varphi(x) = 1$ if $|x| \leq \epsilon/2$,
$0 \leq \varphi(x) \leq 1$ if $\epsilon/2 \leq |x| \leq \epsilon$,
$\varphi(x) = 0$ if $\epsilon \leq |x| \leq 1$.

Integrate φ from 0 to x, n times to get $f(x)$. Then let ϵ_n be chosen so that $\sum |a_n| \epsilon_n$ converges. Put

$$g(x) = \sum_{n=0}^{\infty} a_n f_{n,\epsilon_n}(x).$$

For $k \geq 0$ the series

$$\sum_{n=0}^{\infty} a_n D^k f_{n,\epsilon_n}$$

converges uniformly on $|x| \leq 1$, as one sees by decomposing into the sum from 0 to k and from $k+1$ to ∞, because for $n > k$ we have

$$|a_n D^k f_{n,\epsilon_n}(x)| \leq |a_n| \epsilon_n.]$$

Solution. We construct the C^∞ function as follows: Let $a = \epsilon/2$ and $b = \epsilon$, and let h be the function (bump function) defined to be equal to 0 if $t \leq a$ or $t \geq b$ and

$$h(t) = e^{-1/(t-a)(t-b)}$$

otherwise. Then h is a C^∞ function (Exercise 6, §1, of Chapter IV) with all derivatives equal to 0 at both a and b. Let

$$g(x) = 1 - C \int_{-\infty}^{x} h(t)dt,$$

where $1/C = \int_a^b h(t)dt$. Then $g(a) = 1, g(b) = 0, 0 \leq g(x) \leq 1$, and g is C^∞ with all derivatives equal to 0 at both a and b. For $0 \leq x \leq 1$ define

$$\varphi(x) = \begin{cases} 1 & \text{if } 0 \leq x \leq \epsilon/2, \\ g(x) & \text{if } \epsilon/2 \leq x \leq \epsilon, \\ 0 & \text{if } \epsilon \leq x. \end{cases}$$

For $-1 \leq x \leq 0$ define φ by $\varphi(x) = \varphi(-x)$.

To get f, let $\psi_0(x) = \varphi(x)$ and $\psi_{k+1}(x) = \int_0^x \psi_k(t)dt$. Then let $f = \psi_n$. If $\epsilon_n = 1/(n^2|a_n| + 1)$, then $\sum |a_n| \epsilon_n$ converges. Finally, let

$$g(x) = \sum_{n=0}^{\infty} a_n f_{n,\epsilon_n}(x).$$

For $k \geq 0$, the series $\sum a_n D^k f_{n,\epsilon_n}$ converges uniformly on $|x| \leq 1$ because as we split the sum

$$\sum a_n D^k f_{n,\epsilon_n} = \sum_{n \leq k} a_n D^k f_{n,\epsilon_n} + \sum_{n \geq k+1} a_n D^k f_{n,\epsilon_n}$$

and if $n \geq k+1$, we have $|D^k f_{n,\epsilon_n}| \leq \epsilon_n$ by construction. Hence

$$D^k g(x) = \sum a_n D^k f_{n,\epsilon_n}.$$

When $k < n$ we have $D^k(f_{n,\epsilon_n}) = 0$ by (1) in the hint and if $k > n$, then we also have $D^k f_{n,\epsilon_n}(0) = 0$ because φ is locally constant near the origin. Since $D^n f_{n,\epsilon_n}(0) = 1$ we see that $D^k g(x) = a_k$, as was to be shown.

Exercise IX.7.6 *Given a C^∞ function $g : [a, b] \to \mathbf{R}$ from a closed interval, show that g can be extended to a C^∞ function defined on an open interval containing $[a, b]$.*

Solution. Let $a_n = g^{(n)}(a)$ and $b_n = g^{(n)}(b)$. Then by the previous exercise we can extend g is some open neighborhood of a and in some open neighborhood of b such that this extension is C^∞ on an interval containing $[a, b]$.

Exercise IX.7.7 *Let $n \geq 0$ be an integer. Show that the series*

$$J_n(x) = \sum_{k=0}^{\infty} \frac{(-1)^k x^{n+2k}}{2^{n+2k} k! (n+k)!}$$

converges for all x. Prove that $y = J_n(x)$ is a solution to Bessel's equation

$$y'' + \frac{1}{x} y' + \left(1 - \frac{n^2}{x^2}\right) y = 0.$$

Solution. The absolute value of the k-th term is $\leq |x|^{n+2k}/k!$ so the series converges absolutely for every x and the convergence is uniform on every closed and bounded interval. A similar argument shows that we can differentiate term by term to get y' and y''. Let

$$a_k = \frac{(-1)^k}{2^{n+2k} k! (n+k)!}.$$

Then the coefficient of the term x^{n+2k-2} in the sum

$$y'' + \frac{1}{x} y' + \left(1 - \frac{n^2}{x^2}\right) y = 0$$

is equal to

$$a_k(n+2k)(n+2k-1) + a_k(n+2k) + a_{k-1} - n^2 a^k = (4kn + 4k^2)a_k + a^{k-1}.$$

But $a_{k-1}/a_k = -4k(n+k)$ so we conclude that $y = J_n(x)$ is a solution to Bessel's equation.

X

The Integral in One Variable

X.3 Approximation by Step Maps

Exercise X.3.1 *If f is a continuous real valued function on $[a, b]$, show that one can approximate f uniformly by step functions whose values are less than or equal to those of f, and also by step functions whose values are greater than or equal to those of f. The integrals of these step functions are then the standard lower and upper Riemann sums.*

Solution. Let $\epsilon > 0$. Choose a partition of $[a, b]$ as in Theorem 3.1, that is, a partition of size $< \delta$ where δ is chosen so that $|f(x) - f(y)| < \epsilon$ whenever $|x - y| < \delta$. If $a_{i-1} < t < a_i$ let

$$g(t) = \max_{a_{i-1} \le t \le a_i} f(t) \quad \text{and} \quad h(t) = \min_{a_{i-1} \le t \le a_i} f(t),$$

and if $t = a_i$ for some i let

$$g(t) = \max_{a \le t \le b} f(t) \quad \text{and} \quad h(t) = \min_{a \le t \le b} f(t).$$

Then g and h are step maps which approximate f uniformly and for all t we have $h(t) \le f(t) \le g(t)$.

Exercise X.3.2 *Show that the product of two regulated maps is regulated. The product of two piecewise continuous maps is piecewise continuous.*

Solution. (i) Suppose that $f, g \in \mathrm{Reg}([a, b], G)$, and let M be a common bound for f and g. Given $\epsilon > 0$, choose step maps f_1 and g_1 such that

$\|f - f_1\| < \epsilon$ and $\|g - g_1\| < \epsilon$. Considering a refinement of the partitions associated to f_1 and g_1, one sees at once that $f_1 g_1$ is a step map. We then have

$$\|fg - f_1 g_1\| = \|fg - f_1 g + f_1 g - f_1 g_1\| \leq \|g\|\|f - f_1\| + \|f_1\|\|g - g_1\|$$
$$\leq M\epsilon + (\|f_1 - f\| + \|f\|)\epsilon$$
$$\leq \epsilon(2M + \epsilon).$$

(ii) Given two piecewise continuous functions f and g with their partitions, consider a refinement of both partitions. Since the product of continuous functions is continuous we conclude that the product fg is also a piecewise continuous function.

Exercise X.3.3 *On the space of regulated maps $f \colon [a, b] \to \mathbf{C}$, show that $|f|$ is regulated, and define*

$$\|f\|_1 = \int_a^b |f|.$$

Show that this is a seminorm (all the properties of a norm except that $\|f\|_1 \geq 0$ but $\|f\|_1$ may be 0 without f itself being 0).

Solution. Given $\epsilon > 0$ choose a step map h such that $\|f - h\| < \epsilon$. Then $|h|$ is also a step map and the inequality

$$\||f(t)| - |h(t)|\| \leq |f(t) - h(t)|,$$

implies that $\||f| - |h|\| < \epsilon$ so $|f|$ is regulated. From this analysis we extract the fact that $|f|$ can be uniformly approximated by positive step maps, so that

$$\|f\|_1 = \int_0^{2\pi} |f| \geq 0.$$

Furthermore, if we consider the step map which is 0 on $(a, b]$ and 1 at a, then its norm is 0. If c is a number and $\{h_n\}$ is a sequence of step maps converging to f, then the sequence $\{|h_n|\}$ converges to $|f|$ and the sequence $\{|ch_n|\}$ converges to $|cf|$. The integral of step maps is a linear function so

$$I_a^b(|ch_n|) = |c| I_a^b(|h_n|)$$

hence $\|cf\|_1 = |c|\|f\|_1$. Finally, if f and g are regulated and $\{f_n\}$, $\{g_n\}$ are sequences of step maps converging to f and g, respectively, then $\{f_n + g_n\}$ converges to $f + g$ and $\{|f_n + g_n|\}$ converges to $|f + g|$. Since

$$|f_n(t) + g_n(t)| \leq |f_n(t)| + |g_n(t)|$$

we conclude that $\|f + g\|_1 \leq \|f\|_1 + \|g\|_1$, thus $\| \cdot \|_1$ is a seminorm.

Exercise X.3.4 *Let F be the vector space of real valued regulated functions on an interval $[a, b]$. We have the sup norm on F. We have the seminorm of Exercise 3. It is called the L^1-seminorm. Prove that the continuous functions are dense in F, for the L^1-seminorm. In other words, prove that given $f \in F$, there exists a continuous function g on $[a, b]$ such that $\|f - g\|_1 < \epsilon$. [Hint: First approximate f by a step function. Then approximate a step function by a continuous function obtained by changing a step function only near its discontinuities.]*

Solution. Given a regulated function $f \in F$ there exists a step function φ such that $\|f - \varphi\|_1 < \epsilon$ (cf. Exercise 6).

Now approximate φ by a continuous function g in the following way. We may assume after changing the values of $\varphi(a)$ and $\varphi(b)$ that φ is continuous at a and b. Let $a_1 < a_2 < \cdots < a_n$ be the points where φ is discontinuous, and let $a_0 = a$ and $a_{n+1} = b$. Let $\delta > 0$ be a number such that

$$\delta < \min_{0 \le j \le n} (a_{j+1} - a_j)/2 \quad \text{and} \quad \delta < \epsilon/(2Bn),$$

where B is a bound for $|\varphi|$. For $1 \le j \le n$ let

$$I_j = \left[a_j - \frac{\delta}{2}, a_j + \frac{\delta}{2}\right].$$

Then on each interval I_j replace φ by the line segment having the same values as φ at the end points of the interval.

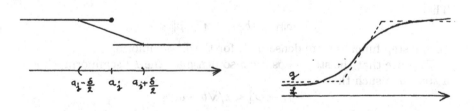

Define g to be equal to these linear functions on the intervals I_j and equal to φ otherwise (i.e. on $[a, b] - \bigcup_j I_j$). Then g is continuous and on $\bigcup_j I_j$ we have $\|\varphi - g\| \le 2B$. On the complement, $[a, b] - \bigcup_j I_j$ we have $g = \varphi$ by construction. Thus

$$\int_a^b |\varphi - g| \le 2Bn\delta,$$

which implies $\|\varphi - g\|_1 < \epsilon$ because of our choice for δ. Putting everything together we get $\|f-g\|_1 < 2\epsilon$, thereby proving that the continuous functions are dense in F for the L^1-seminorm.

Exercise X.3.5 *On the space of regulated functions as in Exercise 4, define the scalar product*

$$\langle f, g \rangle = \int_a^b f(x)g(x)dx.$$

The seminorm associated with this scalar product is called the L^2-seminorm. (Cf. Exercise 11 of Chapter VI, §2.) Show that the continuous functions are dense in F for the L^2-seminorm.

Solution. Given a regulated function f, and $\epsilon > 0$ there exists a step function φ such that $\|f - \varphi\|_2 < \epsilon$ (see Exercise 6). Then we argue as in Exercise 4 with $\delta < \epsilon^2/4B^2n$ instead of $\delta < \epsilon/2Bn$. The resulting continuous function g satisfies

$$\int_a^b |\varphi - g|^2 \leq 4B^2n\delta < \epsilon^2,$$

so that $\|\varphi - f\|_2 < \epsilon$. Then $\|f - g\|_2 < 2\epsilon$.

Exercise X.3.6 *The space F still being as in Exercise 4 or 5, show that the step functions are dense in F for the L^1-seminorm and the L^2-seminorm.*

Solution. Given a regulated function f and $\epsilon > 0$ there exists a step map φ such that

$$\|f - \varphi\| < \epsilon/(b - a).$$

Then

$$\|f - \varphi\|_1 \leq (b - a)\|f - \varphi\| < \epsilon.$$

So the step functions are dense in F for the L^1-seminorm.

To prove that the step maps are also dense for the L^2-seminorm, choose a step map such that

$$\|f - \varphi\| < \epsilon/\sqrt{b - a}.$$

Then

$$\int_a^b |f - \varphi|^2 \leq (b - a)\|f - \varphi\|^2 < \epsilon^2,$$

and therefore $\|f - \varphi\|_2 < \epsilon$.

Exercise X.3.7 *Let F be the space of regulated functions on $[a, b]$ once more. Let $C^\infty = C^\infty([a, b])$ be the space of infinitely differentiable real valued functions on $[a, b]$. Prove that C^∞ is (a) L^1-dense and (b) L^2-dense in F. [Hint: First approximate by step functions, then smooth out the corners using bump functions which are 0 in a δ-interval around a corner and 1 outside a 2δ-interval around each corner. Pick δ sufficiently small.]*

Solution. We give two methods of smoothing out the corners. The first method follows the hint and is important, for example, in the context of Stokes' theorem with singularities. The second method is the one we use in Exercise 1, §2, of Chapter V.

(a) We use the same notation as in Exercise 4. We approximate f by a step function φ and again, we let B be a bound for $|\varphi|$. Now we use bump functions to smooth out the corners on I_j.

Method 1. We multiply φ on I_j by the bump function which is 0 on $[a_j - \delta/4, a_j + \delta/4]$ and 1 outside $[a_j - \delta/2, a_j + \delta/2]$, so that we get the following picture:

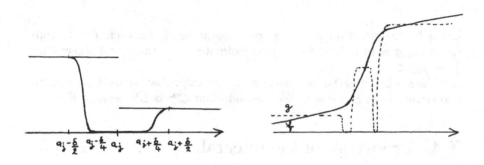

Let g be the function resulting from modifying φ on each I_j. By construction, g is in C^∞ and the same estimate as in Exercise 4 shows that $\|\varphi - g\|_1 < \epsilon$.

Method 2. Let a_j be a discontinuity point of φ. We assumed that $a < a_j < b$. Pick δ as in Exercise 4. Then choose a C^∞ function g_j which is defined on I_j and which has the following properties:

g_j is equal to $\varphi(a_j - \delta/2)$ on $[a_j - \delta/2, a_j - \delta/4]$;

g_j is monotone on $[a_j - \delta/4, a_j + \delta/4]$; and

g_j is equal to $\varphi(a_j + \delta/2)$ on $[a_j + \delta/4, a_j + \delta/2]$.

To do so, use Exercise 6, §1, of Chapter IV (multiply and translate if necessary). We have the following picture:

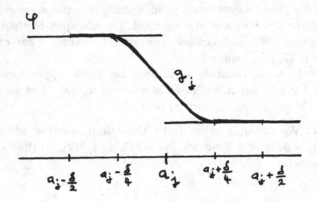

Let g be the function resulting from modifying φ on each I_j. By construction, g is in C^∞ and the same estimate as in Exercise 4 shows that $\|\varphi - g\|_1 < \epsilon$.

(b) The argument is the same as in part (a) except that we use the notation and estimates of Exercise 5. We conclude that C^∞ is L^2-dense in F.

X.4   Properties of the Integral

Exercise X.4.1 *Let $a \leq t \leq b$ be a closed interval and let*

$$P = \{a = t_0 \leq t_1 \leq \cdots \leq t_n\}$$

*be a partition of this interval. By the **size** of P we mean*

$$\text{size } P = \max_k (t_{k+1} - t_k).$$

Let f be a continuous function on $[a, b]$, or even a regulated function. Given numbers c_k with

$$t_k \leq c_k \leq t_{k+1},$$

*form the **Riemann sum***

$$S(P, c, f) = \sum_{k=0}^{n-1} f(c_k)(t_{k+1} - t_k).$$

Let

$$L = \int_a^b f(t)dt.$$

Show that given $\epsilon > 0$, there exists δ such that if size $(P) < \delta$, then

$$|S(P, c, f) - L| < \epsilon.$$

Solution. Given $\epsilon > 0$ and a regulated function f, choose a step function φ such that

$$\|f - \varphi\| < \frac{\epsilon}{b-a} \quad \text{so} \quad \int_a^b |f - \varphi| < \epsilon.$$

Then we have

$$\left| S(P, c, f) - \int_a^b f \right| \leq \left| S(P, c, f) - \int_a^b (f - \varphi) - \int_a^b \varphi \right| \leq \left| S(P, c, f) - \int_a^b \varphi \right| + \epsilon.$$

Now we estimate the term $\left| S(P, c, f) - \int_a^b \varphi \right|$. Let $a = a_0 \leq a_1 \leq \cdots \leq a_p = b$ be a partition associated to φ. Let $P = \{a = t_0 \leq t_1 \leq \cdots \leq t_n = b\}$ be a partition of $[a, b]$ and let B be a bound for f which is also a bound for φ. By linearity of the integral we can write

$$S(P, c, f) - \int_a^b \varphi = \sum_{k=0}^{n-1} f(c_k)(t_{k+1} - t_k) - \int_{t_k}^{t_{k+1}} \varphi$$

$$= \sum_{k=0}^{n-1} \int_{t_k}^{t_{k+1}} f(c_k) - \varphi(t) dt.$$

For each k $(k = 0, \ldots, n-1)$, look at the interval $[t_k, t_{k+1}]$. If no a_i $(i = 0, \ldots, p)$ belongs to $[t_k, t_{k+1}]$, then φ is constant on this interval and we get

$$\left| \int_{t_k}^{t_{k+1}} f(c_k) - \varphi(t) dt \right| \leq \frac{\epsilon}{b-a}(t_{k+1} - t_k).$$

If for some integer i $(i = 1, \ldots, p)$ the point a_i belongs to $[t_k, t_{k+1}]$, then

$$|\varphi(t) - f(c_k)| \leq 2B$$

for all $t \in [t_k, t_{k+1}]$ and therefore

$$\left| \int_{t_k}^{t_{k+1}} f(c_k) - \varphi(t) dt \right| \leq 2B(t_{k+1} - t_k) \leq 2B\text{size}(P).$$

Since each a_i belongs to at most two intervals $[t_k, t_{k+1}]$ we conclude that

$$\left| S(P, c, f) - \int_a^b \varphi \right| \leq \epsilon + 4B(p+1)(\text{size } P).$$

Let $\delta = \epsilon/(4B(p+1))$. Then if size $P < \delta$, we see that

$$|S(P, c, f) - L| < 3\epsilon$$

and the proof is complete. If f is continuous, we can also argue as in the next exercise with $h(x) = x$.

Exercise X.4.2 (The Stieltjes Integral) *Let f be a continuous function on an interval $a \leq t \leq b$. Let h be an increasing function on this interval, and assume that h is bounded. Given a partition*

$$P = \{a = t_0 \leq t_1 \leq \cdots \leq t_n = b\}$$

*of the interval, let c_k be a number, $t_k \leq c_k \leq t_{k+1}$, and define the **Riemann-Stieltjes** sum relative to h to be*

$$S(P, c, f) = \sum_{k=0}^{n-1} f(c_k)[h(t_{k+1}) - h(t_k)].$$

Prove that the limit

$$L = \lim_{P,c} S(P, c, f)$$

exists as the size of the partition approaches 0. This means that there exists a number L having the following property. Given ϵ there exists δ such that for any partition P of size $< \delta$ we have

$$|S(P, c, f) - L| < \epsilon.$$

[Hint: Selecting values for c_k such that $f(c_k)$ is a maximum (resp. minimum) on the interval $[t_k, t_{k+1}]$, use upper and lower sums, and show that the difference is small when the size of the partition is small.] The above limit is usually denoted by

$$L = \int_a^b f \, dh.$$

Solution. Given a partition P of $[a, b]$ we define the upper Riemann-Stieltjes sum to be

$$\overline{S}(P, f) = \sum_{k=0}^{n-1} f(c_k)[h(t_{k+1}) - h(t_k)],$$

where $c_k = \max_{t_k \leq t \leq t_{k+1}} f(t)$. Similarly, we define the lower Riemann-Stieltjes sum to be

$$\underline{S}(P, f) = \sum_{k=0}^{n-1} f(c_k')[h(t_{k+1}) - h(t_k)],$$

where $c_k' = \min_{t_k \leq t \leq t_{k+1}} f(t)$. Clearly, we always have

$$\underline{S}(P, f) \leq \overline{S}(P, f).$$

In fact, if P' is a refinement of P, that is, if P' is obtained from P by adding finitely many points, we see that

$$\underline{S}(P,f) \leq \underline{S}(P',f) \leq \overline{S}(P',f) \leq \overline{S}(P,f).$$

This is obvious when P' is obtained by adding one point, hence the above inequalities follow by induction on the number of points added to P. Now if P_1 and P_2 are partitions of $[a,b]$, then

$$\underline{S}(P_1,f) \leq \overline{S}(P_2,f)$$

because if P' is a refinement of both P_1 and P_2, then

$$\underline{S}(P_1,f) \leq \underline{S}(P',f) \leq \overline{S}(P',f) \leq \overline{S}(P_2,f).$$

This proves that

$$\inf_{P}\{\overline{S}(P,f)\} \quad \text{and} \quad \sup_{P}\{\underline{S}(P,f)\}$$

exist, where the infimum and supremum are taken over all partitions of $[a,b]$.

Let B be a bound for h. Given $\epsilon > 0$ choose a positive number α such a that $2B\alpha < \epsilon$. Since f is uniformly continuous, there exists $\delta > 0$ such that $|x-y| < \delta$ implies $|f(x) - f(y)| < \alpha$. If the size of P is $< \delta$, then

$$\overline{S}(P,f) - \underline{S}(P,f) \leq 2B\alpha < \epsilon.$$

Together with the results obtained above we conclude that both limits

$$\lim_{P,\ \text{size}(P)\to 0} \overline{S}(P,f) \quad \text{and} \quad \lim_{P,\ \text{size}(P)\to 0} \underline{S}(P,f)$$

exist, and that they are equal, say to a number L. Then, given any partition P of size $< \delta$ we have

$$\underline{S}(P,f) \leq S(P,c,f) \leq \overline{S}(P,f)$$

so that

$$|S(P,c,f) - L| < \epsilon.$$

Exercise X.4.3 *Suppose that h is of class C^1 on $[a,b]$, that is h has a derivative which is continuous. Show that*

$$\int_a^b f\,dh = \int_a^b f(t)h'(t)\,dt.$$

Solution. Let $I = \int_a^b f(t)h'(t)\,dt$ and $J = \int_a^b f\,dh$. The integral I exists because fh' is continuous. Using the notation of Exercises 1 and 2, given $\epsilon > 0$ choose a partition $P = \{a = t_0 \leq \cdots \leq t_n = b\}$ of size $< \delta$. By the mean value theorem we know that

$$h'(c_k)(t_{k+1} - t_k) = h(t_{k+1}) - h(t_k),$$

for some $c_k \in [t_{k+1}, t_k]$. So, with this choice of c, we get

$$
S_{RS}(P, c, f) = \sum_{k=0}^{n-1} f(c_k)(h(t_{k+1}) - h(t_k)) = \sum_{k=0}^{n-1} f(c_k)h'(c_k)(t_{k+1} - t_k)
$$
$$
= S_R(P, c, fh'),
$$

where S_{SR} and S_R denote the Riemann-Stieltjes sum and the Riemann sum, respectively. We also have as a consequence of our choice the following inequalities

$$
|S_{RS}(P, c, f) - J| < \epsilon \quad \text{and} \quad |S_R(P, c, fh') - I| < \epsilon.
$$

Thus

$$
|I - J| \leq |I - S_R(P, c, fh')| + |J - S_{RS}(P, c, f)| < 2\epsilon.
$$

Since this is true for all $\epsilon > 0$ we conclude that $I = J$.

Exercise X.4.4 (The Total Variation.) *Let*

$$
f \colon [a, b] \to \mathbf{C}
$$

*be a complex valued function. Let $P = \{t_0 \leq t_1 \leq \cdots \leq t_n\}$ be a partition of $[a, b]$. Define the **variation** $V_P(f)$ to be*

$$
V_P(f) = \sum_{k=0}^{n-1} |f(t_{k+1}) - f(t_k)|.
$$

*Define the **variation***
$$
V(f) = \sup_P V_P(f),
$$

*where the sup (least upper bound if it exists, otherwise ∞) is taken over all partitions. If $V(f)$ is finite, then f is called of **bounded variation**.*
(a) Show that if f is real valued, increasing, and bounded on $[a, b]$, then f is of bounded variation, in fact bounded by $f(b) - f(a)$.
(b) Show that if f is differentiable on $[a, b]$ and f' is bounded, then f is of bounded variation. This is so in particular if f has a continuous derivative.
(c) Show that the set of functions of bounded variations on $[a, b]$ is a vector space, and that if f, g are of bounded variation, so is the product fg.

Solution. (a) Under the assumptions of (a) we have

$$
V_P(f) = \sum_{k=0}^{n-1} f(t_{k+1}) - f(t_k) = f(t_n) - f(t_0) = f(b) - f(a).
$$

(b) Let B be a bound for f'. By the mean value theorem, there exists for each k a number c_k which belongs to $[t_k, t_{k+1}]$ and which verifies

$$f(t_{k+1}) - f(t_k) = f'(c_k)(t_{k+1} - t_k),$$

so that

$$V_P(f) = \sum_{k=0}^{n-1} |f'(c_k)(t_{k+1} - t_k)| \le B(b - a),$$

whence $V(f)$ is finite.

(c) Suppose that f and g are of bounded variation on $[a, b]$. Since

$$V_P(f + g) = \sum_{k=0}^{n-1} |f(t_{k+1}) + g(t_{k+1}) - f(t_k) - g(t_k)|$$

we see that

$$V_P(f + g) \le \sum_{k=0}^{n-1} |f(t_{k+1}) - f(t_k)| + \sum_{k=0}^{n-1} |g(t_{k+1}) - g(t_k)| \le V(f) + V(g),$$

hence $f + g$ is of bounded variation on $[a, b]$. If c is a scalar, then

$$V_P(cf) = \sum_{k=0}^{n-1} |cf(t_{k+1}) - cf(t_k)| = |c|V_P(f) \le |c|V(f),$$

so cf is of bounded variation on $[a, b]$, and we see that the set of mappings of bounded variation on $[a, b]$ forms a vector space.

In order to show that if f and g are of bounded variation on $[a, b]$, then so is the product fg, we first prove that f and g are bounded. For f we have

$$\begin{aligned}
|f(x)| &\le |f(a)| + |f(x) - f(a)| \\
&\le |f(a)| + |f(x) - f(a)| + |f(x) - f(b)| \\
&\le |f(a)| + V(f),
\end{aligned}$$

so f is bounded, and a similar argument shows that g is bounded. Then since

$$f(x)g(x) - f(y)g(y) = f(x)g(x) - f(x)g(y) + f(x)g(y) - f(y)g(y)$$

putting absolute values and using the triangle inequality we obtain

$$|f(x)g(x) - f(y)g(y)| = \|f\||g(x) - g(y)| + |f(x) - f(y)|\|g(y)\|,$$

where $\| \cdot \|$ denotes the sup norm. Consequently

$$V_P(fg) \le \|f\|V_P(g) + V_P(f)\|g\| \le \|f\|V(g) + V(f)\|g\|$$

which proves that fg is of bounded variation whenever f and g are of bounded variation.

The notation for the variation really should include the interval, and we should write

$$V(f, a, b).$$

Define

$$V_f(x) = V(f, a, x),$$

so V_f is a function of x, called the **variation function** *of f.*

Exercise X.4.5 *(a) Show that V_f is an increasing function.*
(b) If $a \leq x \leq y \leq b$ show that

$$V(f, a, y) = V(f, a, x) + V(f, x, y).$$

Solution. (a) Suppose $x < y$, and let P be a partition of $[a, x]$. Then consider the partition, $P' = P \cup \{y\}$ of $[a, y]$. Then

$$V_P(f, a, x) \leq V_{P'}(f, a, y),$$

so $V_f(x) \leq V_f(y)$.
(b) Let P_1 be a partition of $[a, x]$ and P_2 a partition of $[x, y]$. Then

$$V_{P_1}(f, a, x) + V_{P_2}(f, x, y) = V_{P_1 \cup P_2}(f, a, y) \leq V(f, a, y).$$

So $V(f, a, x) + V(f, x, y) \leq V(f, a, y)$. It suffices to prove the reverse inequality. If P is a partition of $[a, y]$ we consider the refinement P' obtained by including the number x to P. Then the triangle inequality implies

$$V_P(f, a, y) \leq V_{P'}(f, a, y)$$

so if P_1 and P_2 are the restrictions of P' to $[a, x]$ and $[x, y]$ we get

$$V_P(f, a, y) \leq V_{P_1}(f, a, x) + V_{P_2}(f, x, y) \leq V(f, a, x) + V(f, x, y)$$

hence $V(f, a, y) \leq V(f, a, x) + V(f, x, y)$, as was to be shown.

Exercise X.4.6 Theorem. *If f is continuous, then V_f is continuous.*
Sketch of Proof: By Exercise 5(b), it suffices to prove (say for continuity on the right) that

$$\lim_{y \to x} V(f, x, y) = 0.$$

If the limit is not 0 (or does not exist), then there exists $\delta > 0$ such that

$$V(f, x, y) > \delta$$

for y arbitrarily close to x, and hence by Exercise 5(b), such that

$$V(f, x, y) > \delta$$

for all y with $x < y \leq y_1$ with some fixed y_1. Let

$$P = \{x_0 = x < x_1 < \cdots < x_n = y_1\}$$

be a partition such that $V_P(f) > \delta$. By continuity of f at x, we can select y_2 such that $x < y_2 < x_1$ and such that $f(y_2)$ is very close to $f(x)$. Replace the term

$$|f(x_1) - f(x)| \quad by \quad |f(x_1) - f(y_2)|$$

in the sum expressing $V_P(f)$. Then we have found y_2 such that $V(f, y_2, y_1) > \delta$. Now repeat this procedure, with a descending sequence

$$\cdots < y_n < y_{n-1} < \cdots < y_1.$$

Using Exercise 5(b), we find that

$$V(f, x, y_1) \geq V(f, y_n, y_{n-1}) + V(f, y_{n-1}, y_{n-2}) + \cdots + V(f, y_2, y_1)$$
$$\geq (n-1)\delta.$$

This is a contradiction for n sufficiently large, thus concluding the proof.

Solution. We justify some details in the almost complete proof. We can choose a partition P because of the definition of the sup. Then we can write $V_P(f) = \delta + \epsilon$ for some $\epsilon > 0$. Then we select y_2 such that $x < y_2 < x_1$ and $|f(y_2) - f(x)| < \epsilon/2$. Then

$$-\frac{\epsilon}{2} + |f(x_1) - f(x)| \leq |f(x_1) - f(x)| - |f(y_2) - f(x)| \leq |f(x_1) - f(y_2)|$$

so

$$\delta < \delta + \epsilon - \frac{\epsilon}{2} \leq V(f, y_2, y_1).$$

Exercise X.4.7 *Prove the following theorem.*

Theorem. *Let f be a real valued function on $[a, b]$, of bounded variation. Then there exist increasing functions g, h on $[a, b]$ such that $g(a) = h(a) = 0$ and*

$$f(x) - f(a) = g(x) - h(x),$$
$$V_f(x) = g(x) + h(x).$$

[Hint: Define g, h by the formulas

$$2g = V_f + f - f(a) \quad and \quad 2h = V_f - f + f(a).]$$

Solution. Define the functions g, h as in the hint. Then we have $2g(a) = 2h(a) = 0$ because $V(f, a, a) = 0$. We also have

$$(2(g - h)) = 2(f - f(a))$$

and

$$2(g+h) = 2V_f.$$

The function g is increasing because if $a \le x \le y \le b$, then part (b) of Exercise 5 implies

$$2(g(y) - g(x)) = V(f, x, y) + f(y) - f(x).$$

But the simple partition $\{x \le y\}$ of $[x, y]$ shows that

$$V(f, x, y) \ge |f(x) - f(y)|,$$

so g is increasing. Similarly for h we have

$$2(h(y) - h(x)) = V(f, x, y) - f(y) + f(x) \ge 0$$

and therefore h is also increasing.

Exercise X.4.8 *Let f be a real valued function of bounded variation on $[a, b]$. Let $c \in [a, b]$. Prove that the limits*

$$\lim_{\substack{h \to 0 \\ h > 0}} f(c+h) \quad and \quad \lim_{\substack{h \to 0 \\ h < 0}} f(c+h)$$

exist if $c \ne a, b$. If $c = a$ or $c = b$, then one has to deal with the right limit with $h > 0$, respectively, the left limit with $h < 0$. [Hint: First prove the result if f is an increasing function.]

Solution. Suppose f is increasing and $c \in (a, b)$. We prove that the first limit exists. By assumption, $f(c) \le f(c+h)$ for all $h > 0$, hence

$$\alpha = \inf_{h > 0} \{f(c+h)\}$$

exists. We contend that α is the desired limit. Given $\epsilon > 0$, there exists $h_1 > 0$ such that $0 \le f(c+h_1) - \alpha < \epsilon$. But f is increasing, so $0 < h < h_1$ implies $f(c+h) \le f(c+h_1)$, so for all $0 < h < h_1$ we have

$$0 \le f(c+h) - \alpha < \epsilon,$$

which proves our contention, namely

$$\lim_{\substack{h \to 0 \\ h > 0}} f(c+h) = \alpha.$$

The same argument with

$$\alpha' = \sup_{h < 0} \{f(c+h)\}$$

shows that

$$\lim_{\substack{h \to 0 \\ h<0}} f(c+h) = \alpha'.$$

If c is one of the end points of the interval $[a, b]$ we only have to investigate the limit which makes sense and the same argument shows that in all cases, the desired limit exists.

Finally, Exercise 7 implies that the same result is true for a function of bounded variation. Indeed, using the notation of Exercise 7, we know that the result is true for g and h. Hence the limit exists for g and $-h$, and therefore the limits also exist for f.

X.6 Relation Between the Integral and the Derivative

Exercise X.6.1 *Let J be an interval and let $f : J \to \mathbf{C}$ be a complex valued differentiable function. Assume that $f(t) \neq 0$ for all $t \in J$. Show that $1/f$ is differentiable, and that its derivative is $-f'/f^2$ as expected.*

Solution. Let $g = 1/f$. Then

$$\frac{g(t_0 + h) - g(t_0)}{h} = \frac{1/f(t_0 + h) - 1/f(t_0)}{h}.$$

Multiply the numerator and denominator by $f(t_0 + h)f(t_0)$ to obtain

$$\frac{g(t_0 + h) - g(t_0)}{h} = -\frac{f(t_0 + h) - f(t_0)}{h} \frac{1}{f(t_0 + h)f(t_0)}.$$

Letting $h \to 0$ shows that g is differentiable and that $g' = -f'/f^2$.

Exercise X.6.2 *Let $f : [a, b] \to E$ be a regulated map. Let $\lambda : E \to G$ be a continuous linear map. Prove that $\lambda \circ f$ is regulated. Prove that*

$$\int_a^b \lambda \circ f = \lambda \left(\int_a^b f \right).$$

Solution. We must show that $\lambda \circ f$ can be uniformly approximated by step maps. Since λ is continuous, there exists a number $C > 0$ such that $|\lambda(x)| \leq C|x|$ for all x. Given $\epsilon > 0$, there exists a step map φ on $[a, b]$ such that $\|f - \varphi\| < \epsilon/C$. Then $\lambda \circ \varphi$ is also a step map on $[a, b]$ and for all $x \in [a, b]$ we have

$$|\lambda \circ f(x) - \lambda \circ \varphi(x)| \leq C|f(x) - \varphi(x)| < \epsilon,$$

so $\|\lambda \circ f - \lambda \circ \varphi\| < \epsilon$, thereby proving that $\lambda \circ f$ is regulated.

Consider a sequence $\{\varphi_n\}$ of step maps which converges uniformly to f. Then

$$\int_a^b \varphi_n \to \int_a^b f$$

and since λ is continuous we get

$$\lambda\left(\int_a^b \varphi_n\right) \to \lambda\left(\int_a^b f\right),$$

We have shown that $\{\lambda \circ \varphi_n\}$ also converges uniformly to $\lambda \circ f$ so

$$\int_a^b \lambda \circ \varphi_n \to \int_a^b \lambda \circ f.$$

Furthermore, if

$$\int_a^b \varphi_n = \sum_{i=0}^{m-1} w_i(a_{i+1} - a_i),$$

then

$$\int_a^b \lambda \circ \varphi_n = \sum_{i=0}^{m-1} \lambda(w_i)(a_{i+1} - a_i) = \lambda\left[\sum_{i=0}^{m-1} w_i(a_{i+1} - a_i)\right] = \lambda\left(\int_a^b \varphi_n\right),$$

because λ is linear. So

$$\lambda\left(\int_a^b \varphi_n\right) \to \int_a^b \lambda \circ f.$$

Therefore we conclude that

$$\int_a^b \lambda \circ f = \lambda\left(\int_a^b f\right),$$

as was to be shown.

Exercise X.6.3 *Prove: Let f be a regulated real valued function on $[a, b]$. Assume that there is a differentiable funtion F on $[a, b]$ such that $F' = f$. Prove that*

$$\int_a^b f = F(b) - F(a).$$

[Hint: For a suitable partition $(a_0 < a_1 < \cdots < a_n)$ use the mean value theorem

$$F(a_{i+1}) - F(a_i) = F'(c_i)(a_{i+1} - a_i) = f(c_i)(a_{i+1} - a_i)$$

and the fact that f is uniformly approximated by a step map on the partition.]

Solution. Given $\epsilon > 0$ there exists a step map φ such that $\|f - \varphi\| < \epsilon/(b-a)$. Then

$$\left| \int_a^b f - \int_a^b \varphi \right| < \epsilon.$$

Moreover, if $P = \{a = a_0 < a_1 < \cdots < a_n = b\}$ is the partition associated to φ and w_i is the value of φ on (a_i, a_{i+1}) we have

$$\int_a^b \varphi - F(b) + F(a) = \sum_{i=0}^{n-1} w_i(a_{i+1} - a_i) - \sum_{i=0}^{n-1} (F(a_{i+1}) - F(a_i)).$$

The mean value theorem implies that there exists $c_i \in (a_i, a_{i+1})$ such that $F(a_{i+1}) - F(a_i) = f(c_i)(a_{i+1} - a_i)$, so

$$\left| \int_a^b \varphi - F(b) + F(a) \right| \leq \sum_{i=0}^{n-1} |w_i - f(c_i)|(a_{i+1} - a_i) < \epsilon,$$

because $|w_i - f(c_i)| < \epsilon/(a-b)$. Thus

$$\left| \int_a^b f - F(b) + F(a) \right| \leq \left| \int_a^b f - \int_a^b \varphi \right| + \left| \int_a^b \varphi - F(b) + F(a) \right| < 2\epsilon,$$

and the formula $\int_a^b f = F(b) - F(a)$ follows.

Exercise X.6.4 *Let $f \colon [a, b] \to E$ be a differentiable map with continuous derivative from a closed interval into a complete normed vector space E. Show that*

$$|f(b) - f(a)| \leq (b - a) \sup |f'(t)|,$$

the sup being, taken for $t \in [a, b]$. This result can be used to replace estimates given by the mean value theorem.

Solution. By Theorem 6.2 we have $f(b) - f(a) = \int_a^b f'$, so we conclude that

$$|f(b) - f(a)| = \left| \int_a^b f' \right| \leq (b - a)\|f'\|,$$

where $\|f'\| = \sup |f'(t)|$.

Exercise X.6.5 *Let f be as in Exercise 4. Let $t_0 \in [a, b]$. Show that*

$$|f(b) - f(a) - f'(t_0)(b - a)| \leq (b - a) \sup |f'(t) - f'(t_0)|,$$

the sup being again taken for t in the interval. [Hint: Apply Exercise 4 to the map $g(t) = f(t) - f'(t_0)t$. We multiply vectors on the right to fit later notation.]

Solution. Let $g(t) = f(t) - f'(t_0)t$. Then $g'(t) = f'(t) - f'(t_0)$ because

$$\frac{f'(t_0)(t + h) - f'(t_0)t}{h} = f'(t_0).$$

Since $g(b) - g(a) = f(b) - f(a) - f'(t_0)(b - a)$, Exercise 4 implies that

$$|f(b) - f(a) - f'(t_0)(b - a)| \le (b - a) \sup |f'(t) - f'(t_0)|,$$

as was to be shown.

XI

Approximation with Convolutions

XI.1   Dirac Sequences

Exercise XI.1.1 *Let K be a real function of a real variable such that $K \geq 0$, K is continuous, zero outside some bounded interval, and*

$$\int_{-\infty}^{\infty} K(t)dt = 1.$$

Define $K_n(t) = nK(nt)$. Show that $\{K_n\}$ is a Dirac sequence.

Solution. Suppose K is 0 outside $[-c, c]$. Clearly we have $K_n \geq 0$. Since K is continuous, so is each K_n and the change of variables formula $t \to nu$ gives

$$1 = \int_{-c}^{c} K(t)dt = \int_{-c/n}^{c/n} nK(nu)du = \int_{-\infty}^{\infty} K_n(u)du.$$

Finally, we see that K_n vanishes outside $[-c/n, c/n]$ so given $\epsilon, \delta > 0$, select N such that $|c|/N < \delta$. Then for all $n > N$ we have

$$\int_{-\infty}^{-\delta} K_n + \int_{\delta}^{\infty} K_n = 0 < \epsilon.$$

Exercise XI.1.2 *Show that one can find a function K as in Exercise 1 which is infinitely differentiable (cf. Exercise 6 of Chapter IV, §1), even, and zero outside the interval $[-1, 1]$.*

Solution. Let $K(x) = e^{-1/(x+1)(1-x)}$, if $x \in (-1, 1)$ and 0 otherwise. Then K satisifies all the desired properties.

Exercise XI.1.3 *Let K be infinitely differentiable, and such that $K(t) = 0$ if t is outside some bounded interval. Let f be a piecewise continuous function, and bounded. Show that $K * f$ is infinitely differentiable, and in fact, $(K * f)' = K' * f$.*

Solution. Suppose K is 0 outside $[-c, c]$. Then we can write

$$K * f(x) = \sum_{i=0}^{n-1} \int_{a_i}^{a_{i+1}} f(t)K(x - t)dt.$$

We can make f continuous on each $[a_i, a_{i+1}]$ without changing the value of the integral. Since K is C^∞, we can differentiate under the integral sign, so it follows that $K * f$ is C^∞ and

$$(K * f)'(x) = \sum_{i=0}^{n-1} \int_{a_i}^{a_{i+1}} f(t)K'(x - t)dt = (K' * f)(x).$$

Exercise XI.1.4 *Let f, g, h be piecewise continuous (or even continuous if it makes you more comfortable), and bounded, and such that g is zero outside some bounded interval. Define*

$$f * g = \int_{-\infty}^{\infty} f(t)g(x - t)dt.$$

Show that $(f * g) * h = f * (g * h)$. With suitable assumptions on f_1, f_2, show that $(f_1 + f_2) * g = f_1 * g + f_2 * g$. Show that $f * g = g * f$.

Solution. We assume for simplicity that f, g and h are continuous. Suppose g is 0 outside the interval $[-c, c]$. The change of variable formula implies

$$(g * f) = \int_{-c}^{c} g(t)f(x - t)dt$$

$$= \int_{x+c}^{x-c} -g(x - u)f(u)du$$

$$= \int_{-\infty}^{\infty} f(u)g(x - u)du = f * g.$$

We then have

$$(f * g) * h(x) = \int_{-\infty}^{\infty} (f * g)(x - u)h(u)du$$

$$= -\int_{-\infty}^{\infty} \left[\int_{-\infty}^{\infty} f(t)g(x - u - t)h(u)dt \right] du,$$

and

$$f * (g * h)(x) = \int_{-\infty}^{\infty} f(t)(g * h)(x - t)dt$$

$$= -\int_{-\infty}^{\infty} \left[\int_{-\infty}^{\infty} f(t)g(x - t - u)h(u)du \right] dt.$$

Interchanging integral signs we see that $(f * g) * h = f * (g * h)$. Furthermore,

$$(f_1 + f_2) * g = \int_{-\infty}^{\infty} (f_1 + f_2)(t)g(x - t)dt$$

$$= \int_{-\infty}^{\infty} f_1(t)g(x - t)dt + \int_{-\infty}^{\infty} f_2(t)g(x - t)dt$$

$$= f_1 * g(x) + f_2 * g(x).$$

XI.2 The Weierstrass Theorem

Exercise XI.2.1 Let f be continuous on $[0, 1]$. Assume that

$$\int_0^1 f(x)x^n dx = 0$$

for every integer $n = 0, 1, 2 \ldots$. Show that $f = 0$. [Hint: Use the Weierstrass theorem to approximate f by a polynomial and show that the integral of f^2 is equal to 0.]

Solution. Let $B > 0$ be a bound for $|f|$ on $[0, 1]$. Given $\epsilon > 0$ there exists a polynomial P such that for all $x \in [0, 1]$ we have $|f(x) - P(x)| < \epsilon/B$. So

$$|f^2(x) - f(x)P(x)| = |f(x)||f(x) - P(x)| < B\frac{\epsilon}{B} = \epsilon.$$

But since $\int_0^1 P(x)f(x) = 0$ we obtain

$$\int_0^1 f^2(x)dx = \int_0^1 f^2(x) - f(x)P(x)dx \leq \int_0^1 |f^2(x) - f(x)P(x)|dx < \epsilon.$$

This is true for all $\epsilon > 0$ so $\int_0^1 f^2(x)dx = 0$. Since f^2 is continuous and ≥ 0, we must have $f^2 = 0$ and therefore $f = 0$.

Exercise XI.2.2 Prove that if f is a continuous function, then

$$\lim_{\substack{h \to 0 \\ h > 0}} \int_{-1}^1 \frac{h}{h^2 + x^2} f(x)dx = \pi f(0).$$

Solution. Suppose that f is continuous on $[-1, 1]$. Then we can extend f to a continuous and bounded function on $\mathbf{R}$. Call this extension f again, and let M be a bound for f. Then Exercise 10, §3, of Chapter XIII implies

$$\lim_{\substack{h \to 0 \\ h>0}} \frac{1}{\pi} \int_{-\infty}^{\infty} \frac{h}{h^2 + x^2} f(x)dx = f(0),$$

because

$$\left\{ \frac{1}{\pi} \frac{h}{h^2 + x^2} \right\}_{h>0}$$

is a Dirac family. To conclude the proof, it suffices to show that

$$\lim_{\substack{h \to 0 \\ h>0}} \int_{-\infty}^{-1} \frac{h}{h^2 + x^2} f(x)dx = \lim_{\substack{h \to 0 \\ h>0}} \int_{1}^{\infty} \frac{h}{h^2 + x^2} f(x)dx = 0.$$

The second limit follows from

$$\left| \int_{1}^{B} \frac{h}{h^2 + x^2} f(x)dx \right| \le M \int_{1}^{B} \frac{h}{h^2 + x^2} dx$$

$$= M \int_{1}^{B} \frac{1}{h} \frac{1}{1 + (x/h)^2} dx$$

$$= M \int_{1/h}^{B/h} \frac{1}{1 + u^2} dx$$

$$= M(\arctan(B/h) - \arctan(1/h)).$$

Let $B \to \infty$ and then $h \to 0$ to conclude.

Exercise XI.2.3 (An Integral Operator) *Let $K = K(x,y)$ be a continuous function on the rectangle defined by the inequalities*

$$a \le x \le b \quad and \quad c \le y \le d.$$

For $f \in C^0([c,d])$, define the function $Tf = T_K f$ by the formula

$$T_K f(x) = \int_{c}^{d} K(x,y)f(y)dy.$$

(a) Prove that T_K is a continuous linear map

$$C^0([c,d]) \to C^0([a,b]),$$

with the sup norm on both spaces.
(b) Prove that T_K is a continuous linear map with the L^2-norm on both spaces.

Solution. (a) The linearity property of the integral implies at once that T_K is linear. We now show that T_K is continuous with the sup norm on both spaces. Since the function K is continuous on a compact set, it is bounded, say by M. If f belongs to $C^0([c, d])$, then for all $x \in [a, b]$ we have

$$|T_K f(x)| \leq \int_c^d |K(x, y)| |f(y)| dy$$
$$\leq M(d - c) \|f\|.$$

Hence $\|T_K f\| \leq M(d - c) \|f\|$, which concludes the proof of the continuity of the linear map T_K in this particular case.

(b) We now show that T_K is also continuous with the L^2-norm on both spaces. We have

$$\|T_K f\|_2^2 = \int_a^b |T_K f(x)|^2 dx$$
$$= \int_a^b \left(\int_c^d K(x, y) f(y) dy \right)^2 dx.$$

Let M be a bound for K. Then the last expression is

$$\leq \int_a^b \left(\int_c^d M |f(y)| dy \right)^2 dx = M^2(b - a) \left(\int_c^d |f(y)| dy \right)^2.$$

By the Schwarz inequality applied to the scalar product defined by the integral (Exercises 4 and 5, §2, of Chapter V) we see that

$$\left(\int_c^d |f(y)| dy \right)^2 \leq (d - c) \int_c^d |f(y)|^2 dy,$$

therefore

$$\|T_K f\|_2^2 \leq M^2(b - a)(d - c) \|f\|_2^2,$$

which proves the continuity of the linear map T_K in this case.

XII
Fourier Series

XII.1 Hermitian Products and Orthogonality

Exercise XII.1.1 *Verify the statements about the orthogonality of the functions χ_n, and the functions $\varphi_0, \varphi_n, \psi_n$. That is, prove $\langle \chi_n, \chi_m \rangle = 0$ and $\langle \varphi_n, \varphi_m \rangle = 0$ if $m \neq n$.*

Solution. For the functions χ_n we have

$$\langle \chi_n, \chi_n \rangle = \int_{-\pi}^{\pi} e^{inx-inx} dx = 2\pi,$$

and if $n \neq m$ we have

$$\langle \chi_n, \chi_m \rangle \int_{-\pi}^{\pi} e^{inx-imx} dx = \frac{1}{i(n-m)} \left[e^{i(n-m)x} \right]_{-\pi}^{\pi} = 0.$$

The orthogonality of the functions $\varphi_0, \varphi_n, \psi_n$ was studied in Exercise 3, §7, of Chapter IX.

Exercise XII.1.2 *On the space $\mathbf{C}^n$ consisting of all vectors $z = (z_1, \ldots, z_n)$ and $w = (w_1, \ldots, w_n)$ where $z_i, w_i \in \mathbf{C}$, define the product*

$$\langle z, w \rangle = z_1 \overline{w}_1 + \cdots + z_n \overline{w}_n.$$

Show that this is a hermitian product, and that $\langle z, z \rangle = 0$ if and only if $z = 0$.

Solution. The verification of all the properties is routine. We have

$$\langle z, w \rangle = \sum_{i=1}^{n} z_i \overline{w}_i = \sum_{i=1}^{n} \overline{w_i \overline{z}_i} = \overline{\langle w, z \rangle}.$$

Then we also have

$$\langle u, v + w \rangle = \sum u_i \overline{(v_i + w_i)} = \sum u_i \overline{v}_i + \sum u_i \overline{w}_i = \langle u, v \rangle + \langle u, w \rangle,$$

and

$$\langle \alpha u, v \rangle = \sum \alpha u_i \overline{v}_i = \alpha \sum u_i \overline{v}_i = \alpha \langle u, v \rangle$$

and also

$$\langle u, \alpha v \rangle = \sum u_i \overline{\alpha v_i} = \overline{\alpha} \sum u_i \overline{v}_i = \overline{\alpha} \langle u, v \rangle.$$

Finally, we have

$$\langle v, v \rangle = \sum |v_i|^2 \geq 0.$$

Now suppose $\langle z, z \rangle = 0$. Then $\sum |z_i|^2 = 0$ so $|z_i| = 0$ for all i, hence $z = 0$. Conversely, if $z = 0$, then it is clear that $\langle z, z \rangle = 0$.

Exercise XII.1.3 *Let l^2 be the set of all sequences $\{c_n\}$ of complex numbers such that $\sum |c_n|^2$ converges. Show that l^2 is a vector space, and that if $\{\alpha_n\}$, $\{\beta_n\}$ are elements of l^2, then the product*

$$(\{\alpha_n\}, \{\beta_n\}) \mapsto \sum \alpha_n \overline{\beta}_n$$

is a hermitian product such that $\langle \alpha, \alpha \rangle = 0$ if and only if $\alpha = 0$. (Show that the series on the right converges, using the Schwarz inequality for each partial sum. Use the same method to prove the first statement.) Prove that l^2 is complete.

Solution. (i) Suppose $\{\alpha_n\}$, $\{\beta_n\}$ are in l^2. Then

$$\sum_{i=1}^{n} |\alpha_n + \beta_n|^2 \leq \sum_{i=1}^{n} |\alpha_n|^2 + \sum_{i=1}^{n} |\beta_n|^2 + 2 \sum_{i=1}^{n} |\alpha_n||\beta_n|.$$

The Schwarz inequality implies that

$$\sum_{i=1}^{n} |\alpha_n||\beta_n| \leq \left(\sum |\alpha_n|^2 \right)^{1/2} \left(\sum |\beta_n|^2 \right)^{1/2}.$$

Letting $n \to \infty$ we conclude that $\{\alpha_n + \beta_n\} \in l^2$. If c is a scalar, then $\sum |c\alpha_n|^2 = |c|^2 \sum |\alpha_n|^2$ so $\{c\alpha_n\} \in l^2$. We conclude that l^2 is a vector space.

(ii) First we show that the product makes sense. The Schwarz inequality implies that

$$\sum_{i=1}^{n} |\alpha_n \beta_n| \leq \left(\sum_{i=1}^{n} |\alpha_n|^2 \right)^{1/2} \left(\sum_{i=1}^{n} |\beta_n|^2 \right)^{1/2} \leq \left(\sum |\alpha_n|^2 \right)^{1/2} \left(\sum |\beta_n|^2 \right)^{1/2},$$

so the series $\sum \alpha_n \overline{\beta_n}$ converges. We have verified in Exercise 2 that the partial sums verify the properties of a hermitian product. So these properties are also true for the product defined in this exercise.

Now we prove that l^2 is complete. Let $X_n = \{\alpha_{n,j}\}_{j=1}^{\infty}$ and suppose $\{X_n\}_{n=1}^{\infty}$ is a Cauchy sequence. Given $0 < \epsilon < 1$ there exists a positive integer N such that for all $n, m > N$ we have $\|X_n - X_m\| < \epsilon$. Then for all $n, m > N$ we have

$$\sum_{j} |\alpha_{n,j} - \alpha_{m,j}|^2 < \epsilon^2.$$

Hence, for each j, $|\alpha_{n,j} - \alpha_{m,j}| < \epsilon$ and therefore $\{\alpha_{n,j}\}_{n=1}^{\infty}$ is a Cauchy sequence in $\mathbf{C}$ for each j. Let α_j be its limit and let $X = \{\alpha_j\}_{j=1}^{\infty}$. We assert that X belongs to l^2 and that $X_n \to X$ as $n \to \infty$ with respect to the norm given by the hermitian product studied at the beginning of the exercise. For each positive integer M and all $n, m > N$ we have

$$\sum_{j=1}^{M} |\alpha_{n,j} - \alpha_{m,j}|^2 < \epsilon^2,$$

so letting $m \to \infty$ and then $M \to \infty$ we see that X belongs to l^2 and that $\|X_n - X\| < \epsilon$ for all $n > N$, thus proving our assertions.

Exercise XII.1.4 *If f is periodic of period 2π, and $a, b \in \mathbf{R}$, then*

$$\int_{a}^{b} f(x)dx = \int_{a+2\pi}^{b+2\pi} f(x)dx = \int_{a-2\pi}^{b-2\pi} f(x)dx.$$

(Change variables, letting $u = x - 2\pi$, $du = dx$.) Also,

$$\int_{-\pi}^{\pi} f(x + a)dx = \int_{-\pi}^{\pi} f(x)dx = \int_{-\pi+a}^{\pi+a} f(x)dx.$$

(Split the integral over the bounds $-\pi+a$, $-\pi$, π, $\pi+a$ and use the preceding statement.)

Solution. The change of variable $u = x - 2\pi$ and the periodicity of f imply

$$\int_{a}^{b} f(x) = \int_{a-2\pi}^{b-2\pi} f(u + 2\pi)du = \int_{a-2\pi}^{b-2\pi} f(u)du.$$

The change of variable $u = x + 2\pi$ and the periodicity of f imply

$$\int_{a}^{b} f(x) = \int_{a+2\pi}^{b+2\pi} f(u - 2\pi)du = \int_{a+2\pi}^{b+2\pi} f(u)du.$$

The change of variable $u = x + a$ implies

$$\int_{-\pi}^{\pi} f(x+a)dx = \int_{-\pi+a}^{\pi+a} f(u)du.$$

Moreover we have

$$\int_{-\pi+a}^{\pi+a} f(u)du = \int_{-\pi+a}^{-\pi} f(u)du + \int_{-\pi}^{\pi} f(u)du + \int_{\pi}^{\pi+a} f(u)du,$$

and

$$\int_{-\pi+a}^{-\pi} f(u)du = \int_{-\pi+a+2\pi}^{-\pi+2\pi} f(u)du = -\int_{\pi}^{\pi+a} f(u)du$$

so

$$\int_{-\pi+a}^{\pi+a} f(u)du = \int_{-\pi}^{\pi} f(u)du$$

as was to be shown.

Exercise XII.1.5 *Let f be an even function (that is $f(x) = f(-x)$). Show that all its Fourier coefficients with respect to $\sin nx$ are 0. Let g be an odd function (that is $g(-x) = -g(x)$). Show that all its Fourier coefficients with respect to $\cos nx$ are 0.*

Solution. If h is odd, then $\int_{-\pi}^{\pi} h(x)dx = 0$. To prove this, split the integral and change variables $u = -x$ in the first integral, so that

$$\int_{-\pi}^{\pi} h(x)dx = \int_{-\pi}^{0} h(x)dx + \int_{0}^{\pi} h(x)dx = -\int_{\pi}^{0} h(-u)du + \int_{0}^{\pi} h(x)dx$$

$$= \int_{\pi}^{0} h(u)du + \int_{0}^{\pi} h(x)dx = 0.$$

The exercise is a consequence of the following observations. If f is even, then the functions $f(x)\sin nx$ are odd, and if g is odd, then the functions $g(x)\cos nx$ are odd.

Exercise XII.1.6 *Compute the real Fourier coefficients of the following functions: (a) x; (b) x^2; (c) $|x|$; (d) $\sin^2 x$; (e) $|\sin x|$; and (f) $|\cos x|$.*

Solution. (a) Since $x \mapsto x$ is odd, Exercise 5 implies that $a_0 = a_k = 0$. Integrating by parts we find

$$b_n = \frac{1}{\pi} \int_{-\pi}^{\pi} x \sin(nx)dx = \left[-\frac{x}{\pi n} \cos(nx) \right]_{-\pi}^{\pi} + \frac{1}{\pi n} \int_{-\pi}^{\pi} \cos(nx)dx$$

$$= (-1)^{n+1} \frac{2}{n}.$$

(b) The function $x \mapsto x^2$ is even, so Exercise 5 implies that $b_k = 0$. We have

$$a_n = \frac{1}{\pi} \int_{-\pi}^{\pi} x^2 \cos(nx)dx = \left[\frac{x^2}{\pi n}\sin(nx)\right]_{-\pi}^{\pi} - \frac{2}{\pi n}\int_{-\pi}^{\pi} x\sin(nx)dx$$

$$= (-1)^{n+1}\frac{4}{n^2}.$$

(c) We find that the Fourier series of the function $x \mapsto |x|$ is

$$|x| = \frac{\pi}{2} - \frac{4}{\pi}\left(\cos x + \frac{\cos 3x}{3^2} + \cdots + \frac{\cos(2n+1)x}{(2n+1)^2} + \cdots\right).$$

(d) A simple trigonometric identity gives

$$\sin^2 x = \frac{1}{2} - \frac{\cos 2x}{2}.$$

(e)

$$|\sin x| = \frac{4}{\pi}\left(\frac{1}{2} - \frac{\cos 2x}{3} - \cdots - \frac{\cos 2nx}{4n^2 - 1} - \cdots\right).$$

(f)

$$|\cos x| = \frac{4}{\pi}\left(\frac{1}{2} + \frac{\cos 2x}{3} + \cdots + (-1)^{n-1}\frac{\cos 2nx}{4n^2 - 1} + \cdots\right).$$

Exercise XII.1.7 *Let $f(x)$ be the function equal to $(\pi - x)/2$ in the interval $[0, 2\pi]$, and extended by periodicity to the whole real line. Show that the Fourier series of f is $\sum(\sin nx)/n$.*

Solution. We see that the function is odd, so $a_n = 0$ by Exercise 5. Exercise 4 implies

$$b_n = \frac{1}{\pi}\int_0^{2\pi}\frac{(\pi - x)}{2}\sin(nx)dx = \frac{1}{2}\int_0^{2\pi}\sin(nx)dx - \frac{1}{2\pi}\int_0^{2\pi} x\sin(nx)dx.$$

The first integral is equal to 0 and the second integral is

$$= \frac{-1}{2\pi}\left\{\left[-\frac{x}{n}\cos(nx)\right]_0^{2\pi} + \frac{1}{n}[\sin(nx)]_0^{2\pi}\right\} = \frac{1}{n}.$$

So the Fourier series of f is indeed $\sum(\sin nx)/n$.

Exercise XII.1.8 *Let f be periodic of period 2π, and of class C^1. Show that there is a constant $C > 0$ such that all Fourier coefficients c_n $(n \neq 0)$ satisfy the bound $|c_n| \leq C/|n|$. [Hint: Integrate by parts.]*

Solution. Let C be a bound for f', and suppose that $n \neq 0$. Integrating by parts and using the fact that f is periodic we find that

$$2\pi c_n = \int_{-\pi}^{\pi} f(x)e^{-inx}dx = \frac{-1}{in}\left[f(x)e^{-inx}\right]_{-\pi}^{\pi} + \frac{1}{in}\int_{\pi}^{\pi} f'(x)e^{-inx}dx$$

$$= \frac{1}{in}\int_{-\pi}^{\pi} f'(x)e^{-inx}dx.$$

Hence the estimate

$$2\pi|c_n| \le \frac{1}{|n|}\int_{-\pi}^{\pi} C|e^{-inx}|dx = \frac{2\pi C}{|n|}.$$

Therefore $|c_n| \le C/|n|$ for all $n \ne 0$.

Exercise XII.1.9 *Let f be periodic of period 2π, and of class C^2 (twice continuously differentiable). Show that there is a constant $C > 0$ such that all Fourier coefficients c_n $(n \ne 0)$ satisfy the bound $|c_n| \le C/n^2$. Generalize.*

Solution. We prove the general result. Suppose $f \in C^p$, $p \ge 1$, and that f is 2π periodic. Then $f', f'', \ldots, f^{(p)}$ are all 2π periodic. Let $n \ne 0$. Then integrating by parts we get the following string of equalities

$$2\pi c_n = \frac{1}{in}\int_{-\pi}^{\pi} f'(x)e^{-inx}dx = \cdots = \frac{1}{(in)^p}\int_{-\pi}^{\pi} f^{(p)}(x)e^{-inx}dx.$$

If B_p is a bound for $f^{(p)}$ we get

$$2\pi|c_n| \le \frac{2\pi B_p}{|n|^p}$$

and therefore $|c_n| \le B_p/|n|^p$ for all $n \ne 0$.

Exercise XII.1.10 *Let t be real and not equal to an integer. Determine the Fourier series for the functions $f(x) = \cos tx$ and $g(x) = \sin tx$.*

Solution. (i) Since f is even conclude that $b_k = 0$ for all $k \ge 1$. For the other coefficients we have

$$a_0 = \frac{1}{2\pi}\int_{-\pi}^{\pi}\cos(tx)dx = \frac{1}{2\pi}\left[\frac{1}{t}\sin(tx)\right]_{-\pi}^{\pi} = \frac{\sin(t\pi)}{t\pi},$$

and for $n \ge 1$,

$$a_n = \frac{2}{\pi}\int_0^{\pi}\cos(tx)\cos(nx)dx = \frac{1}{\pi}\int_0^{\pi}\cos(tx+nx)+\cos(tx-nx)dx$$

$$= \frac{1}{\pi}\left[\frac{1}{t+n}\sin(t\pi+n\pi) - \frac{1}{t-n}\sin(t\pi-n\pi)\right]$$

$$= (-1)^n\frac{2t\sin(t\pi)}{\pi(t^2-n^2)}.$$

So the Fourier series of f is

$$\frac{\sin(t\pi)}{t\pi} + \frac{2t\sin(t\pi)}{\pi}\sum_{n=1}^{\infty}\frac{(-1)^n}{t^2-n^2}\cos(nx).$$

(ii) The function g is odd, so $a_n = 0$ for all $n \geq 0$. The other coefficients are obtained by integration

$$
\begin{aligned}
b_n &= \frac{2}{\pi} \int_0^\pi \sin(tx)\sin(nx)dx = \frac{1}{\pi} \int_0^\pi \cos(tx - nx) - \cos(tx + nx)dx \\
&= \frac{1}{\pi}\left[\frac{1}{t-n}\sin(t\pi - n\pi) - \frac{1}{t-n}\sin(t\pi + n\pi)\right] \\
&= (-1)^n \frac{2n\sin(t\pi)}{\pi(t^2 - n^2)}.
\end{aligned}
$$

So the Fourier series of g is

$$
\frac{\sin(t\pi)}{\pi} \sum_{n=1}^\infty (-1)^n \frac{2n}{t^2 - n^2} \sin(nx).
$$

Exercise XII.1.11 *Let E be a vector space over $\mathbf{C}$ with a hermitian product. Prove the* **parallelogram law**: *For all $v, w \in E$ we have*

$$
\|v + w\|^2 + \|v - w\|^2 = 2\|v\|^2 + 2\|w\|^2.
$$

Solution. Using the properties of the hermitian product we find

$$
\|v + w\|^2 = \langle v + w, v + w \rangle = \langle v, v \rangle + \langle v, w \rangle + \langle w, v \rangle + \langle w, w \rangle,
$$

and

$$
\|v - w\|^2 = \langle v - w, v - w \rangle = \langle v, v \rangle - \langle v, w \rangle - \langle w, v \rangle + \langle w, w \rangle,
$$

hence $\|v + w\|^2 + \|v - w\|^2 = 2\|v\|^2 + 2\|w\|^2$, as was to be shown.

Exercise XII.1.12 *Let E be a vector space with a hermitian product which is positive definite, that is if $\|v\| = 0$, then $v = 0$. Let F be a complete subspace of E. Let $v \in E$ and let*

$$
a = \inf_{x \in F} \|x - v\|.
$$

Prove that there exists an element $x_0 \in F$ such that $a = \|v - x_0\|$. [Hint: Let $\{y_n\}$ be a sequence in F such that $\|y_n - v\|$ converges to a. Prove that $\{y_n\}$ is Cauchy, using the parallelogram law on

$$
y_n - y_m = (y_n - x) - (y_m - x).]
$$

Solution. Let $\{y_n\}$ be a sequence in F such that $\|y_n - v\| \to a$ as $n \to \infty$. The parallelogram law implies

$$
2\|y_n - v\|^2 + 2\|y_m - v\|^2 = \|y_n - y_m\|^2 + \|y_n + y_m - 2v\|^2.
$$

But

$$\|y_n + y_m - 2v\|^2 = 4\left\|\frac{1}{2}(y_n + y_m) - v\right\|^2 \geq 4a^2$$

because F is a subspace. So

$$\|y_n - y_m\|^2 \leq 2\|y_n - v\|^2 + 2\|y_m - v\|^2 - 4a^2.$$

Given $\epsilon > 0$ there exists N such that if $n \geq N$, then $\|y_n - v\|^2 < a^2 + \epsilon$. Then for all $n, m \geq N$ we have

$$\|y_n - y_m\|^2 < 4\epsilon,$$

so the sequence $\{y_n\}$ is Cauchy and therefore has a limit $x_0 \in E$. Since F is closed, $x_0 \in F$, and $\|x_0 - v\| = a$.

Remark. In this exercise we assume that F is a subspace. From this assumption, we concluded that $\frac{1}{2}(y_n + y_m) \in F$ whenever y_n, y_m in F. So if F is only assumed to be a closed convex set, the conclusion of the exercise still holds. For a similar result, see Exercise 16, §2, of Chapter VII.

Exercise XII.1.13 *Notation as in the preceding exercise, assume that $F \neq E$. Show that there exists a vector $z \in E$ which is perpendicular to F and $z \neq 0$. [Hint: Let $v \in E$, $v \notin F$. Let x_0 be as in Exercise 12, and let $z = v - x_0$. Changing v by a translation, you may assume that $z = v$, so that*

$$\|z\|^2 \leq \|z + x\|^2 \quad \text{for all } x \in F.$$

You can use two methods. One of them is to consider $z + t\alpha x$, with small positive values of t, and suitable $\alpha \in \mathbf{C}$. The other is to use Pythagoras' theorem.]

Solution. We wish to show that z is perpendicular to all x in F.

Method 1. Suppose $\|x\| \neq 0$ and let c be the component of z along x. Write $z = z - cx + cx$ so by Pythagoras

$$\|z\|^2 = \|z - cx\|^2 + \|cx\|^2 \geq \|z - cx\|^2.$$

By the choice of z we have $\|z\|^2 \leq \|z + y\|^2$ for all $y \in F$. Hence we obtain

$$\|z\|^2 = \|z - cx\|^2$$

which implies that $\|cx\|^2 = 0$ and therefore $c = 0$, as was to be shown.

Method 2. Consider $\|z\|^2 \leq \|z + tx\|^2$ for small values of t. Expanding yields

$$\langle z, z \rangle \leq \langle z, z \rangle + 2t\mathrm{Re}\langle z, x \rangle + t^2\langle x, x \rangle$$

which implies

$$0 \leq 2t\mathrm{Re}\langle z, x \rangle + t^2\langle x, x \rangle.$$

We pick t small of the opposite sign of $\mathrm{Re}\langle z, x \rangle$ to get a contradiction when $\mathrm{Re}\langle z, x \rangle \neq 0$. Using ix instead of x proves that $\mathrm{Im}\langle z, x \rangle = 0$. This concludes the proof.

Exercise XII.1.14 *Notation as in Exercises 12 and 13, let* $\lambda\colon E \to \mathbf{C}$ *be a continuous linear map. Show that there exists* $y \in E$ *such that* $\lambda(x) = \langle x, y \rangle$ *for all* $x \in E$. *[Hint: Let F be the subspace of all $x \in E$ such that $\lambda(x) = 0$. Show that F is closed. If $F \neq E$, use Exercise 13 to get an element $z \in E$, $z \notin F$, $z \neq 0$, such that z is perpendicular to F. Show that there exists some complex α such that $\alpha z = y$ satisfies the requirements, namely $\alpha = \overline{\lambda(z)}/\|z\|^2$.]*

Solution. The subspace F is closed because if $x_n \in F$ and $x_n \to x$, then $0 = \lambda(x_n) \to \lambda(x)$ by continuity, so $\lambda(x) = 0$ and $x \in F$. Suppose $F \neq E$ and choose z as in Exercise 13. Then let $y = \alpha z$ where $\alpha = \overline{\lambda(z)}/\|z\|^2$. Any x in E can be expressed as

$$x = x - \frac{\lambda(x)}{\lambda(z)}z + \frac{\lambda(x)}{\lambda(z)}z.$$

But $u = x - (\lambda(x)/\lambda(z))z$ belongs to F because

$$\lambda(u) = \lambda(x) - \frac{\lambda(x)}{\lambda(z)}\lambda(z) = 0.$$

So

$$\langle x, y \rangle = \left\langle u + \frac{\lambda(x)}{\lambda(z)}z, \alpha z \right\rangle = \overline{\alpha}\frac{\lambda(x)}{\lambda(z)}\langle z, z \rangle = \lambda(x).$$

Exercise XII.1.15 *Let E be a vector space over $\mathbf{C}$ with a hermitian product which is positive definite. Let $v_1, \ldots, v_n$ be elements of E, and assume that they are linearly independent. This means: if $c_1 v_1 + \cdots + c_n v_n = 0$ with $c_i \in \mathbf{C}$, then $c_i = 0$ for all i. Prove that for each $k = 1, \ldots, n$ there exist elements $w_1, \ldots, w_k$ which are of length 1, mutually perpendicular (that is $\langle w_i, w_j \rangle = 0$ if $i \neq j$), and generate the same subspace as $v_1, \ldots, v_k$. These elements are unique up to multiplication by complex numbers of absolute value 1. [Hint: For the existence, use the usual orthogonalization process: Let*

$$u_1 = v_1,$$
$$u_2 = v_2 - c_1 v_1,$$
$$\vdots$$
$$u_k = v_k - c_{k-1}v_{k-1} - \cdots - c_1 v_1,$$

where c_i are chosen to orthogonalize. Divide each u_i by its length to get w_i. Put in all the details and complete this proof.]

Solution. Let $u_1 = v_1$ and $w_1 = u_1/\|u_1\|$. Then for $k = 1$ we see that $\|w_1\| = 1$ and that $\{v_1\}$ and $\{w_1\}$ generate the same subspace. If w_1 and w_1' generate the same subspace and both have norm 1, then $w_1 = \lambda w_1'$ where $|\lambda| = 1$.

Let $u_2 = v_2 - c_1 w_1$ where $c_1 = \langle v_2, w_1 \rangle$. Then

$$\langle u_2, w_1 \rangle = \langle v_2, w_1 \rangle - \langle v_2, w_1 \rangle = 0.$$

The vector u_2 is non-zero for otherwise $\{v_1, v_2\}$ are linearly dependent. Let $w_2 = u_2/\|u_2\|$. The vectors w_1 and w_2 are both linear combinations of v_1 and v_2. Conversely, v_1 and v_2 are both linear combinations of w_1 and w_2, so $\{v_1, v_2\}$ and $\{w_1, w_2\}$ generate the same space. The definition shows that this vector is unique up to multiplication by a complex number of absolute value 1.

Now proceed by induction. For $k \geq 2$ we let

$$u_k = v_k - c_{k-1} w_{k-1} - \cdots - c_1 w_1,$$

where $c_j = \langle v_k, w_j \rangle$. Then $\langle u_k, w_j \rangle = 0$ for all $i \leq j \leq k-1$. The vector u_k is non-zero, otherwise the vectors, $v_1, \ldots, v_k$ would be linearly dependent. Let $w_k = u_k/\|u_k\|$. Here we see that w_k is unique up to multiplication by a complex number of norm 1. The induction hypothesis implies that $\{w_1, \ldots, w_{k-1}\}$ and $\{v_1, \ldots, v_{k-1}\}$ generate the same space. So w_k is a linear combination of the vectors $v_1, \ldots, v_k$, hence the space generated by $\{w_1, \ldots, w_k\}$ is contained in the space generated by $\{v_1, \ldots, v_k\}$. Since $\{w_1, \ldots, w_k\}$ is a set of k linearly independent vectors we conclude that the space generated by $\{w_1, \ldots, w_k\}$ is the same as the space generated by $\{v_1, \ldots, v_k\}$. This concludes the proof.

Exercise XII.1.16 *In this exercise, take all functions to be real valued, and all vector spaces over the reals. Let $K(x, y)$ be a continuous function of two variables, defined on the square $a \leq x \leq b$ and $a \leq y \leq b$. A continuous function f on $[a, b]$ is said to be an* **eigenfunction** *for K, with respect to a real number λ, if*

$$\int_a^b K(x, y) f(y) dy = \lambda f(x).$$

Use the L^2-norm on the space E of continuous functions on $[a, b]$. Prove that if $f_1, \ldots, f_n$ are in E, mutually orthogonal, and of L^2-norm equal to 1, and are eigenfunctions with respect to the same number $\lambda \neq 0$, then n is bounded by a number depending only on K and λ. [Hint: Use Theorem 1.5.]

Solution. Fix x with $a \leq x \leq b$. Let

$$c_k = \frac{\langle K, f_k \rangle}{\langle f_k, f_k \rangle} = \langle K, f_k \rangle.$$

Theorem 1.5 implies that $\sum_{k=1}^n |c_k|^2 \leq \|K\|^2$ so

$$\sum_{k=1}^n \lambda^2 |f_k(x)|^2 \leq \|K\|^2.$$

If B is a bound for K on the square $[a, b] \times [a, b]$, we see that

$$\sum_{k=1}^{n} \lambda^2 |f_k(x)|^2 \le B^2(b - a).$$

This last inequality is true for all x with $a \le x \le b$, so integrating with respect to x and using the fact that $\|f_k\|_2^2 = 1$, we get

$$\lambda^2 n \le B^2(b - a)^2.$$

XII.2   Trigonometric Polynomials as a Total Family

Exercise XII.2.1 *Let α be an irrational number. Let f be a continuous function (complex valued, of a real variable), periodic of period 1. Show that*

$$\lim_{N \to \infty} \frac{1}{N} \sum_{n=1}^{N} f(n\alpha) = \int_0^1 f(x)dx.$$

[Hint: First, let $f(x) = e^{2\pi i k x}$ for some integer k. If $k \ne 0$, then you can compute explicitly the sum on the left, and one sees at once that the geometric sums

$$\left| \sum_{n=1}^{N} e^{2\pi i k n \alpha} \right|$$

are bounded, whence the assertion follows. If $k = 0$, it is even more trivial. Second, prove that if the relationship is true for two functions, then it is true for a linear combination of these functions. Hence if the relationship is true for a family of generators of a vector space of functions, then it is true for all elements of this vector space. Third, prove that if the relationship is true for a sequence of functions $\{f_k\}$, and these functions converge uniformly to a function f, then the relationship is true for f.]

Solution. Suppose $f(x) = e^{2\pi i k x}$. First, suppose $k \ne 0$. Then since α is irrational, we see that

$$\left| \frac{1}{N} \sum_{n=1}^{N} e^{2\pi i k n \alpha} \right| = \frac{1}{N} \left| \frac{e^{2\pi i k \alpha} - e^{2\pi i (N+1) k \alpha}}{1 - e^{2\pi i k \alpha}} \right| \le \frac{1}{N} \frac{2}{|1 - e^{2\pi i k \alpha}|} \to 0$$

as $N \to \infty$. Since $\int_0^1 e^{2\pi i k x} dx = 0$, the formula holds. Suppose $k = 0$. Then

$$\frac{1}{N} \sum_{n=1}^{N} 1 = 1 = \int_0^1 1 dx,$$

so in this case, the formula is also true.

If f and g verify the relationship, then so does $af + bg$ for all complex numbers a and b because the integral is linear and

$$\frac{1}{N} \sum_{n=1}^{N} af(n\alpha) + bg(n\alpha) = a\frac{1}{N} \sum_{n=1}^{N} f(n\alpha) + b\frac{1}{N} \sum_{n=1}^{N} g(n\alpha).$$

Finally, suppose that $\{f_k\}$ is a sequence of continuous functions, of period 1, which satisfy the formula and which converges uniformly to a function f. We contend that f verifies the formula. Given $\epsilon > 0$, choose k such that $\|f - f_k\| < \epsilon$, where $\|\cdot\|$ denotes the sup norm. Then

$$\left| \frac{1}{N} \sum_{n=1}^{N} f(n\alpha) - \int_0^1 f \right| \le \frac{1}{N} \sum_{n=1}^{N} |f(n\alpha) - f_k(n\alpha)| + \left| \frac{1}{N} \sum_{n=1}^{N} f_k(n\alpha) - \int_0^1 f_k \right|$$
$$+ \int_0^1 |f_k - f|.$$

The first and third term are $< \epsilon$ for all N. For all large N the second term is $< \epsilon$, so for all large N we have

$$\left| \frac{1}{N} \sum_{n=1}^{N} f(n\alpha) - \int_0^1 f \right| < 3\epsilon,$$

thereby proving our contention.

Now we give the concluding argument. The first and second step show that the relationship is true for all trigonometric polynomials. Since any continuous and periodic function is the uniform limit or trigonometric polynomials, the third step shows that the relationship is true for all continuous functions with period 1.

Exercise XII.2.2 *Prove that the limit of the preceding exercise is valid if f is an arbitrary real valued periodic (period 1) regulated function (or Riemann integrable function) by showing that given ϵ, there exist continuous functions g, h periodic of period 1, such that*

$$g \le f \le h \quad and \quad \int_0^1 (h - g) < \epsilon.$$

In particular, the limit is valid if f is the characteristic function of a subinterval of $[0, 1]$. In probabilistic terms, this means that the probability that $2\pi k\alpha$ (with a positive integer k), up to addition of some integral multiple of 2π, lies in a subinterval $[a, b]$, is exactly the length of the interval $b - a$. This result provides a quantitative continuation of Chapter I, §4, Exercise 6. It is also called the **equidistribution** *of the numbers $\{k\alpha\}$ modulo* **Z***.*

Solution. Given $\epsilon > 0$, there exists a step map φ such that for $x \in [0, 1]$ we have $|f(x) - \varphi(x)| < \epsilon$. We can assume $\varphi(0) = \varphi(1)$ because f is periodic.

We extend φ on $\mathbf{R}$ by defining $\varphi(x) = \varphi(x - n)$ if $x \in [n, n+1]$, so that φ is now periodic of period 1, and the inequality $|f(x) - \varphi(x)| < \epsilon$ holds for all x.

First, we build g and h on the interval $[0, 1]$. Suppose φ has discontinuities at the points of the partition

$$0 = a_0 < a_1 < \cdots < a_n < 1.$$

We call δ-intervals, intervals of length 2δ centered at points of the partition. Suppose that a_i is not an end point. If x is not in one of the δ-intervals, we set $h(x) = \varphi(x) + \epsilon$ and $g(x) = \varphi(x) - \epsilon$. Let M be a bound for f and select B such that $B > M + \epsilon$. We now define h and g on the δ-intervals. Let h be the linear function having the value $\varphi(a_i - \delta) + \epsilon$ at $a_i - \delta$ and B at a_i. Let h have the value $\varphi(a_i + \delta) + \epsilon$ at $a_i + \delta$ and be linear on $[a_i, a_i + \delta]$. Similarly, for g, which is defined to be the linear function having value $\varphi(a_i - \delta) - \epsilon$ at $a_i - \delta$, $-B$ at a_i, and $\varphi(a_i + \delta) - \epsilon$ at $a_i + \delta$. We also modify the end points, so that $g(0) = g(1)$ and $h(0) = h(1)$, so we can extend g and h to be periodic of period 1 on $\mathbf{R}$. Then g and h are continuous, periodic, and $g \le f \le h$.

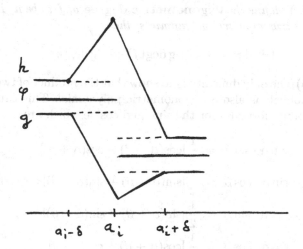

Furthermore,

$$\int_0^1 h - g \le 2\epsilon + 2n\delta B,$$

hence we can choose δ so that the integral is $< 3\epsilon$. Then we have

$$\frac{1}{N}\sum_{n=1}^{N} g(n\alpha) \le \frac{1}{N}\sum_{n=1}^{N} f(n\alpha) \le \frac{1}{N}\sum_{n=1}^{N} h(n\alpha).$$

But by the previous exercise we know that

$$\frac{1}{N}\sum_{n=1}^{N}g(n\alpha) \to \int_0^1 g \quad \text{and} \quad \frac{1}{N}\sum_{n=1}^{N}h(n\alpha) \to \int_0^1 h,$$

and

$$\int_0^1 g \le \int_0^1 f \le \int_0^1 h \quad \text{and} \quad \int_0^1 h - g \le 3\epsilon$$

so we conclude that the desired limit exists and

$$\lim_{N\to\infty}\frac{1}{N}\sum_{n=1}^{N}f(n\alpha) = \int_0^1 f.$$

Exercise XII.2.3 *(a) Let P, Q be trigonometric polynomials. Show that $P + Q$ and PQ is also a trigonometric polynomial. If c is a constant, then cP is a trigonometric polynomial.*
(b) Suppose a trigonometric polynomial P is written in the form

$$P(x) = a_0 + \sum_{k=1}^{n}(a_k \cos kx + b_k \sin kx).$$

If a_n or $b_n \neq 0$, define the **trigonometric degree** *of f to be n. Prove that if f, g are two trigonometric polynomials, then*

$$\text{trig deg}(fg) = \text{trig deg}(f) + \text{trig deg}(g).$$

Solution. (a) The only difficulty is to show that the product of two trigonometric polynomials is also a trigonometric polynomial. This can be done using the product formulas for the sine and cosine which are

$$\sin\alpha \cdot \sin\beta = \frac{1}{2}\left[\cos(\alpha - \beta) - \cos(\alpha + \beta)\right],$$

$$\sin\alpha \cdot \cos\beta = \frac{1}{2}\left[\sin(\alpha + \beta) + \sin(\alpha - \beta)\right],$$

$$\cos\alpha \cdot \sin\beta = \frac{1}{2}\left[\sin(\alpha + \beta) - \sin(\alpha - \beta)\right],$$

$$\cos\alpha \cdot \cos\beta = \frac{1}{2}\left[\cos(\alpha + \beta) + \cos(\alpha - \beta)\right].$$

These formulas show that any expression of the form $\cos^i x \cdot \sin^j x$ can be written as a linear combination of sine and cosine.
(b) The fact that $\text{trig deg}(fg) = \text{trig deg}(f) + \text{trig deg}(g)$ follows from the product formulas.

Exercise XII.2.4 *Let C_0^∞ be the space of C^∞ functions with are periodic and vanish at $-\pi, \pi$. Show that C_0^∞ is L^2-dense in E (the space of piecewise continuous periodic functions). [Hint: Approximate the function by step functions, and use bump functions.]*

Solution. The hint gives the proof away. Let φ be a step map that approximates f, that is $\|f - \varphi\|_2^2 \leq \epsilon$. The C^∞ bump functions smooth out the corners and eliminate the singularities of φ. For the construction of such functions, see Exercise 6, §1, of Chapter IV and Exercise 7, §3, of Chapter X. Doing this construction at each discontinuity of φ and making sure that at the end-points we let the bump function be equal to 0, we get a function $g \in C_0^\infty$ which approximates φ. We have the following pictures:

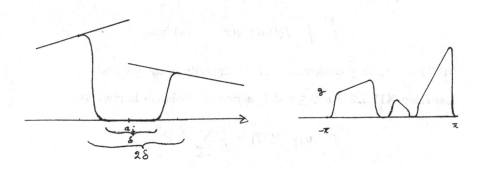

The smaller δ, the better the approximation. Indeed, let n be the number points of the partition which defines the step map φ. Then if B is a bound for φ, we see that estimating the integral on the δ-intervals, we get

$$\|\varphi - g\|_2^2 \leq 2\delta n B^2.$$

So for small δ, we will get $\|f - g\|_2^2 < 2\epsilon$.

XII.3   Explicit Uniform Approximation

Exercise XII.3.1 *Let E be as in the text, the vector space of piecewise continuous periodic functions. If $f, g \in E$, define*

$$f * g(x) = \int_{-\pi}^{\pi} f(t)g(x - t)dt.$$

Prove the following properties:
(a) $f * g = g * f$.
(b) If $h \in E$, then $f * (g + h) = f * g + f * h$.
(c) $(f * g) * h = f * (g * h)$.
(d) If α is a number, then $(\alpha f) * g = \alpha(f * g)$.

Solution. (a) The change of variable $u = x - t$ implies

$$(f * g)(x) = \int_{-\pi}^{\pi} f(t)g(x-t)dt = -\int_{x+\pi}^{x-\pi} f(x-u)g(u)du$$

$$= \int_{-\pi}^{\pi} g(u)f(x-u)du = (g * f)(x).$$

(b) The linearity of the integral implies that $f * (g+h) = f * g + f * h$.
(c) Both $(f * g) * h(x)$ and $f * (g * h)(x)$ are equal to

$$\int_{-\pi}^{\pi} \int_{-\pi}^{\pi} f(t)h(u)g(x-t-u)dudt.$$

(d) This is an easy consequence of the linearity of the integral.

Exercise XII.3.2 *For $0 \leq r < 1$, define the **Poisson kernel** as*

$$P(r,\theta) = P_r(\theta) = \frac{1}{2\pi} \sum_{-\infty}^{\infty} r^{|n|} e^{in\theta}.$$

Show that

$$P_r(\theta) = \frac{1}{2\pi} \frac{1-r^2}{1-2r\cos\theta + r^2}.$$

Solution. We use the formula to sum a geometric series

$$2\pi P_r(\theta) = 1 + \sum_{n=1}^{\infty} (re^{-i\theta})^n + \sum_{n=1}^{\infty} (re^{i\theta})^n = 1 + \frac{re^{-i\theta}}{1-re^{-i\theta}} + \frac{re^{i\theta}}{1-re^{i\theta}}$$

$$= 1 + \frac{re^{-i\theta} - r^2 + re^{i\theta} - r^2}{1-re^{i\theta} - re^{-i\theta} + r^2} = \frac{1-r^2}{1-2r\cos\theta + r^2}.$$

Exercise XII.3.3 *Prove that $P_r(\theta)$ satisfies the three conditions **DIR 1, 2, 3**, where n is replaced by r and $r \to 1$ instead of $n \to \infty$. In other words:*
DIR 1. *We have $P_r(\theta) \geq 0$ for all r and all θ.*
DIR 2. *Each P_r is continuous and*

$$\int_{-\pi}^{\pi} P_r(\theta)d\theta = 1.$$

DIR 3. *Given ϵ and δ, there exists $r_0, 0 < r_0 < 1$ such that if $r_0 < r < 1$, then*

$$\int_{-\pi}^{-\delta} P_r + \int_{\delta}^{\pi} P_r < \epsilon.$$

Solution. We use the formula obtained in the previous exercise. First, **DIR 1** holds because $-1 \leq \cos x \leq 1$ hence

$$1 - 2r \cos \theta + r^2 \geq 1 - 2r + r^2 = (1 - r)^2 > 0,$$

and the numerator is > 0 because $0 \leq r < 1$.

DIR 2 is also verified. Indeed, each P_r is continuous because the series converges uniformly (comparison to the geometric series). Furthermore, we can integrate the series term by term, so

$$\int_{-\pi}^{\pi} P_r(\theta) d\theta = \frac{1}{2\pi} \sum_{-\infty}^{\infty} \int_{-\pi}^{\pi} r^{|n|} e^{in\theta} d\theta.$$

But if $n \neq 0$, $\int_{-\pi}^{\pi} e^{in\theta} d\theta = 0$ so

$$\int_{-\pi}^{\pi} P_r(\theta) d\theta = \frac{1}{2\pi} \int_{-\pi}^{\pi} 1 d\theta = 1.$$

Finally, we verify **DIR 3**. If $\delta \leq \theta \leq \pi$, then $\cos \theta \leq \cos \delta$ so

$$\int_{\delta}^{\pi} P_r(\theta) d\theta \leq \int_{\delta}^{\pi} \frac{1 - r^2}{1 - 2r \cos \delta + r^2} d\theta.$$

But $1 - 2r \cos \delta + r^2 \to 2(1 - \cos \delta)$ and $1 - r^2 \to 0$ as $r \to 1$, and since P_r is even, there exists $0 < r_0 < 1$ such that if $r_0 < r < 1$, then

$$\int_{-\pi}^{-\delta} P_r + \int_{\delta}^{\pi} P_r < \epsilon.$$

Exercise XII.3.4 *Show that Theorem 1.1 concerning Dirac sequences applies to the Poisson kernels, again letting $r \to 1$ instead of $n \to \infty$. In other words: Let f be a piecewise continuous function on $\mathbf{R}$ which is periodic. Let S be a compact set on which f is continuous. Let*

$$f_r = P_r * f.$$

Then f_r converges to f uniformly on S as $r \to 1$.

Solution. Let $\epsilon > 0$. Then **DIR 2** implies that

$$(P_r * f)(x) - f(x) = \int_{-\pi}^{\pi} [f(x - \theta) - f(x)] P_r(\theta) d\theta.$$

The function f is uniformly continuous on S so there exists $\delta > 0$ such that $|f(x - \theta) - f(x)| < \epsilon$ whenever $|\theta| < \delta$. Let B be a bound for f, and select r_0, $0 < r_0 < 1$ such that if $r_0 < r < 1$, then

$$\int_{-\pi}^{-\delta} P_r + \int_{\delta}^{\pi} P_r < \frac{\epsilon}{2B}.$$

Then we have

$$|(P_r * f)(x) - f(x)| \leq \int_{-\pi}^{-\delta} + \int_{-\delta}^{\delta} + \int_{\delta}^{\pi} |f(x - \theta) - f(x)| P_r(\theta) d\theta.$$

The middle integral is estimated by

$$\int_{-\delta}^{\delta} |f(x - \theta) - f(x)| P_r(\theta) d\theta \leq \int_{-\delta}^{\delta} \epsilon P_r(\theta) d\theta \leq \epsilon \int_{-\pi}^{\pi} P_r(\theta) d\theta = \epsilon.$$

The sum of the first and third integral is

$$\leq \int_{-\pi}^{-\delta} + \int_{-\delta}^{\pi} 2BP_r(\theta) d\theta < \epsilon,$$

so if $r_0 < r < 1$, then for all $x \in S$ we have $|(P_r * f)(x) - f(x)| \leq 2\epsilon$.

Exercise XII.3.5 *In this exercise we use partial derivatives which you should know from more elementary courses. See Chapter XV, §1, for a systematic treatment.*

Let $x = r \cos \theta$ and $y = r \sin \theta$ where (r, θ) are the usual polar coordinates. Prove that in terms of polar coordinates, we have the relation

$$\frac{\partial^2}{\partial x^2} + \frac{\partial^2}{\partial y^2} = \frac{\partial^2}{\partial r^2} + \frac{1}{r} \frac{\partial}{\partial r} + \frac{1}{r^2} \frac{\partial^2}{\partial \theta^2}.$$

This means that if $f(x, y)$ is a function of the rectangular coordinates x, y, then

$$f(x, y) = f(r \cos \theta, r \sin \theta) = u(r, \theta)$$

is also a function of (r, θ), and if we apply the left-hand side to f, that is

$$\frac{\partial^2 f}{\partial x^2} + \frac{\partial^2 f}{\partial y^2},$$

*then we get the same thing as if we apply the right-hand side to $u(r, \theta)$. The above relation gives the expression for the **Laplace operator** $(\partial/\partial x)^2 + (\partial/\partial y)^2$ in terms of polar coordinates. The Laplace operator is denoted by Δ.*

*A function f is called **harmonic** if $\Delta f = 0$.*

Solution. We simply differentiate. For the first term $\partial^2/\partial r^2$ we have

$$\frac{\partial u}{\partial r} = \frac{\partial f}{\partial x} \cos \theta + \frac{\partial f}{\partial y} \sin \theta$$

and therefore, we find that $(\partial^2 u/\partial r^2)$ is equal to

$$= \cos \theta \left(\frac{\partial^2 f}{\partial x^2} \cos \theta + \frac{\partial^2 f}{\partial y \partial x} \sin \theta \right) + \sin \theta \left(\frac{\partial^2 f}{\partial y \partial x} \cos \theta + \frac{\partial^2 f}{\partial y^2} \sin \theta \right)$$

$$= \frac{\partial^2 f}{\partial x^2} \cos^2 \theta + \frac{\partial^2 f}{\partial x \partial y} \sin \theta \cos \theta + \frac{\partial^2 f}{\partial y \partial x} \sin \theta \cos \theta + \frac{\partial^2 f}{\partial y^2} \sin^2 \theta.$$

For the last term $\partial^2/\partial\theta^2$ we have

$$\frac{\partial u}{\partial \theta} = -r\frac{\partial f}{\partial x}\sin\theta + r\frac{\partial f}{\partial y}\cos\theta,$$

so we find that

$$\frac{\partial^2 u}{\partial \theta^2} = \quad -r\frac{\partial f}{\partial x}\cos\theta - r\frac{\partial f}{\partial y}\sin\theta + r^2\frac{\partial^2 f}{\partial x^2}\sin^2\theta - r^2\frac{\partial^2 f}{\partial x\partial y}\sin\theta\cos\theta$$
$$-r^2\frac{\partial^2 f}{\partial y\partial x}\sin\theta\cos\theta + r^2\frac{\partial^2 f}{\partial y^2}\cos^2\theta.$$

The formula that gives the expression of the Laplace operator in terms of polar coordinates drops out.

Exercise XII.3.6 *(a) Show that the functions $r^{|k|}e^{ik\theta}$ are harmonic, for every integer k.*
(b) Show that $\Delta P = 0$. In other words, the function

$$P(r,\theta) = \frac{1}{2\pi}\sum_{k=-\infty}^{\infty} r^{|k|}e^{ik\theta}$$

is harmonic for $0 \le r < 1$. Justify the term by term differentiations.

Solution. (a) Let $u(r,\theta) = r^{|k|}e^{ik\theta}$. If $k = 0$ there is nothing to prove. Assume $k \ne 0$. Then

$$\frac{\partial u}{\partial r} = |k|r^{|k|-1}e^{ik\theta}, \quad \frac{\partial^2 u}{\partial r^2} = |k|(|k|-1)r^{|k|-2}e^{ik\theta} \quad \text{and} \quad \frac{\partial^2 u}{\partial \theta^2} = -k^2 r^{|k|}e^{ik\theta},$$

so Exercise 5 implies

$$\Delta u = r^{|k|-2}e^{ik\theta}(k^2 - |k| + |k| - k^2) = 0.$$

(b) Differentiating with respect to r we get the following two series:

$$\sum_{-\infty}^{\infty} |k|r^{|k|-1}e^{ik\theta} \quad \text{and} \quad \sum_{-\infty}^{\infty} |k|(|k|-1)r^{|k|-2}e^{ik\theta}.$$

The ratio test implies that these two series converge absolutely and uniformly on every closed interval $[0,c]$ with $0 < c < 1$ and with $\theta \in \mathbf{R}$. Differentiating with respect to θ we get

$$\sum_{-\infty}^{\infty} ikr^{|k|}e^{ik\theta} \quad \text{and} \quad \sum_{-\infty}^{\infty} -k^2 r^{|k|}e^{ik\theta}.$$

If $r \in [0,c]$ with $0 < c < 1$ and $\theta \in \mathbf{R}$, the ratio test and the fact that $|e^{ik\theta}| = 1$ implies that the series converges absolutely and uniformly. So we can differentiate term by term the given function, and part (a) implies that $P_r(\theta)$ is harmonic for $0 \le r < 1$.

Exercise XII.3.7 *Let g be a continuous function of θ, periodic of period 2π. Define*

$$u(r, \theta) = (P_r * g)(\theta) \quad \text{for } 0 \le r < 1.$$

(a) Show that u(r, θ) is harmonic. (You will need to differentiate under an intergral sign.) In fact, $\Delta(P * g) = (\Delta P) * g$.
(b) Show that

$$\lim_{r \to 1} u(r, \theta) = g(\theta)$$

uniformly in θ, as a special case of approximation of Dirac families.

Solution. (a) The functions g and P_r are periodic of period 2π, so u is also periodic of period 2π. Suppose $r \in [0, c]$ with $0 < c < 1$. The convolution is equal to

$$u(r, \theta) = \int_{-\pi}^{\pi} P_r(t) g(\theta - t) dt.$$

All the functions being continuous, we differentiate under the integral sign to conclude that

$$\Delta u = (\Delta P_r) * g.$$

In Exercise 6 we proved that $\Delta P_r = 0$ so $\Delta u = 0$. This result holds for any c with $0 < c < 1$, thus $u(r, \theta)$ is harmonic.
(b) For $\theta \in [0, 2\pi]$, Exercise 4 implies that $u(r, \theta)$ converges uniformly to g as $r \to 1$. Then since all the functions are periodic 2π, we conclude that

$$\lim_{r \to 1} u(r, \theta) = g(\theta)$$

uniformly in θ.

XII.4   Pointwise Convergence

Exercise XII.4.1 *(a) Carry out the computation of the Fourier series of $(\pi - x)^2/4$ on $[0, 2\pi]$. Show that this Fourier series can be differentiated term by term in every interval $[\delta, 2\pi - \delta]$ and deduce that*

$$\frac{\pi - x}{2} = \sum_{k=1}^{\infty} \frac{\sin kx}{k}, \quad 0 < x < 2\pi.$$

(b) Deduce the same identity from Theorem 4.5.

Solution. The function is odd, so $b_n = 0$ for all n. For $n \ge 1$ we integrate by parts to obtain

$$a_n = \frac{1}{\pi} \int_0^{2\pi} \frac{(\pi - x)^2}{4} \cos nx dx = \frac{1}{\pi n} \frac{(\pi - x)}{2} \sin nx.$$

Exercise 7, §1, implies that $a_n = 1/n^2$. When $n = 0$ we get

$$a_0 = \frac{1}{2\pi} \int_0^{2\pi} \frac{(\pi - x)^2}{4} dx = \frac{\pi^2}{12},$$

so the Fourier series of $(\pi - x)^2/4$ on $[0, 2\pi]$ is

$$\frac{\pi^2}{12} + \sum_{n=1}^{\infty} \frac{\cos nx}{n^2}.$$

The derived series is

$$-\sum_{n=1}^{\infty} \frac{\sin nx}{n}.$$

In Exercise 6, §5, of Chapter IX, we proved the uniform convergence of this series in every interval $[\delta, 2\pi - \delta]$, so we can differentiate the Fourier series term by term. The derivative of $(\pi - x)^2/4$ is $-(\pi - x)/2$, so for $0 < x < 2\pi$ we have

$$\frac{\pi - x}{2} = \sum_{n=1}^{\infty} \frac{\sin nx}{n}.$$

(b) Let $0 < x < 2\pi$. Since $u \mapsto (\pi - u)/2$ is differentiable at x, Theorem 4.5 implies that the Fourier series at x converges to $(\pi - x)/2$. In Exercise 7, §1, we proved that this Fourier series is $\sum_{n=1}^{\infty} (\sin nx)/n$. The result follows.

Exercise XII.4.2 *Let f be a C^∞ periodic function (period 2π). Prove that given a positive integer k, one has*

$$\int_{-\pi}^{\pi} f(x) e^{iAx} dx = O(1/|A|^k) \quad for \ A \to \pm\infty.$$

Solution. We assume that $A \in \mathbf{Z}$. Integrating by parts once we get

$$\int_{-\pi}^{\pi} f(x) e^{iAx} dx = \left[\frac{f(x) e^{iAx}}{iA} \right]_{-\pi}^{\pi} - \frac{1}{iA} \int_{-\pi}^{\pi} f'(x) e^{iAx} dx$$

$$= -\frac{1}{iA} \int_{-\pi}^{\pi} f'(x) e^{iAx} dx.$$

So integrating by parts k times and putting absolute values we get

$$\left| \int_{-\pi}^{\pi} f(x) e^{iAx} dx \right| \leq \frac{1}{|A|^k} \left| \int_{-\pi}^{\pi} f^{(k)}(x) e^{iAx} dx \right|$$

$$\leq \frac{1}{|A|^k} \int_{-\pi}^{\pi} |f^{(k)}(x)| dx$$

$$\leq \frac{2\pi}{|A|^k} \sup_{-\pi \leq x \leq \pi} |f^{(k)}(x)|.$$

The last sup exists because we assume that $f \in C^\infty$ and $[-\pi, \pi]$ is compact. Conclude.

Exercise XII.4.3 *Show that the convergence of the Fourier series to $f(x)$ at a given point x depends only on the behavior of f near x. In other words, if $g(t) = f(t)$ for all t in some open interval containing x, then the Fourier series of g converges to $g(x)$ at x if and only if the Fourier series of f converges to $f(x)$ at x.*

Solution. Suppose that $f = g$ in the open ball of radius r centered at x. We can write $D_n * g(x) - g(x)$ and $D_n * f(x) - f(x)$ as the sum of three integrals as in Theorem 4.5. Suppose the Fourier series of f converges to $f(x)$ at x. Given $\epsilon > 0$ for all large n we have

$$|D_n * f(x) - f(x)| < \epsilon.$$

Let δ be also $< r$. So the middle integral in $D_n * g(x) - g(x)$ and $D_n * f(x) - f(x)$ are equal. The Riemann-Lebesgue lemma implies that the first and third integral converge to 0, so we conclude that for large n, $D_n * g(x) - g(x)$ is small. Therefore, the Fourier series of g converges to $g(x)$. Conversely, the Fourier series of f converges to $f(x)$ at x if the Fourier series of g converges to $g(x)$ at x.

Exercise XII.4.4 *Let F be the complete normed vector space of continuous periodic functions on $[-\pi, \pi]$ with the sup norm. Let l^1 be the vector space of all real sequences $\alpha = \{a_n\}$ $(n = 1, 2, \ldots)$ such that $\sum |a_n|$ converges. We define, as in Exercise 8 of Chapter IX, §5, the norm*

$$\|\alpha\|_1 = \sum_{n=1}^{\infty} |a_n|.$$

Let $L\alpha(x) = \sum a_n \cos nx$, so that $L \colon l^1 \to F$ is a linear map, satisfying

$$\|L(\alpha)\| \leq \|\alpha\|_1.$$

Let B be the closed unit ball of radius 1 centered at the origin in l^1. Show that $L(B)$ is closed in F. [Hint: Let $\{f_k\}$ $(k = 1, 2, \ldots)$ be a sequence of elements of $L(B)$ which converges uniformly to a function f in F. Let $f_k = L(\alpha^k)$ with $\alpha^k = \{a_n^k\}$ in l^1. Show that

$$a_n^k = \frac{1}{\pi} \int_{-\pi}^{\pi} f_k(x) \cos nx\, dx.$$

Let $b_n = 1/\pi \int_{-\pi}^{\pi} f(x) \cos nx\, dx$. Note that $|b_n - a_n^k| \leq 2\|f - f_k\|_\infty$. Let $\beta = \{b_n\}$. Show first that β is an element of l^1, proceeding as follows. If $\beta \notin l^1$, then for some N and $c > 0$ we have $\sum_{n=1}^{N} |b_n| \geq 1 + c$. Taking k large enough, show that $\sum_{n=1}^{N} |a_n^k| > 1$, which is a contradiction. Why can you now conclude that $L(\beta) = f$?]

Solution. We use the notation of the chapter and of the hint. Since $|\cos mx|$ ≤ 1, the series $\sum_{n=1}^{\infty} a_n^k \varphi_n \varphi_m$ converges uniformly to $f_k \varphi_m$. We can integrate term by term so that

$$\int_{-\pi}^{\pi} f_k(x) \cos mx \, dx = \sum_{n=1}^{\infty} a_n^k \langle \varphi_n, \varphi_m \rangle = \pi a_m^k.$$

Let $b_n = 1/\pi \int_{-\pi}^{\pi} f(x) \cos nx \, dx$. Then

$$|b_n - a_n^k| \leq \frac{1}{\pi} \int_{-\pi}^{\pi} |f(x) - f_k(x)| |\cos nx| \, dx \leq \frac{2\pi}{\pi} \|f - f_k\|.$$

Now we prove that $\beta \in B$. If not, then for some N and $c > 0$, we have

$$\sum_{n=1}^{N} |b_n| \geq 1 + c.$$

The hypotheses implies that for some k we have $\|f - f_k\| < c/(4N)$. Then

$$|a_n^k| = |a_n^k - b^n + b_n| \geq |b_n| - |a_n^k - b_n| \geq |b_n| - 2\|f - f_k\| \geq |b_n| - \frac{c}{2N},$$

so

$$\sum_{n=1}^{N} |a_n^k| \geq \sum_{n=1}^{N} |b_n| - N\frac{c}{2N} \geq 1 + \frac{c}{2}.$$

But $\|\alpha^k\|_1 \leq 1$ so we get a contradiction which proves that $\beta \in B$. Now we prove that $L(\beta) = f$. For each k, the function f_k is even and continuous, so f is even and continuous. Hence the Fourier coefficients of f with respect to the sine are 0. Furthermore, the 0-th Fourier coefficient of f is 0 because

$$\int_{-\pi}^{\pi} f(x) dx = 0,$$

and this follows from the fact that $\int_{-\pi}^{\pi} f_k(x) dx = 0$ for all k, by symmetry. So the Fourier series of f simply is $\sum_{n=1}^{\infty} b_n \cos nx = L(\beta)$. By Theorem 4.1 we conclude that $f = L(\beta)$. This proves that $L(B)$ is closed.

Exercise XII.4.5 *Determine the Fourier series for the function whose values are e^x for*

$$0 < x < 2\pi.$$

Solution. We compute the Fourier series of the function e^{ax} where a is not of the form in where n is an integer. The complex Fourier coefficients are given by

$$c_k = \frac{1}{2\pi} \int_0^{2\pi} e^{ax} e^{-ikx} dx = \frac{1}{2\pi(a - ik)} \left[e^{2\pi a} - 1 \right].$$

So the Fourier series of the function e^{ax} is

$$= \frac{e^{2\pi a} - 1}{2\pi a} + (e^{2\pi a} - 1) \sum_{k \geq 1} \frac{a + ik + a - ik}{2\pi(a^2 + k^2)} \cos kx + i \frac{a + ik - a + ik}{2\pi(a^2 + k^2)} \sin kx$$

$$= \frac{e^{2\pi a} - 1}{2\pi a} + \frac{e^{2\pi a} - 1}{\pi} \sum_{k \geq 1} \frac{a \cos kx - k \sin kx}{a^2 + k^2}.$$

Exercise XII.4.6 *For $0 < x < 2\pi$ and $a \neq 0$ we have*

$$\pi e^{ax} = (e^{2a\pi} - 1) \left(\frac{1}{2a} + \sum_{k=1}^{\infty} \frac{a \cos kx - k \sin kx}{k^2 + a^2} \right).$$

Solution. By factoring $e^{2\pi a} - 1$ and multiplying by π the expression of the Fourier series computed in Exercise 5, we get the right-hand side of the formula. Theorem 4.5 shows that the Fourier series converges to the function. This yields the desired formula.

Exercise XII.4.7 *For $0 < x < 2\pi$ and a not an integer, we have*

$$\pi \cos ax = \frac{\sin 2a\pi}{2a} + \sum_{k=1}^{\infty} \frac{a \sin 2a\pi \cos kx + k(\cos 2a\pi - 1) \sin kx}{a^2 - k^2}.$$

Solution. The formula obtained in Exercise 6 is valid for any complex number a not equal to in for some integer n. Suppose α is a real number not equal to an integer and let $i\alpha = a$. Then we have

$$\pi e^{ax} = \pi(\cos \alpha x + i \sin \alpha x),$$

so $\pi \cos \alpha x$ is simply the real part of the right-hand side of the formula derived in Exercise 6. The expression in the first parentheses equals

$$e^{2i\alpha\pi} - 1 = \cos(2\alpha\pi) - 1 + i \sin(2\alpha\pi) = u_1 + iv_1,$$

and the expression in the second parentheses equals

$$\frac{-i}{2\alpha} + \sum_{k=1}^{\infty} \frac{i\alpha \cos kx - k \sin kx}{k^2 - \alpha^2} = u_2 + iv_2.$$

Therefore, $\pi \cos \alpha x = u_1 u_2 - v_1 v_2$ and the desired formula follows from a simple computation.

Exercise XII.4.8 *Letting $x = \pi$ in Exercise 7, conclude that*

$$\frac{a\pi}{\sin a\pi} = 1 + 2a^2 \sum_{k=1}^{\infty} \frac{(-1)^k}{a^2 - k^2}$$

when a is not an integer.

Solution. Letting $x = \pi$ in Exercise 7, we get

$$\pi \cos a\pi = \sin(2a\pi) \left(\frac{1}{2a} + a \sum_{k=1}^{\infty} \frac{(-1)^k}{a^2 - k^2} \right),$$

thus

$$2a\pi \cos a\pi = 2 \sin a\pi \cos a\pi \left(1 + 2a^2 \sum_{k=1}^{\infty} \frac{(-a)^k}{a^2 - k^2} \right).$$

Exercise XII.4.9 (Elkies) *Let B be the periodic function with period 1 defined on $[0, 1]$ by*

$$B(x) = x^2 - x + \frac{1}{6}.$$

(a) Prove that $B(x) = \frac{1}{2\pi^2} \sum_{n \neq 0} \frac{1}{n^2} e^{2\pi i n x}$.

(b) Prove the polynomial identity for every positive integer M:

$$\frac{1}{M+1} \left(\sum_{n=1}^{M+1} z^n \right) \left(\sum_{k=1}^{M+1} z^{-k} \right) = \sum_{m=1}^{M+1} \left(1 - \frac{m}{M+1} \right) (z^m + z^{-m}) + 1.$$

(c) Prove that for all integers $M \geq 1$ we have:

$$\sum_{m=1}^{M} \left(1 - \frac{m}{M+1} \right) B(mu) \geq -\frac{1}{12}.$$

(d) More generally, let $A = (a_1, \ldots, a_r)$ be an r-tuple of positive numbers. Let $X = (x_1, \ldots, x_r)$ be an r-tuple of real numbers. Define

$$E(A, X) = \sum_{i \neq j} a_i a_j \frac{1}{2} B(x_i - x_j).$$

Prove that

$$E(A, X) \geq -\frac{1}{12} \sum_{j=1}^{r} a_j^2.$$

Solution. (a) Let $u = 2\pi x$ and consider the function B_2 defined by $B_2(u) = B(u/2\pi)$, and which is periodic of period 2π. For $n \neq 0$ the complex Fourier coefficients of B_2 are given by

$$2\pi c_n = \frac{1}{4\pi^2} \int_0^{2\pi} u^2 e^{-inu} du - \frac{1}{2\pi} \int_0^{2\pi} u e^{-inu} du + \frac{1}{6} \int_0^{2\pi} e^{-inu} du.$$

The last integral equals 0, and integration by parts shows that the second term equals $1/(in)$. Integrating by parts also shows that the first integral is equal to $-1/(in) + 1/(\pi n^2)$, so

$$c_n = \frac{1}{2\pi^2 n^2}.$$

If $n = 0$, we have

$$2\pi c_0 = \int_0^{2\pi} B_2(u)du = \pi\left(\frac{8}{12} - 1 + \frac{1}{3}\right) = 0.$$

Theorem 4.5 implies

$$B_2(u) = \frac{1}{2\pi^2}\sum_{n\neq 0}\frac{1}{n^2}e^{inx},$$

but $B_2(2\pi x) = B(x)$ so the formula for $B(x)$ drops out.

(b) Induction. The formula is true when $M = 1$. Assume that the formula is true for some positive integer M. Set

$$A = \frac{1}{M+2}\left(\sum_{n=1}^{M+2}z^n\right)\left(\sum_{k=1}^{M+2}z^{-k}\right).$$

Then we see that

$$A = \frac{1}{M+2}\left(\sum_{n=1}^{M+1}z^n\cdot\sum_{k=1}^{M+1}z^{-k} + z^{M+2}\sum_{k=1}^{M+1}z^{-k} + z^{-M-2}\sum_{n=1}^{M+1}z^n + 1\right).$$

The induction hypothesis implies that

$$\begin{aligned}
A &= \frac{1}{M+2}\left[\left(\sum_{m=1}^{M+1}(M+1-m)(z^m + z^{-m}) + M+1\right)\right.\\
&\quad\left. + \left(\sum_{m=1}^{M+1}z^{-m} + \sum_{n=1}^{M+1}z^n + 1\right)\right]\\
&= \sum_{m=1}^{M+2}\left(1 - \frac{m}{M+2}\right)(z^m + z^{-m}) + 1,
\end{aligned}$$

as was to be shown.

(c) Part (a) implies

$$\sum_{m=1}^{M+1}\left(1 - \frac{m}{M+1}B(mu)\right)$$

is equal to

$$\frac{1}{2\pi^2}\sum_{n>0}\frac{1}{n^2}\sum_{m=1}^{M+1}\left(1 - \frac{m}{M+1}\right)(e^{2\pi inmu} + e^{-2\pi inmu}),$$

and by part (b) we see that this last expression is

$$= \frac{1}{2\pi^2} \sum_{n>0} \frac{1}{n^2(M+1)} \left(\sum_{k=1}^{M+1} e^{2\pi i n k u} \right) \left(\sum_{k=1}^{M+1} e^{-2\pi i n k u} \right) - \frac{1}{n^2}.$$

But $\sum e^{-2\pi i n k u}$ is the complex conjugate of $\sum e^{2\pi i n k u}$ so we see that the above expression is

$$\geq \frac{-1}{2\pi^2} \sum_{n>0} \frac{1}{n^2} = -\frac{1}{12},$$

where we have used the fact that $\sum_{n=1}^{\infty} 1/n^2 = \pi^2/6$. This proves the inequality of (c).

(d) We have

$$E(A, X) = \frac{1}{4\pi^2} \sum_{n \neq 0} \frac{1}{n^2} \sum_{i \neq j} a_i a_j e^{2\pi i n (x_i - x_j)}$$

$$= \frac{1}{4\pi^2} \sum_{n \neq 0} \frac{1}{n^2} \left(\left| \sum_j a_j e^{2\pi i n x_j} \right|^2 - \sum_j a_j^2 \right)$$

$$\geq -\frac{1}{4\pi^2} \sum_{n \neq 0} \frac{1}{n^2} \sum_j a_j^2 = -\frac{1}{12} \sum_j a_j^2.$$

XIII

Improper Integrals

XIII.1 Definition

Exercise XIII.1.1 *Let f be complex valued, $f = f_1 + if_2$ where f_1, f_2 are real valued, and piecewise continuous.*
(a) Show that

$$\int_a^\infty f \ \text{converges if and only if} \ \int_a^\infty f_1 \ \text{and} \ \int_a^\infty f_2 \ \text{converge.}$$

(b) The function f is absolutely integrable on $\mathbf{R}$ if and only if f_1 and f_2 are absolutely integrable.

Solution. (a) Suppose $\int_a^\infty f$ converges to $u_1 + iu_2$, then since

$$\left| \int_a^b f_i - u_i \right| \leq \left| \int_a^b f - (u_1 + iu_2) \right|,$$

for $i = 1, 2$ we conclude that $\int_a^\infty f_1$ and $\int_a^\infty f_2$ converge to u_1 and u_2, respectively.

Conversely, suppose that $\int_a^\infty f_1$ and $\int_a^\infty f_2$ converge to u_1 and u_2, respectively. Then

$$\left| \int_a^b f - (u_1 + iu_2) \right| \leq \left| \int_a^b f_1 - u_1 \right| + \left| \int_a^b f_2 - u_2 \right|,$$

so $\int_a^\infty f$ converges to $u_1 + iu_2$.

(b) Since $|f_1| \leq |f|$, $|f_2| \leq |f|$, and $|f| \leq |f_1| + |f_2|$, f is absolutely integrable on $\mathbf{R}$ if and only if f_1 and f_2 are absolutely integrable on $\mathbf{R}$.

Exercise XIII.1.2 *Integrating by parts, show that the following integrals exist and evaluate them:*

$$\int_0^\infty e^{-x} \sin x \, dx$$

and

$$\int_0^\infty e^{-x} \cos x \, dx.$$

Solution. Integrating by parts twice we find that

$$\int_0^\infty e^{-x} \sin x \, dx = 1 - \int_0^\infty e^{-x} \sin x \, dx$$

so that

$$\int_0^\infty e^{-x} \sin x \, dx = \frac{1}{2},$$

and similarly for the other integral. Note that we can compute both integrals simultaneously. Compute $\int_0^b e^{-x} e^{ix} \, dx$ and use Exercise 1. You will get

$$\int_0^b e^{-x} e^{ix} \, dx = \frac{1}{i-1} \left[e^{-x} e^{ix} \right]_0^b = \frac{e^{-b} e^{ib} - 1}{i-1},$$

but $|e^{-b} e^{ib}| = e^{-b} \to 0$ as $b \to \infty$, so that

$$\int_0^\infty e^{-x} e^{ix} \, dx = \frac{-1}{i-1} = \frac{1+i}{2}.$$

Therefore the integrals $\int_0^\infty e^{-x} \sin x \, dx$ and $\int_0^\infty e^{-x} \cos x \, dx$ both converge to $1/2$.

Exercise XIII.1.3 *Let f be a continuous function on $\mathbf{R}$ which is absolutely integrable.*
(a) Show that

$$\int_{-\infty}^\infty f(-x) \, dx = \int_{-\infty}^\infty f(x) \, dx.$$

(b) Show that for every real number a we have

$$\int_{-\infty}^\infty f(x+a) \, dx = \int_{-\infty}^\infty f(x) \, dx.$$

(c) Assume that the function $f(t)/|t|$ is continuous and absolutely integrable. Use the symbols

$$\int_{\mathbf{R}^*} f(t) \, d^*t = \int_{-\infty}^\infty f(t) \frac{1}{|t|} \, dt.$$

If a is any real number $\neq 0$, show that

$$\int_{\mathbf{R}^*} f(at)d^*t = \int_{\mathbf{R}^*} f(t)d^*t.$$

This is called the **invariance of the integral under multiplicative translations with respect to** dt/t.

Solution. The change of variable $u = -x$ implies

$$\int_{-a}^{0} f(x)dx = \int_{0}^{a} f(-u)du \quad \text{and} \quad \int_{0}^{b} f(x)dx = \int_{-b}^{0} f(-u)du.$$

Taking limits yields the desired result.
(b) The change of variable $u = x + a$ implies

$$\int_{-\alpha}^{0} f(x+a)dx = \int_{-\alpha+a}^{a} f(u)du \quad \text{and} \quad \int_{0}^{\beta} f(x+a)dx = \int_{a}^{\beta+a} f(u)du.$$

Taking limits yields the desired result.
(c) Suppose $a > 0$. Then

$$\int_{-\alpha}^{c} f(at)\frac{1}{|at|}dt = \int_{-a\alpha}^{ac} f(u)\frac{1}{|u|}du \quad \text{and} \quad \int_{c}^{\beta} f(at)\frac{1}{|at|}dt = \int_{ac}^{a\beta} f(u)\frac{1}{|u|}du.$$

Taking limits yields the desired result. If $a < 0$, part (a) implies the result.

XIII.2   Criteria for Convergence

Exercise XIII.2.1 *Show that the following integrals converge absolutely. We take $a > 0$, and P is a polynomial.*

(a) $\int_0^\infty P(x)e^{-x}dx$.   (b) $\int_0^\infty P(x)e^{-ax}dx$.
(c) $\int_0^\infty P(x)e^{-ax^2}dx$.   (d) $\int_{-\infty}^\infty P(x)e^{-a|x|}dx$.
(e) $\int_0^\infty (1+|x|)^n e^{-ax}dx$ for every positive integer n.

Solution. (a) Let $a = 1$ in (b).
(b) It is sufficient to prove the absolute convergence in the case where P is a monomial, i.e. $P(x) = x^n$. We can write

$$x^n e^{-ax} = x^n e^{-ax/2} e^{-ax/2}.$$

For all large $x > 0$ we have $|x^n e^{-ax/2}| \leq 1$. Clearly, the integral $\int_0^\infty e^{-ax/2}dx$ converges because

$$\int_0^B e^{-ax/2}dx = \frac{-2}{a}\left[e^{-aB/2} - 1\right] \to \frac{2}{a}$$

as $B \to \infty$. Therefore, the integral $\int_0^\infty P(x)e^{-ax}dx$ converges.

(c) For all large x, $e^{-ax^2} \le e^{-ax}$ so (b) guarantees the absolute convergence of the integral $\int_0^\infty P(x)e^{-ax^2}dx$.

(d) If $A, B > 0$, we can write

$$\int_0^B P(x)e^{-a|x|}dx = \int_0^B P(x)e^{-ax}dx \quad \text{and}$$

$$\int_{-A}^0 P(x)e^{-a|x|}dx = \int_0^A P(-x)e^{-ax}dx.$$

Conclude using part (b).

(e) Since $x \ge 0$, $|x| = x$ and expanding $(1 + x)^n$ we get a polynomial, we conclude from part (b) that $\int_0^\infty (1 + |x|)^n e^{-ax}dx$ converges absolutely.

Exercise XIII.2.2 *Show that the integrals converge.*

(a) $\int_0^{\pi/2} \frac{1}{|\sin x|^{1/2}}dx.$ (b) $\int_{\pi/2}^\pi \frac{1}{|\sin x|^{1/2}}dx.$

Solution. (a) Write

$$\frac{1}{|\sin x|^{1/2}} = \frac{x^{1/2}}{|\sin x|^{1/2}} \frac{1}{x^{1/2}}.$$

The function $x \mapsto x^{1/2}/|\sin x|^{1/2}$ is continuous on $[0, \pi/2]$ and bounded because $u/\sin u \to 1$ as $u \to 0$. The result follows from the fact that the integral

$$\int_0^{\pi/2} \frac{dx}{x^{1/2}}$$

converges. Indeed,

$$\int_0^{\pi/2} \frac{dx}{x^{1/2}} = [2\sqrt{x}]_0^{\pi/2} = 2\sqrt{\pi/2}.$$

(b) Suppose $\delta > 0$ is small. The change of variable $u = \pi - x$ implies

$$\int_{\pi/2}^{\pi - \delta} \frac{dx}{|\sin x|^{1/2}}dx = \int_{\pi/2}^\delta \frac{-du}{|\sin(\pi - u)|^{1/2}} = \int_\delta^{\pi/2} \frac{du}{|\sin u|^{1/2}}.$$

Let $\delta \to 0$ and use part (a) to conclude.

Exercise XIII.2.3 *Interpret the following integral as a sum of integrals between $n\pi$ and $(n + 1)\pi$, and then show that it converges:*

$$\int_0^\infty \frac{1}{(x^2 + 1)|\sin x|^{1/2}}dx.$$

Solution. Let

$$a_n = \int_{n\pi}^{(n+1)\pi} \frac{1}{(x^2+1)|\sin x|^{1/2}} dx.$$

The change of variables $u = x - n\pi$ shows that

$$\int_{n\pi+\delta_1}^{n\pi+\pi/2} \frac{1}{(x^2+1)|\sin x|^{1/2}} dx = \int_{\delta_1}^{\pi/2} \frac{1}{((u+n\pi)^2+1)|\sin u|^{1/2}} du$$

$$\leq \frac{1}{n^2\pi^2} \int_0^{\pi/2} \frac{du}{|\sin u|^{1/2}},$$

and a similar estimate holds for the integral from $n\pi + \pi/2$ to $(n+1)\pi - \delta_2$. Using Exercise 2, we see that a_n is finite and since $\sum 1/n^2$ converges and $a_n \geq 0$ we conclude that the series $\sum a_n$ converges. Since the integrand is positive it follows that the integral

$$\int_0^\infty \frac{1}{(x^2+1)|\sin x|^{1/2}} dx$$

converges.

Exercise XIII.2.4 *Show that the following integrals converge:*

(a) $\int_0^\infty \frac{1}{\sqrt{x}} e^{-x} dx.$ *(b)* $\int_0^\infty \frac{1}{x^s} e^{-x} dx$ *for* $s < 1.$

Solution. (a) Put $s = 1/2$ in (b).

(b) Because of Exercise 1, we may assume that $0 < s < 1$. We split the integral

$$\int_\delta^B \frac{1}{x^s} e^{-x} dx = \int_\delta^1 \frac{1}{x^s} e^{-x} dx + \int_1^B \frac{1}{x^s} e^{-x} dx.$$

The first integral converges because for $x \geq 0$

$$0 \leq \frac{e^{-x}}{x^s} \leq \frac{1}{x^s}$$

and

$$\int_\delta^1 \frac{dx}{x^s} = \frac{1}{-s+1} - \frac{\delta^{-s+1}}{-s+1}$$

tends to $1/(1-s)$ as $\delta \to 0$. The second integral also converges because for all large x we have

$$0 \leq \frac{e^{-x}}{x^s} \leq e^{-x}$$

and

$$\int_1^\infty e^{-x}$$

converges. This concludes our argument.

Exercise XIII.2.5 *Assume that f is continuous for $x \geq 0$. Prove that if $\int_1^\infty f(x)dx$ exists, then*

$$\int_a^\infty f(x)dx = a \int_1^\infty f(ax)dx \quad for \ a \geq 1.$$

Solution. Changing variable, $u = x/a$ we see that

$$\int_a^B f(x)dx = \int_1^{B/a} f(au)a\,du.$$

Let $B \to \infty$ and conclude.

Exercise XIII.2.6 *Let E be the set of functions f (say real valued, of one variable, defined on $\mathbf{R}$) which are continuous and such that*

$$\int_{-\infty}^\infty |f(x)|dx$$

converges.
(a) Show that E is a vector space.
(b) Show that the association

$$f \mapsto \int_{-\infty}^\infty |f(x)|dx$$

is a norm on this space.
(c) Give an example of a Cauchy sequence in this space which does not converge (in other words, this space is not complete).

Solution. Suppose $f, g \in E$. Then $f + g$ is continuous and $|f(x) + g(x)| \leq |f(x)| + |g(x)|$ so $\int_{-\infty}^\infty |f(x) + g(x)|dx$ converges. If α is a number, then $\int_a^b |\alpha f| = |\alpha| \int_a^b |f|$, so $\alpha f \in E$. Thus E is a vector space.
(b) Let

$$\|f\|_1 = \int_{-\infty}^\infty |f(x)|dx.$$

Then if $f = 0$ we clearly have $\|f\|_1 = 0$, and conversely, if $\|f\|_1$, we must have $f = 0$ because f is continuous. Indeed, if $|f(y)| > 0$ for some y, then $|f| > 0$ in some open ball centered at y and therefore the integral would be > 0. Clearly, $\|\alpha f\|_1 = |\alpha|\|f\|_1$ whenever α is a real number. Finally, the triangle inequality follows from the fact that $|f(x) + g(x)| \leq |f(x)| + |g(x)|$ for every x.
(c) Consider the function f_n equal to 0 outside $[-1/n, 1+1/n]$ which takes the value 1 on $[0, 1]$ and which is linear on the intervals $[-1/n, 0]$ and $[1, 1+1/n]$. Consider the sequence $\{f_n\}_{n=1}^\infty$. Then if $n > m$ we have

$$\|f_n - f_m\|_1 \leq \frac{2}{m},$$

so $\{f_n\}$ is a Cauchy sequence. Suppose that there exists a continuous function $f \in E$ such that $f_n \to f$ in the $\| \cdot \|_1$ norm. Then

$$\int_0^1 |1 - f(x)|dx = \int_0^1 |f_n(x) - f(x)|dx \leq \|f_n - f\|_1.$$

But $\|f_n - f\|_1 \to 0$ so we conclude that $f(x) = 1$ if $x \in [0, 1]$ because f is continuous. The same argument shows that $f(x) = 0$ if $x \notin [0, 1]$. Since we assumed f continuous, we get a contradiction and this proves that E is not complete under $\| \cdot \|_1$.

In the following exercise, you may assume that

$$\int_{-\infty}^{\infty} e^{-t^2} dt = \sqrt{\pi}.$$

Exercise XIII.2.7 *(a) Let k be an integer ≥ 0. Let $P(t)$ be a polynomial, and let c be the coefficient of its term of highest degree. Integrating by parts, show that the integral*

$$\int_{-\infty}^{\infty} \left(\frac{d^k}{dt^k} e^{-t^2} \right) P(t)dt$$

is equal to 0 if $\deg P < k$, and is equal to $(-1)^k k! c \sqrt{\pi}$ if $\deg P = k$.
(b) Show that

$$\frac{d^k}{dt^k}(e^{-t^2}) = P_k(t)e^{-t^2},$$

where P_k is a polynomial of degree k, and such that the coefficient of t^k in P_k is equal to

$$a_k = (-1)^k 2^k.$$

(c) Let m be an integer ≥ 0. Let H_m be the function defined by

$$H_m(t) = e^{t^2/2} \frac{d^m}{dt^m}(e^{-t^2}).$$

Show that

$$\int_{-\infty}^{\infty} H_m(t)^2 dt = (-1)^m m! a_m \sqrt{\pi},$$

and that if $m \neq n$, then

$$\int_{-\infty}^{\infty} H_m(t)H_n(t)dt = 0.$$

Solution. (a) Let I be the integral we wish to compute. When integrating by parts, the first term equals 0 because of (b) and because if R is a polynomial, then

$$\lim_{|t|\to\infty} R(t)e^{-t^2} = 0.$$

So if $r = \deg P < k$ we see that integrating by parts r times we get

$$I = (-1)^r \int_{-\infty}^{\infty} \left(\frac{d^{k-r}}{dt^{k-r}} e^{-t^2} \right) P^{(r)}(t)dt = (-1)^r r! c \left[\frac{d^{k-r-1}}{dt^{k-r-1}} e^{-t^2} \right]_{-\infty}^{\infty} = 0.$$

If $\deg P = k$, then

$$I = (-1)^k k! c \int_{-\infty}^{\infty} e^{-t^2} dt = (-1)^k k! c \sqrt{\pi}.$$

(b) We use induction. If $k = 0$, then $P_0 = 1$, so the result is true. Suppose

$$\frac{d^k}{dt^k}(e^{-t^2}) = \left[(-2)^k t^k + Q_k(t) \right] e^{-t^2}$$

with $\deg Q_k \leq k - 1$. Then differentiating the above expression we see that

$$\frac{d^{k+1}}{dt^{k+1}}(e^{-t^2}) = P_{k+1}(t)e^{-t^2},$$

where

$$P_{k+1}(t) = \left[k(-2)^k t^{k-1} + Q'_k(t) \right] + \left[(-2)^{k+1} t^{k+1} - 2t Q_k(t) \right].$$

This proves (b).
(c) Note that

$$H_m(t)^2 = e^{t^2/2} P_m(t) e^{-t^2} e^{t^2/2} \frac{d^m}{dt^m}(e^{-t^2}) = P_m(t) \frac{d^m}{dt^m}(e^{-t^2}),$$

so part (a) implies

$$\int_{-\infty}^{\infty} H_m(t)^2 dt = (-1)^m m! a_m \sqrt{\pi}.$$

For the second case, assume without loss of generality that $n < m$. Then

$$H_n(t) H_m(t) = P_n(t) \frac{d^m}{dt^m}(e^{-t^2}).$$

Part (a) implies

$$\int_{-\infty}^{\infty} H_n(t) H_m(t)dt = 0.$$

Exercise XIII.2.8 (a) *Let f be a real valued continuous function on the positive real numbers, and assume that f is monotone decreasing to 0. Show that the integrals*

$$\int_A^B f(t)\sin t\,dt, \quad \int_A^B f(t)\cos t\,dt, \quad \int_A^B f(t)e^{it}\,dt$$

are bounded uniformly for all numbers $B \geq A \geq 0$.
(b) Show that the improper integrals exist:

$$\int_0^\infty f(t)\sin t\,dt, \quad \int_0^\infty f(t)\cos t\,dt, \quad \int_0^\infty f(t)e^{it}\,dt.$$

The integrals of this exercise are called the **oscillatory integrals**.

Solution. (a) The function f is bounded, so it is sufficient to show that the integral of cosine, sine, and e^{it} are uniformly bounded. We have

$$\left|\int_A^B e^{it}\,dt\right| = \left|\frac{e^{iB} - e^{iA}}{i}\right| \leq 2.$$

Since $e^{it} = \cos t + i\sin t$ we have

$$\left|\int_A^B \cos t\,dt\right| \leq 2 \quad \text{and} \quad \left|\int_A^B \sin t\,dt\right| \leq 2.$$

(b) Theorem 2.6 implies the convergence of the three oscillatory integrals.

XIII.3   Interchanging Derivatives and Integrals

Exercise XIII.3.1 *Show that the integral*

$$g(x) = \int_0^\infty \frac{\sin t}{t} e^{-tx}\,dt$$

converges uniformly for $x \geq 0$ but does not converge absolutely for $x = 0$.

Solution. We isolate the key step.

Theorem 1 *Let f be a differentiable, non-negative decreasing function and let h be a continuous function. Suppose that there exists $M > 0$ such that*

$$\left|\int_A^B h(x)\,dx\right| \leq M$$

for all $A \leq B$. Then

$$\left|\int_A^B f(x)h(x)\,dx\right| \leq f(A)M.$$

The proof goes as follows. Let $H(x) = \int_A^x h(t)dt$. Integrating by parts we obtain

$$\int_A^B f(x)g(x)dx = [f(x)H(x)]_A^B - \int_A^B f'(x)H(x)dx.$$

Putting absolute values and using the triangle inequality we get

$$\left| \int_A^B f(x)h(x)dx \right| \leq f(B)M + M \int_A^B -f'(x)dx$$

$$= f(B)M + M(-f(B) + f(A)) = f(A)M.$$

We apply this result to $f(t) = e^{-tx}/t$ and $h(t) = \sin t$. Then $f(t) \geq 0$ and $f'(t) \leq 0$, and

$$\left| \int_A^B \sin t\, dt \right| \leq 2$$

by Exercise 8 of §2. So we can apply the theorem. Since

$$f(A) = \frac{e^{-Ax}}{A} \leq \frac{1}{A}$$

when $x \geq 0$, we get

$$\left| \int_A^B \frac{\sin t}{t} e^{-tx} dt \right| \leq \frac{1}{A}$$

and the convergence is uniform for $x \geq 0$, as was to be shown.

Now we prove that the integral

$$\int_0^\infty \left| \frac{\sin t}{t} \right| dt$$

diverges. On the interval $[(n-1)\pi, n\pi]$ we have

$$\left| \frac{\sin t}{t} \right| \geq \frac{1}{n\pi} |\sin t|,$$

and the change of variable $t - (n-1)\pi = u$ implies

$$\int_{(n-1)\pi}^{n\pi} |\sin t| dt = 2.$$

The series $\sum 1/n$ diverges, so the integral is not absolutely convergent.

Exercise XIII.3.2 *Let g be as in Exercise 1. (a) Show that you can differentiate under the integral sign with respect to x. Integrating by parts and justifying all the steps, show that for x > 0,*

$$g(x) = -\arctan x + \text{const}.$$

(b) Taking the limit as x → ∞, show that the above constant is π/2.
(c) Justifying taking the limit for x → 0, conclude that

$$\int_0^\infty \frac{\sin t}{t} \, dt = \frac{\pi}{2}.$$

Solution. (a) The argument in the text justifies differentiating under the integral sign

$$g'(x) = -\int_0^\infty e^{-tx} \sin t \, dt.$$

Now we can integrate by parts the integral from 0 to B,

$$-\int_0^B e^{-tx} \sin t \, dt = \left[e^{-tx} \cos t\right]_0^B + \int_0^B x e^{-tx} \cos t \, dt.$$

Now we integrate by parts the integral on the right

$$-\int_0^B e^{-tx} \sin t \, dt = \left[e^{-tx} \cos t\right]_0^B + \left[x e^{-tx} \sin t\right]_0^B + x^2 \int_0^B e^{-tx} \sin t \, dt.$$

Letting $B \to \infty$ we get

$$g'(x) = -1 - x^2 g'(x)$$

hence

$$g'(x) = \frac{-1}{1 + x^2}.$$

Integrating we see that for $x > 0$ we have

$$g(x) = -\arctan x + \text{const}.$$

(b) To find the limit of $g(x)$ as $x \to \infty$ we must estimate the integral. The mean value theorem implies $|\sin t| \leq |t|$ so

$$|g(x)| \leq \int_0^\infty e^{-tx} dt = \frac{1}{x},$$

and therefore $\lim_{x \to \infty} g(x) = 0$. The value of the constant is $\pi/2$ because $\lim_{x \to \infty} -\arctan x = -\pi/2$.
(c) The uniform convergence proved in Exercise 1 implies the continuity of g at 0, and since $\lim_{x \to 0} \arctan x = 0$ we conclude that

$$g(0) = \int_0^\infty \frac{\sin t}{t} \, dt = \frac{\pi}{2}.$$

Exercise XIII.3.3 *Show that for any number b > 0 we have*

$$\int_0^\infty \frac{\sin bt}{t} \, dt = \frac{\pi}{2}.$$

Solution. The change of variable $u = bt$ implies

$$\int_\delta^B \frac{\sin bt}{t} dt = \int_{b\delta}^{bB} b \frac{\sin u}{u} \frac{1}{b} du,$$

so taking the limits $\delta \to 0$ and $B \to \infty$ we get

$$\int_0^\infty \frac{\sin bt}{t} dt = \int_0^\infty \frac{\sin u}{u} du = \frac{\pi}{2}.$$

Exercise XIII.3.4 *Show that there exists a constant C such that*

$$\int_0^\infty e^{-t^2} \cos tx\, dt = Ce^{-x^2/4}.$$

[Hint: Let $f(x)$ be the integral. Show that $f'(x) = -xf(x)/2$. See the proof of Theorem 1.3 of the next chapter. Using the value

$$\int_0^\infty e^{-t^2} dt = \frac{\sqrt{\pi}}{2},$$

one sees that $C = \sqrt{\pi}/2$.]

Solution. Let $g(t, x) = e^{-t^2} \cos tx$. Then g and $D_2 g$ are both continuous and we have the estimates

$$|g(t, x)| \leq e^{-t^2} \quad \text{and} \quad |D_2 g(t, x)| \leq te^{-t^2}$$

so we can differentiate under the integral sign. We obtain

$$f'(x) = \int_0^\infty -te^{-t^2} \sin tx\, dt.$$

Integrating by parts we get

$$f'(x) = \left[\frac{e^{-t^2}}{2} \sin tx \right]_0^\infty - \frac{x}{2} \int_0^\infty e^{-t^2} \cos tx\, dt = -\frac{x}{2} f(x).$$

Conclude by arguing as in Exercise 1, §1, of Chapter IV.

Exercise XIII.3.5 *Determine the following functions in terms of elementary functions:*

(a) $f(x) = \int_{-\infty}^\infty e^{-t^2} \sin tx\, dt$.   (b) $f(x) = \int_{-\infty}^\infty e^{-t^2} e^{itx} dt$.

Solution. (a) Let $g(t, x) = e^{-t^2} \sin tx$. We have $f(x) = 0$ because $g(-t, x) = g(t, x)$, and putting absolute values we see that the integral converges.
(b) The function g defined in Exercise 4 is even in t, so

$$\int_{-\infty}^{\infty} e^{-t^2} \cos txdt = 2\int_0^{\infty} e^{-t^2} \cos txdt.$$

Exercise 4 implies that

$$\int_{-\infty}^{\infty} e^{-t^2} \cos txdt = 2Ce^{-x^2/4},$$

where $C = \sqrt{\pi}/2$. To conclude, remember that $e^{ix} = \sin x + i\cos x$.

Exercise XIII.3.6 *Determine whether the following integrals converge:*

(a) $\int_0^{\infty} \frac{1}{x\sqrt{1+x^2}}dx$. (b) $\int_0^1 \sin(1/x)dx$.

Solution. (a) We must check convergence at 0 and ∞. For $0 < x < 1$ we have $\sqrt{1+x^2} \leq 2$, so for $\delta > 0$ close to 0 we have

$$\int_{\delta}^1 \frac{dx}{x\sqrt{1+x^2}} \geq \int_{\delta}^1 \frac{dx}{2x} = \frac{-\log \delta}{2}.$$

Since $-\log\delta \to \infty$ as $\delta \to 0$, the integral does not converge. Notice however that

$$\int_1^{\infty} \frac{dx}{x\sqrt{1+x^2}}$$

converges because

$$\frac{1}{x\sqrt{1+x^2}} \leq \frac{1}{x^2}.$$

(b) We must check convergence near the origin. Let $\epsilon > 0$. If $0 < a < b < \epsilon$ then

$$\left| \int_a^b \sin(1/x)dx \right| \leq \int_a^b |\sin(1/x)|dx \leq b - a \leq \epsilon,$$

so the Cauchy criterion is verified, hence $\int_0^1 \sin(1/x)dx$ converges. One could also change variables $x = 1/u$ and use the fact that $\int_1^{\infty} (\sin u)/u^2 du$ converges.

Exercise XIII.3.7 *Show that $\int_0^{\infty} \sin(x^2)dx$ converges. [Hint: Use the substitution $x^2 = t$.]*

Solution. We change variables $x^2 = t$. Then $dt/(2\sqrt{t}) = dx$ so

$$\int_0^B \sin(x^2)dx = \int_0^{B^2} \frac{\sin t}{2\sqrt{t}}dt.$$

But we know from Exercise 8, §2, that $\left| \int_a^b \sin tdt \right| \leq 2$. Since $t \mapsto 1/\sqrt{t}$ is monotone decreasing to 0, we conclude that $\int_0^{\infty} \sin(x^2)dx$ converges.

Exercise XIII.3.8 *Evaluate the integrals*

$$\int_1^\infty \int_1^\infty \frac{t-x}{(x+t)^3} \, dt \, dx \quad and \quad \int_1^\infty \int_1^\infty \frac{t-x}{(x+t)^3} \, dx \, dt$$

to see that they are not equal. Some sort of assumption has to be made to make the interchange of Theorem 3.5 possible.

Solution. Direct computations show that the integrals converge. For the first integral, change variables $u = x + t$,

$$\int_1^B \frac{t-x}{(x+t)^3} \, dt = \int_{1+x}^{B+x} \frac{u-2x}{u^3} \, du = \left[\frac{-1}{u}\right]_{1+x}^{B+x} + x \left[\frac{1}{u^2}\right]_{1+x}^{B+x}.$$

Taking the limit as $B \to \infty$ we obtain

$$\int_1^\infty \frac{t-x}{(x+t)^3} \, dt = \frac{1}{1+x} - \frac{x}{(1+x)^2} = \frac{1}{(1+x)^2}.$$

So

$$\int_1^\infty \int_1^\infty \frac{t-x}{(x+t)^3} \, dt \, dx = \int_1^\infty \frac{dx}{(1+x)^2} = \frac{1}{2}.$$

For the second integral, a similar argument shows that

$$\int_1^\infty \frac{t-x}{(x+t)^3} \, dx = \frac{-1}{(1+t)^2} \quad and \quad \int_1^\infty \int_1^\infty \frac{t-x}{(x+t)^3} \, dx \, dt = \frac{-1}{2}.$$

Exercise XIII.3.9 *For $x \geq 0$ let*

$$g(x) = \int_0^\infty \frac{\log(u^2 x^2 + 1)}{u^2 + 1} \, du$$

so that $g(0) = 0$. Show that g is continuous for $x \geq 0$. Show that g is differentiable for $x > 0$. Differentiate under the integral sign and use a partial fraction decomposition to show that

$$g'(x) = \frac{\pi}{1+x} \quad for \ x > 0,$$

and thus prove that $g(x) = \pi \log(1+x)$. (This proof is due to Seeley.)

Solution. Let

$$f(u, x) = \frac{\log(u^2 x^2 + 1)}{u^2 + 1}.$$

Suppose $x \in [0, c]$. There exists a number $M > 0$ such that for all $u > M$ we have

$$\frac{\log(u^2 x^2 + 1)}{(u^2 + 1)^{1/4}} \leq 1 \quad and \quad 1 \leq u^2,$$

for all $x \in [0, c]$. Then for all $B > M$ we have

$$\int_M^B \frac{\log(u^2 x^2 + 1)}{u^2 + 1} du \le \int_M^B \frac{(2u^2)^{1/4}}{u^2 + 1} \le K \int_M^B \frac{du}{u^{3/2}},$$

for some fixed constant K. The convergence of $\int_1^\infty 1/u^{3/2} du$ implies the uniform convergence of $\int_0^\infty f(u, x) du$. So g is continuous on $[0, c]$ and since c was arbitrary we conclude that g is continuous for $x \ge 0$.

Now we prove differentiability. Let $0 < a < b$ and suppose $x \in [a, b]$. Then we have

$$|D_2 f(u, x)| = \frac{2u^2 x}{(u^2 + 1)(u^2 x^2 + 1)} \le \frac{2bu^2}{u^2 a^2 u^2} \le \frac{K'}{u^2}$$

so the integral $\int_0^\infty D_2 f(u, x) du$ converges uniformly for $x \in [a, b]$. Since a and b are arbitrary, we conclude that g is differentiable for $x > 0$ and

$$g'(x) = \int_0^\infty D_2 f(u, x) du.$$

Suppose $x \ne 1$. Since

$$\frac{u^2}{(u^2 + 1)(u^2 x^2 + 1)} = \frac{1}{(x^2 - 1)} \left[\frac{1}{u^2 + 1} - \frac{1}{u^2 x^2 + 1} \right],$$

we have

$$g'(x) = \frac{2x}{(x^2 - 1)} \left[\frac{\pi}{2} - \frac{\pi}{2x} \right] = \frac{\pi}{x + 1}.$$

If $x = 1$, then we see that

$$g'(1) = 2 \int_0^\infty \frac{u^2}{(u^2 + 1)^2} du,$$

and to evaluate this integral, consider the change of variable $u = \tan t$ so that

$$g'(1) = 2 \int_0^{\pi/2} \sin^2 t \, dt = \frac{\pi}{2}.$$

Thus for all $x > 0$ we have

$$g'(x) = \frac{\pi}{x + 1}.$$

Integrating, we obtain $g(x) = \pi \log(1 + x) + \text{constant}$. The continuity of g at 0 implies that the constant is 0.

Exercise XIII.3.10 *(a) For $y > 0$ let*

$$\varphi_y(x) = \frac{1}{\pi} \frac{y}{x^2 + y^2}.$$

Prove that $\{\varphi_y\}$ is a Dirac family for $y \to 0$.

(b) Let f be continuous on $\mathbf{R}$ and bounded. Prove that $(\varphi_y * f)(x)$ converges to $f(x)$ as $y \to 0$.

(c) Show that $\varphi(x, y) = \varphi_y(x)$ is harmonic. Probably using the Laplace oeprator in polar coordinates makes the computation easier.

Solution. (a) Clearly, φ_y is continuous and positive. Changing variables $x = uy$ we get

$$\frac{1}{\pi} \int_{-\infty}^{\infty} \frac{y}{x^2 + y^2} dx = \frac{1}{\pi} \int_{-\infty}^{\infty} \frac{1}{u^2 + 1} du = 1.$$

Finally, if $\delta > 0$, then

$$\int_{-\infty}^{-\delta} + \int_{\delta}^{\infty} \frac{y}{x^2 + y^2} dx = 2 \int_{\delta}^{\infty} \frac{y}{x^2 + y^2}$$

$$= 2 \int_{\delta/y}^{\infty} \frac{1}{u^2 + 1} du$$

$$= 2 \left(\frac{\pi}{2} - \arctan \frac{\delta}{y} \right),$$

but $\arctan t \to \pi/2$ as $t \to \infty$, so this completes the proof that $\{\varphi_y\}$ is a Dirac family for $y \to 0$.

(b) We give the standard proof. Let $\epsilon > 0$ and fix x. Then we have

$$(\varphi_y * f)(x) - f(x) = \int_{-\infty}^{\infty} [f(x - t) - f(x)]\varphi_y(t) dt.$$

Let B be a bound for f and choose δ such that $|f(x - t) - f(x)| < \epsilon$ whenever $|t| < \delta$. Then we estimate

$$|(\varphi_y * f)(x) - f(x)| \leq \int_{-\infty}^{-\delta} + \int_{-\delta}^{\delta} + \int_{\delta}^{\infty} |f(x - t) - f(x)|\varphi_y(t) dt.$$

$$\text{(XIII.1)}$$

Select y_0 such that if $0 < y < y_0$, then $\int_{-\infty}^{-\delta} + \int_{\delta}^{\infty} \varphi_y < \epsilon/(2B)$. Then we see that in (XIII.1) the sum of the first and second integral is $< \epsilon$, and the middle integral is also $< \epsilon$. This proves that $(\varphi_y * f)(x) \to f(x)$ as $y \to 0$.

(c) The Laplace operator in polar coordinates is

$$\Delta = \frac{\partial^2}{\partial x^2} + \frac{\partial^2}{\partial y^2} = \frac{\partial^2}{\partial r^2} + \frac{1}{r} \frac{\partial}{\partial r} + \frac{1}{r^2} \frac{\partial^2}{\partial \theta^2}.$$

Letting $x = r \cos \theta$ and $y = r \sin \theta$ we find that $\varphi = (\sin \theta)/r$. Hence

$$\Delta \varphi = \frac{2 \sin \theta}{r^3} - \frac{\sin \theta}{r^3} - \frac{\sin \theta}{r^3} = 0,$$

which proves that φ is harmonic.

Exercise XIII.3.11 *For each real number t let [t] be the largest integer ≤ t. Let*

$$P_1(t) = t - [t] - \frac{1}{2}.$$

*(a) Sketch the graph of $P_1(t)$, which is called the **sawtooth function** for the obvious reason.*

(b) Show that the integral

$$\int_0^\infty \frac{P_1(t)}{1+t} dt$$

converges.

(c) Let δ > 0. Show that the integral

$$f(x) = \int_0^\infty \frac{P_1(t)}{x+t} dt$$

converges uniformly for $x \geq \delta$.

(d) Let

$$P_2(t) = \frac{1}{2}(t^2 - t) \quad \text{for } 0 \leq t \leq 1,$$

*and extend $P_2(t)$ by periodicity to all of **R** (period 1). Then $P_2(n) = 0$ for all integers n and P_2 is bounded. Furthermore, $P_2'(t) = P_1(t)$. Show that for x > 0,*

$$\int_0^\infty \frac{P_1(t)}{x+t} dt = \int_0^\infty \frac{P_2(t)}{(x+t)^2} dt.$$

(e) Show that if $f(x)$ denotes the integral in part (d), then $f'(x)$ can be found by differentiating under the integral sign on the right-hand side, for x > 0.

Solution. (a) On $[0,1)$, $P_1(t) = t - \frac{1}{2}$ and since P_1 is periodic of period 1 the graph of P_1 follows at once:

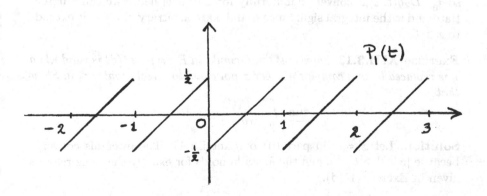

(b) It is clear from the cancellations that the integral $\int_a^b P_1(t)dt$ is bounded for all $0 \le a \le b$. Since $t \mapsto 1/(1+t)$ is monotone decreasing to 0 we conclude that the integral

$$\int_0^\infty \frac{P_1(t)}{1+t}\,dt$$

converges.

(c) The uniform convergence follows from inequality

$$\frac{1}{x+t} \le \frac{1}{\delta+t}$$

which holds for all $x \ge \delta$.

(d) The second integral converges because P_2 is bounded and $\int_0^\infty 1/(x+t^2)dt$ converges. We have

$$\int_0^n \frac{P_1(t)}{x+t} - \frac{P_2(t)}{(x+t)^2}\,dt = \sum_{i=0}^{n-1} \int_i^{i+1} \frac{P_1(t)(x+t) - P_2(t)}{(x+t)^2}\,dt,$$

and since $P_2'(t) = P_1(t)$ we see that

$$\int_i^{i+1} \frac{P_1(t)(x+t) - P_2(t)}{(x+t)^2}\,dt = \int_i^{i+1} \frac{P_2'(t)(x+t) - P_2(t)}{(x+t)^2}\,dt = \left[\frac{P_2(t)}{x+t}\right]_i^{i+1} = 0$$

because $P_2(j) = 0$ for all j. Let $n \to \infty$ and conclude.

(e) Let $g(t,x) = P_2(t)/(x+t)^2$. Then

$$|D_2 g(t,x)| = \left|\frac{P_1(t)(x+t)^2 - 2P_2(t)(x+t)}{(x+t)^4}\right| \le \frac{|P_1(t)|}{(x+t)^2} + 2\frac{|P_2(t)|}{(x+t)^3}.$$

If $x \in [a,b]$ with $0 < a < b$ we have the estimates

$$\frac{|P_1(t)|}{(x+t)^2} \le \frac{|P_1(t)|}{(a+t)^2} \quad \text{and} \quad \frac{|P_2(t)|}{(x+t)^3} \le \frac{|P_2(t)|}{(a+t)^3},$$

so $\int_0^\infty D_2 g(t,x)dt$ converges uniformly for $x \in [a,b]$ hence we can differentiate under the integral sign. Since a and b are arbitrary, the result extends to $x > 0$.

Exercise XIII.3.12 *Show that the formula in Exercise 11(d) is valid when x is replaced by any complex number z not equal to a real number ≤ 0. Show that*

$$\lim_{y \to \infty} \int_0^\infty \frac{P_1(t)}{iy+t}\,dt = 0.$$

Solution. Let $x = z$ in part (d) of Exercise 11. The integrals converge because $|z+t| \ge t - |z|$ and the formula holds for exactly the same reasons given in Exercise 11(d).

Now if $z = iy$ and B is a bound for P_2 we have

$$\left| \int_0^M \frac{P_1(t)}{iy + t} dt \right| = \left| \int_0^M \frac{P_2(t)}{(iy + t)^2} dt \right| \leq B \int_0^M \frac{dt}{(t - |y|)^2} = B \left[\frac{-1}{t - |y|} \right]_0^M.$$

Taking the limit as $M \to \infty$ and then as $y \to \infty$ yields the desired result.

Exercise XIII.3.13 (The Gamma Function) *Define*

$$f(x) = \int_0^\infty t^{x-1} e^{-t} dt = \int_0^\infty e^{-t} t^x \frac{dt}{t}$$

for $x > 0$.
(a) Show that f is continuous.
(b) Integrate by parts to show that $f(x + 1) = x f(x)$. Show that $f(1) = 1$, and hence that $f(n + 1) = n!$ for $n = 0, 1, 2, \ldots$.
(c) Show that for any $a > 0$ we have

$$\int_0^\infty e^{-at} t^{x-1} dt = \frac{f(x)}{a^x}.$$

(d) Sketch the graph of f for $x > 0$, showing that f has one minimum point, and tends to infinity as $x \to \infty$, and as $x \to 0$.
(e) Evaluate $f(\frac{1}{2}) = \sqrt{\pi}$. [Hint: Substitute $t = u^2$ and you are allowed to use the value of the integral in the hint of Exercise 4.]
(f) Evaluate $f(3/2), f(5/2), \ldots, f(n + \frac{1}{2})$.
(g) Show that

$$\sqrt{\pi} f(2n) = 2^{2n-1} f(n) f\left(n + \frac{1}{2} \right).$$

(h) Show that f is infinitely differentiable, and that

$$f^{(n)}(x) = \int_0^\infty (\log t)^n t^{x-1} e^{-t} dt.$$

*For any complex number s with $\operatorname{Re}(s) > 0$ one defines the **gamma function***

$$\Gamma(s) = \int_0^\infty e^{-t} t^s \frac{dt}{t}.$$

Show that the gamma function is continuous as a function of s. If you know about complex differentiability, your proof that it is differentiable should also apply for the complex variable s.

Solution. (a) Let $0 < a < b$ and suppose that $x \in [a, b]$. Split the integral

$$\int_0^\infty t^{x-1} e^{-t} dt = \int_0^1 t^{x-1} e^{-t} dt + \int_1^\infty t^{x-1} e^{-t} dt.$$

For $0 < t \leq 1$ we have $t^{x-1}e^{-t} \leq t^{a-1}e^{-t}$ and since $a > 0$ the first integral on the right converges uniformly for $x \in [a, b]$. If $t \geq 1$ we have $t^{x-1}e^{-t} \leq t^{b-1}e^{-t}$ and since $t^{b-1}e^{-t/2} \to 0$ as $t \to \infty$ we see that the second integral converges uniformly for $x \in [a, b]$. Since the continuity condition on $(t, x) \mapsto t^{x-1}e^{-t}$ is verified, we conclude that f is continuous.

(b) Integrating by parts, we get

$$\int_0^B t^{x-1}e^{-t}dt = \left[\frac{t^x}{x}e^{-t}\right]_0^B + \frac{1}{x}\int_0^B t^x e^{-t}dt,$$

letting $B \to \infty$ we see that $xf(x) = f(x+1)$. Moreover, we have

$$\int_0^B e^{-t}dt = \left[-e^{-t}\right]_0^B,$$

so letting $B \to \infty$ we see that $f(1) = 1$. Argue by induction to prove that $f(n+1) = n!$.

(c) The change of variable $u = at$ implies

$$\int_\delta^1 + \int_1^B e^{-at}t^{x-1}dt = \int_{a\delta}^a + \int_a^{aB} e^{-u}u^{x-1}a^{1-x}a^{-1}du.$$

Let $\delta \to 0$ and $B \to \infty$ to conclude.

(d) First we investigate the limit of f as $x \to 0$. We write

$$f(x) = \int_0^1 t^{x-1}e^{-t}dt + \int_1^\infty t^{x-1}e^{-t}dt.$$

The second integral is positive and the first tends to ∞ as x tends to 0. Indeed, when $0 < t \leq 1$ we have $e^{-t} \geq e^{-1}$ and therefore

$$\int_0^1 t^{x-1}e^{-t}dt \geq e^{-1}\int_0^1 t^{x-1}dt = \frac{e^{-1}}{x},$$

so the desired limit drops out.

By (h) we know that $f^{(2)} \geq 0$ so f' is increasing. We contend that for x close to 0, f' is negative. Suppose $x < 1/2$ and write

$$f'(x) = \int_0^1 + \int_1^\infty (\log t)t^{x-1}e^{-t}dt.$$

The second integral is uniformly bounded, and the first can be estimated as follows:

$$\int_0^1 (\log t)t^{x-1}e^{-t}dt \leq K\int_0^1 (\log t)t^{x-1}dt$$

where K is a positive constant. Integrating by parts we see that

$$\int_0^1 (\log t) t^{x-1} dt = -\int_0^1 \frac{t^{x-1}}{x} dt.$$

This last integral tends to $-\infty$ as $x \to 0$, thus proving our contention. If $x > 2$ we see at once that f is increasing, so f has one minimum point.

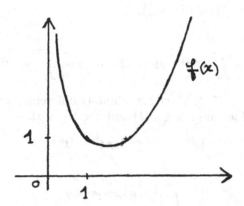

(e) Change variables $t = u^2$ and use finite integrals as in (c) to obtain

$$f\left(\frac{1}{2}\right) = \int_0^\infty e^{-t} t^{\frac{1}{2}-1} dt = 2\int_0^\infty e^{-u^2} u^{-1} u \, du = 2\frac{\sqrt{\pi}}{2} = \sqrt{\pi}.$$

(f) We simply use (b). We find that

$$f\left(\frac{3}{2}\right) = \frac{1}{2} f\left(\frac{1}{2}\right) = \frac{\sqrt{\pi}}{2},$$

and

$$f\left(\frac{5}{2}\right) = \frac{3}{2} f\left(\frac{3}{2}\right) = 3\frac{\sqrt{\pi}}{2^2},$$

and

$$f\left(\frac{7}{2}\right) = \frac{5}{2} f\left(\frac{5}{2}\right) = 5 \times 3\frac{\sqrt{\pi}}{2^3}.$$

By induction we prove that

$$f\left(n+\frac{1}{2}\right) = \frac{(2n-1)(2n-3)\cdots 5 \times 3}{2^n} \sqrt{\pi}.$$

Indeed, by (b)

$$f\left(n+1+\frac{1}{2}\right) = \frac{2n+1}{2} f\left(n+\frac{1}{2}\right) = \frac{(2n+1)(2n-1)(2n-3)\cdots 5 \times 3}{2^{n+1}} \sqrt{\pi}.$$

(g) By (b) we know that $\sqrt{\pi}f(2n) = \sqrt{\pi}(2n-1)!$. By (e) we see that

$$2^{2n-1}f(n)f\left(n+\frac{1}{2}\right) = 2^{2n-1}(n-1)!\frac{(2n-1)(2n-3)\cdots 5 \times 3}{2^n}\sqrt{\pi}$$

$$= (2n-1)!\sqrt{\pi},$$

so $\sqrt{\pi}f(2n) = 2^{2n-1}f(n)f\left(n+\frac{1}{2}\right)$.

(h) Let

$$f_1 : x \mapsto \int_0^1 e^{-t}t^{x-1}dt \quad \text{and} \quad f_2 : x \mapsto \int_1^\infty e^{-t}t^{x-1}dt,$$

and let $g(t,x) = e^{-t}t^{x-1}$. Then g is infinitely differentiable with respect to the second variable, and if $x \in [a,b]$ and $0 < t \leq 1$, then

$$|g^{(n)}(t,x)| \leq |\log t|^n e^{-t}t^{a-1}$$

and since the integral

$$\int_0^1 |\log t|^n e^{-t}t^{a-1}dt$$

converges (to see this, write $|\log t|^n t^{a-1} = |\log t|^n t^{\frac{a}{2}}t^{\frac{a}{2}-1}$), an easy induction shows that f_1 is infinitely differentiable on $[a,b]$ and

$$f_1^{(n)}(x) = \int_0^1 |\log t|^n e^{-t}t^{x-1}dt.$$

For f_2 we see that if $1 \leq t$, then

$$|g^{(n)}(t,x)| \leq |\log t|^n e^{-t}t^{b-1},$$

and just as for f_1 we conclude that f_2 is infinitely differentiable on $[a,b]$ and that

$$f_2^{(n)}(x) = \int_1^\infty |\log t|^n e^{-t}t^{x-1}dt.$$

In the complex case, we see that if $s = x + iy$, then

$$|e^{-t}t^{s-1}| = e^{-t}t^{x-1}$$

so the above argument can be adapted to the complex gamma function. Note that in the complex case, one works with compact rectangles in the complex plane instead of compact intervals on the real line.

Exercise XIII.3.14 *Show that*

$$\int_{-\infty}^\infty \frac{1}{(u^2+1)^s}du = \sqrt{\pi}\frac{\Gamma(s-\frac{1}{2})}{\Gamma(s)} \quad \text{for } \mathrm{Re}(s) > \frac{1}{2}.$$

[Hint: Multiply the desired integral by $\Gamma(s)$ and let $t \mapsto (u^2+1)t.$]

Solution. We have

$$\Gamma(s) \int_{-\infty}^{\infty} \frac{1}{(u^2+1)^s} du = 2 \int_0^{\infty} \int_0^{\infty} \frac{e^{-t} t^{s-1}}{(u^2+1)^s} dt du.$$

If $t = (u^2+1)q$, then $dt = (u^2+1)dq$ so that

$$\Gamma(s) \int_{-\infty}^{\infty} \frac{1}{(u^2+1)^s} du = 2 \int_0^{\infty} \int_0^{\infty} e^{-u^2 q} e^{-q} q^{s-1} dq du.$$

The change of variable $\alpha = u\sqrt{q}$ implies

$$\int_0^{\infty} e^{-u^2 q} du = \frac{1}{\sqrt{q}} \int_0^{\infty} e^{-\alpha^2} d\alpha = \frac{\sqrt{\pi}}{2} q^{-1/2}.$$

For $\mathrm{Re}(s) > 1/2$ the hypotheses of Theorem 3.5 are verified so that

$$\Gamma(s) \int_{-\infty}^{\infty} \frac{1}{(u^2+1)^s} du = 2 \int_0^{\infty} \int_0^{\infty} \frac{e^{-u^2 q} e^{-q} q^{s-1}}{(u^2+1)^s} du dq$$

$$= \sqrt{\pi} \int_0^{\infty} e^{-q} q^{s-1-1/2} \frac{dq}{q} = \sqrt{\pi} \Gamma\left(s - \frac{1}{2}\right).$$

Exercise XIII.3.15 (A Bessel Function) *Let a, b be real numbers > 0. For any complex number s define*

K 1.
$$K_s(a,b) = \int_0^{\infty} e^{-(a^2 t + b^2/t)} t^s \frac{dt}{t}.$$

Show that the integral converges absolutely. For $c > 0$ define

K 2.
$$K_s(c) = \int_0^{\infty} e^{-c(t+1/t)} t^s \frac{dt}{t}.$$

Show that

K 3.
$$K_s(a,b) = \left(\frac{b}{a}\right)^s K_s(ab).$$

Show that

K 4.
$$K_s(c) = K_{-s}(c).$$

K 5.
$$K_{1/2}(c) = \sqrt{\frac{\pi}{c}} e^{-2c}.$$

[Hint: Let

$$g(x) = K_{1/2}(x).$$

Change variables, let $t \mapsto t/x$. Let $h(x) = \sqrt{x}g(x)$. Differentiate under the integral sign and twiddle the integral to find that

$$h'(x) = -2h(x),$$

whence $h(x) = Ce^{-2x}$ for some constant C. Let $x = 0$ in the integral for $h(x)$ to evaluate C, which comes out as $\Gamma(\frac{1}{2}) = \sqrt{\pi}$.]

Solution. If $s = x + iy$, then we want to study the integral

$$\int_0^1 e^{-(a^2t+b^2/t)}t^{x-1}dt \quad \text{and} \quad \int_1^\infty e^{-(a^2t+b^2/t)}t^{x-1}dt$$

separately. The first integral converges because near 0, e^{-a^2t} is bounded and $\int_0^1 e^{-b^2/t}t^{x-1}dt$ converges because for all small t we have $e^{-b^2/t}t^{x-1} \le e^{-b^2/2t}$.

The second integral also converges because for all large t, $e^{-b^2/t}$ is bounded and $\int_0^1 e^{-a^2t}t^{x-1}dt$ converges.

K 3. In the integral $K_s(a,b)$ put $t = (bu)/a$. Then

$$K_s(a,b) = \int_0^\infty e^{-(abu+ab/u)}u^s\left(\frac{b}{a}\right)^s\frac{du}{u} = \left(\frac{b}{a}\right)^s K_s(ab).$$

K 4. We change variables $u = 1/t$, then

$$K_{-s}(c) = \int_\infty^0 e^{-c(1/u+u)}u^s u(-u^{-2})du = \int_0^\infty e^{-c(u+1/u)}u^s\frac{du}{u} = K_s(c).$$

K 5. Changing variables $t = u/x$ we get

$$g(x) = \int_0^\infty e^{-u-x^2/u}u^{-1/2}x^{-1/2}du$$

so

$$h(x) = \int_0^\infty e^{-u-x^2/u}u^{-1/2}du.$$

Since

$$\frac{\partial}{\partial x}(e^{-u-x^2/u}u^{-1/2}) = \frac{-2x}{u^{3/2}}e^{-u-x^2/u},$$

we see that the integral

$$\int_0^\infty D_2(e^{-u-x^2/u}u^{-1/2})du$$

converges uniformly for $x \in [a,b]$ with $0 < a < b$, so we can differentiate under the integral sign and we get

$$h'(x) = \int_0^\infty \frac{-2x}{u^{3/2}}e^{-u-x^2/u}du.$$

The change of variables $q = x^2/u$ gives

$$h'(x) = \int_\infty^0 \frac{-2x}{(x^2/q)^{3/2}} e^{-q-x^2/q}(-x^2/q^2)dq = -2h(x),$$

so $h(x) = Ce^{-2x}$ for some constant C. Moreover, if we let $u = \alpha^2$ and if we use continuity of h at 0 we get

$$C = \int_0^\infty e^{-u}u^{-1/2}du = 2\int_0^\infty e^{-\alpha^2}\alpha^{-1}\alpha d\alpha = \sqrt{\pi},$$

so

$$K_{1/2}(c) = \sqrt{\frac{\pi}{c}}e^{-2c}.$$

XIV
The Fourier Integral

XIV.1 The Schwartz Space

Exercise XIV.1.1 *Let $g \in S$ and define $g_a(x) = g(ax)$ for $a > 0$. Show that*

$$\hat{g}_a(y) = \frac{1}{a} \hat{g}\left(\frac{y}{a}\right).$$

In particular, if $g(x) = e^{-x^2}$, find $\hat{g}_a(x)$.

Solution. Change variables $u = ax$ so that

$$\int_{-B}^{B} g(ax)e^{-ixy}d_1x = \frac{1}{a}\int_{-aB}^{aB} g(u)e^{-iuy/a}d_1u.$$

Let $B \to \infty$ and conclude.

Let $f(x) = e^{-x^2/2}$. Then $g(x) = f(\sqrt{2}x)$, so we have

$$\hat{g}(y) = \frac{1}{\sqrt{2}}f\left(\frac{y}{\sqrt{2}}\right),$$

and therefore

$$\hat{g}_a(x) = \frac{1}{a}\hat{g}\left(\frac{x}{a}\right) = \frac{1}{a\sqrt{2}}f\left(\frac{x}{a\sqrt{2}}\right) = \frac{1}{a\sqrt{2}}e^{-x^2/(4a^2)}.$$

Exercise XIV.1.2 *Normalize the Fourier series differently, for the interval $[0, 1]$. That is, define the scalar product for two functions f, g periodic of period 1 to be*

$$\int_0^1 f(t)\overline{g(t)}dt.$$

The total orthogonal family that corresponds to the one studied in Chapter 12 is then the family of functions

$$\{e^{2\pi inx}\}, \quad n \in \mathbf{Z}.$$

These are already unit vectors, that is these functions form an orthonormal family, which is often convenient because one does not have to divide by 2π. The theorems of Chapter XII go over this situation, of course. In particular, if we deal with a very smooth fucntion g, its Fourier series is uniformly convergent to the function. That's the application we are going to consider now.

Let f be in the Schwartz space. Define a different normalization of the Fourier transform for the present purposes, namely define the **Poisson dual**

$$f^\vee(x) = \int f(t)e^{-2\pi itx}dt.$$

Prove the **Poisson summation formula**:

$$\sum_{n \in \mathbf{Z}} f(n) = \sum_{n \in \mathbf{Z}} f^\vee(n).$$

[Hint: Let

$$g(x) = \sum_{n \in \mathbf{Z}} f(x+n).$$

Then g is periodic of period 1 and infinitely differentiable. Let

$$c_m = \int_0^1 g(x)e^{-2\pi imx}dx = \int_0^1 \sum_{n \in \mathbf{Z}} f(x+n)e^{-2\pi imx}dx.$$

Then

$$\sum_{m \in \mathbf{Z}} c_m = g(0) = \sum_{n \in \mathbf{Z}} f(n).$$

On the other hand, using the integral for c_m, insert the factor $1 = e^{-2\pi imn}$, change variables, and show that $c_m = f^\vee(m)$. The formula drops out.]

Solution. The Fourier series of g at 0 converges to $g(0)$, so we have

$$\sum_{m \in \mathbf{Z}} c_m = g(0) = \sum_{n \in \mathbf{Z}} f(n).$$

But

$$c_m = \int_0^1 \sum_{n \in \mathbf{Z}} f(x+n)e^{-2\pi imx}dx = \int_0^1 \sum_{n \in \mathbf{Z}} f(x+n)e^{-2\pi im(x+n)}dx$$

$$= \sum_{n \in \mathbf{Z}} \int_0^1 f(x+n)e^{-2\pi im(x+n)}dx = \sum_{n \in \mathbf{Z}} \int_n^{n+1} f(u)e^{-2\pi imu}du$$

$$= \int f(u)e^{-2\pi imu}du = f^\vee(m),$$

whence the Poisson summation formula

$$\sum_{n \in \mathbf{Z}} f(n) = \sum_{n \in \mathbf{Z}} f^\vee(n).$$

Exercise XIV.1.3   Functional Equation of the Theta Function
Let θ be the function defined for $x > 0$ by

$$\theta(x) = \sum_{-\infty}^{\infty} e^{-n^2 \pi x}.$$

Prove the functional equation, namely

$$\theta(x^{-1}) = x^{1/2}\theta(x).$$

Solution. Fix x and let $f(y) = e^{-(\pi x)y^2}$. Then using the notation of Exercise 2 and an argument similar to the proof of Theorem 1.3, one finds that

$$D\hat{f}(y) = \left(\frac{-2\pi}{x}y\right)\hat{f}(y),$$

so $\hat{f} = Ce^{-\pi y^2/x}$. We see that $\hat{f}(0) = C$, so changing variables $t\sqrt{\pi x} = u$ we have

$$C = \int_{-\infty}^{\infty} e^{-(\pi x)t^2}dt = \frac{2}{\sqrt{\pi x}}\int_0^\infty e^{-u^2}du = \frac{1}{\sqrt{x}}.$$

The Poisson summation formula applied to f implies the functional equation of the theta function.

Exercise XIV.1.4   (Functional Equation of the Zeta Function (Riemann)) *Let s be a complex number, $s = \sigma + it$ with σ, t real. If $\sigma > 1$, and $a > 1$, show that the series*

$$\zeta(s) = \sum_{n=1}^{\infty} \frac{1}{n^s}$$

converges absolutely, and uniformly in every region $\sigma \geq a > 1$. Let F be the function of s defined for $\sigma > 1$ by

$$F(s) = \pi^{-s/2}\Gamma\left(\frac{s}{2}\right)\zeta(s).$$

Let $g(x) = \sum_{n=1}^{\infty} e^{-n^2 \pi x}$, so that $2g(x) = \theta(x) - 1$. Show that

$$
\begin{aligned}
F(s) &= \int_0^\infty x^{s/2} g(x) \frac{dx}{x} \\
&= \int_1^\infty x^{s/2} g(x) \frac{dx}{x} + \int_1^\infty x^{-s/2} g\left(\frac{1}{x}\right) \frac{dx}{x}.
\end{aligned}
$$

Use the functional equation of the theta function to show that

$$
F(s) = \frac{1}{s-1} - \frac{1}{s} + \int_1^\infty (x^{s/2} + x^{(1-s)/2}) g(x) \frac{dx}{x}.
$$

Show that the integral on the right converges absolutely for all complex s, and uniformly for s in a bounded region of the complex plane. The expression on the right then defines F for all values of $s \neq 0, 1$, and we see that

$$
F(s) = F(1-s).
$$

Solution. We have

$$
\left| \frac{1}{n^s} \right| = \frac{1}{n^\sigma} \le \frac{1}{n^a},
$$

so the series $\zeta(s)$ converges absolutely and uniformly on every region $\sigma \ge a > 1$.
We have

$$
F(s) = \pi^{-s/2} \sum_{n=1}^{\infty} \int_0^\infty \frac{e^{-t} t^{s/2}}{n^s} \frac{dt}{t},
$$

and after the change of variable $t = \pi n^2 x$ we obtain

$$
F(s) = \pi^{-s/2} \sum_{n=1}^{\infty} \int_0^\infty e^{-\pi n^2 x} \pi^{s/2} x^{s/2} \frac{dx}{x} = \int_0^\infty x^{s/2} g(x) \frac{dx}{x}.
$$

We can split the integral from 0 to 1 and from 1 to ∞. In the integral from 0 to 1 we change variables $x = 1/y$ so

$$
\int_0^1 x^{s/2} g(x) \frac{dx}{x} = -\int_\infty^1 y^{-s/2} g\left(\frac{1}{y}\right) y \frac{dy}{y^2} = \int_1^\infty y^{-s/2} g\left(\frac{1}{y}\right) \frac{dy}{y},
$$

and the second formula for F drops out.
From the functional equation of the theta function we get

$$
\begin{aligned}
2g(x^{-1}) &= \theta(x^{-1}) - 1 = x^{1/2}\theta(x) - 1 = x^{1/2}[2g(x) + 1] - 1 \\
&= 2x^{1/2} g(x) + x^{1/2} - 1.
\end{aligned}
$$

From the second formula for F we get

$$F(s) = \int_1^\infty x^{s/2} g(x) + x^{-s/2} \left[x^{1/2} g(x) + \frac{x^{1/2}}{2} - \frac{1}{2} \right] \frac{dx}{x}$$

$$= \int_1^\infty g(x)[x^{s/2} + x^{(1-s)/2}] \frac{dx}{x} + \frac{1}{2} \left[\frac{2}{1-s} x^{(1-s)/2} \right]_1^\infty - \frac{1}{2} \left[\frac{-2}{s} x^{-s/2} \right]_1^\infty ,$$

But $\mathrm{Re}(s) > 1$ so

$$F(s) = \int_1^\infty g(x)[x^{s/2} + x^{(1-s)/2}] \frac{dx}{x} + \frac{1}{s-1} - \frac{1}{s}.$$

We now show that the integral on the right converges uniformly on every bounded region of the complex plane. Suppose there exists a number B such that $|\sigma| \le B$ for all s. We have $|x^{s/2}| = x^{\sigma/2}$ and $|x^{(1-s)/2}| = x^{(1-\sigma)/2}$ and $e^{-n^2 \pi x} \le (e^{-\pi x})^n$ so

$$g(x) \le \sum_{n=1}^\infty (e^{\pi x})^n = \frac{e^{\pi x}}{1 - e^{\pi x}}.$$

For all $x \ge A$ we have $1 - e^{-\pi x} \ge 1/2$ so combined with the triangle inequality we have

$$\int_A^M \left| g(x)[x^{s/2} + x^{(1-s)/2}] \right| \frac{dx}{x} \le 2 \int_A^M (x^{1+B/2} + x^{1+(1+B)/2}) e^{-\pi x} dx.$$

The integral on the right converges as $M \to \infty$, thus proving the uniform convergence on every bounded region of the complex plane, and the absolute convergence for every complex number s.

XIV.2   The Fourier Inversion Formula

Exercise XIV.2.1 *Let T denote the Fourier transform, i.e. $Tf = \hat{f}$. Then $T: S \to S$ is an invertible linear map. If $f \in S$ and $g = f + Tf + T^2 f + T^3 f$, show that $Tg = g$, that is $\hat{g} = g$. This shows how to get a lot of functions equal to their roofs.*

Solution. We simply have

$$Tg = Tf + T^2 f + T^3 f + T^4 f = Tf + T^2 f + T^3 f + f = g.$$

Exercise XIV.2.2 *Show that every infinitely differentiable function which is equal to 0 outside some bounded interval is in S. Show that there exist such functions not identically zero. (Essentially an exercise in the chapter on the exponential function!)*

The **support** of a function f is the closure of the set of points x such that $f(x) \ne 0$. In particular, the support is a closed set. We may say that a C^∞ function with compact support is in the Schwartz space. The support of f is denoted by $\mathrm{supp}(f)$.

Solution. Suppose f is zero outside the closed and bounded interval I. All the derivatives of f are 0 outside I so for all m and d, the function $|x|^m f^{(d)}(x)$ is bounded because this function is continuous and is equal to 0 outside I. Thus $f \in S$.

Exercise 6, §1, of Chapter IV gives an example of a C^∞ function with compact support which is not identically zero.

Exercise XIV.2.3 *Write out in detail the statements and proofs for the theory of Fourier integrals as in the text but in dimension n, following the remark at the end of the section.*

Solution. For a complete exposition of the theory of the Fourier integral in n variables see S. Lang's *Real and Functional Analysis*.

The next exercises are formulated for $\mathbf{R}$, *but you may also do them for* $\mathbf{R}^n$ *in light of Exercise 3.*

Exercise XIV.2.4 *Let* $g \in C_c(\mathbf{R})$, $g \geq 0$, *and* $\int g = 1$. *Show that* $|\hat{g}| \leq 1$.

Solution. We estimate the Fourier transform of g,

$$|\hat{g}(y)| = \frac{1}{\sqrt{2\pi}} \left| \int g(x) e^{-ixy} dx \right| \leq \int |g(x)||e^{-ixy}| dx.$$

But $|e^{-ixy}| = 1$ and $g \geq 0$ so

$$|\hat{g}(y)| \leq \int g = 1.$$

Exercise XIV.2.5 *Suppose that g is even, real valued in S. Let* $f = g * g$. *Show that* $\hat{f} = |\hat{g}|^2$. *How does* $supp(\hat{f})$ *compare with* $supp(\hat{g})$?

Solution. By Theorem 1.2 we know that $\hat{f} = \hat{g}\hat{g}$. The assumption that g is even and real valued implies that $\hat{g} = \bar{\hat{g}}$. Indeed,

$$\bar{\hat{g}}(y) = \overline{\int g(x) e^{-ixy} d_1 x} = \int \overline{g(x)} e^{ixy} d_1 x = \int g(x) e^{ixy} d_1 x$$

and changing variables we get

$$\bar{\hat{g}}(y) = \int g(-x) e^{-ixy} d_1 x = \int g(x) e^{-ixy} d_1 x = \hat{g}(y),$$

which proves the assertion. Hence $\hat{f} = \hat{g}\bar{\hat{g}} = |\hat{g}|^2$. This equality shows that $\hat{f}(x) = 0$ if and only if $\hat{g}(x) = 0$, whence $supp(\hat{f}) = supp(\hat{g})$.

Exercise XIV.2.6 *Given* $\epsilon > 0$, *show that there exists a function* $f \in S$, *real valued, such that:*

$$f \geq 0, \quad \hat{f}(0) = 1, \quad supp(\hat{f}) \subset [-\epsilon, \epsilon].$$

Solution. First we show that there exists a function h such that

$$\hat{h} \geq 0, \quad h(0) = 1, \quad \text{supp}(\hat{h}) \subset [-\epsilon, \epsilon].$$

Let g be a function in $C_c^\infty(\mathbf{R})$ with support in $[-\epsilon/2, \epsilon/2]$ such that g is even, positive and not identically 0. This can be achieved by using bump functions (see Chapter IV). Then let $h = g * g$. By the previous exercise, we know that $\hat{h} \geq 0$. By multiplying g by a positive constant if necessary, we get $h(0) = 1$. Finally,

$$\text{supp}(h) \subset \text{supp}(g) + \text{supp}(g),$$

where $\text{supp}(g) + \text{supp}(g) = \{x + y : x, y \in \text{supp}(g)\}$. Indeed, changing variables we have

$$h(x) = \int_{\mathbf{R}} g(x - t) g(t) dt$$

so if $x \notin \text{supp}(g) + \text{supp}(g)$, then for any $t \in \text{supp}(g)$ we see that we have $x - t \notin \text{supp}(g)$, so $g(x - t)g(t) = 0$ for all t, hence $h(x) = 0$. Therefore $\text{supp}(h) \subset [-\epsilon, \epsilon]$ as was to be shown.

Now let f be the Fourier transform of h, that is $f = \hat{h}$. Then $f \geq 0$, and $\hat{f} = \hat{\hat{h}} = h^-$, so $\hat{f}(0) = h^-(0) = h(0) = 1$ and

$$\text{supp}(\hat{f}) = \text{supp}(h) \subset [-\epsilon, \epsilon].$$

This concludes the exercise.

Exercise XIV.2.7 *As for the Poisson formula, define the **Poisson dual***

$$f^\vee(y) = \int_{\mathbf{R}} f(x) e^{-2\pi i x y} dx.$$

Verify the the formula $f^{\vee\vee} = f^-$, which thus holds also for this normalization of the Fourier transform. You can get this one out of the other one by changes of variables in the integrals. Keep cool, calm, and collected.

Solution. We have

$$f^\vee(y) = \int_{\mathbf{R}} f(x) e^{-2\pi i x y} dx \quad \text{and} \quad f^\wedge(y) = \frac{1}{2\sqrt{\pi}} \int_{\mathbf{R}} f(x) e^{-i x y} dx.$$

Hence $f^\vee(y) = \sqrt{2\pi} f^\wedge(2\pi y)$ and therefore,

$$f^{\vee\vee}(y) = \sqrt{2\pi} \int_{\mathbf{R}} f^\wedge(2\pi x) e^{-2\pi i x y} dx.$$

Changing variables $u = 2\pi x$ we get

$$f^{\vee\vee}(y) = \frac{1}{\sqrt{2\pi}} \int_{\mathbf{R}} f^\wedge(u) e^{-i u y} dx = f^{\wedge\wedge}(y).$$

Therefore $f^{\vee\vee} = f^-$.

XIV.3   An Example of Fourier Transform Not in the Schwartz Space

Exercise XIV.3.1 (The Lattice Point Problem) *Let $N(R)$ be the number of lattice points (that is, elements of $\mathbf{Z}^2$) in the closed disc of radius R in the plane. A famous conjecture asserts that*

$$N(R) = \pi R^2 + O(R^{1/2+\epsilon})$$

for every $\epsilon > 0$. It is known that the error term cannot be $O(R^{1/2}(\log R)^k)$ for any positive integer k (result of Hardy and Landau). Prove the following best know result of Sierpinski-Van der Corput-Vinogradov-Hua:

$$N(R) = \pi R^2 + O(R^{2/3}).$$

[Hint: Let φ be the characteristic function of the unit disc, and put

$$\varphi_R(x) = \varphi\left(\frac{x}{R}\right).$$

Let ψ be a C^∞ function with compact support, positive, and such that

$$\int_{\mathbf{R}^2} \psi(x)dx = 1, \quad \text{and let} \quad \psi_\epsilon(x) = \epsilon^{-2}\psi\left(\frac{x}{\epsilon}\right).$$

Then $\{\psi_\epsilon\}$ is a Dirac family for $\epsilon \to 0$, and we can apply the Poisson summation formula to the convolution $\varphi_R * \psi_\epsilon$ to get

$$\sum_{m\in\mathbf{Z}^2} \varphi_R * \psi_\epsilon(m) = \sum_{m\in\mathbf{Z}^2} \hat{\varphi}_R(m)\hat{\psi}_\epsilon(m).$$

$$= \pi R^2 + \sum_{m\neq 0} R^2\hat{\varphi}(Rm)\hat{\psi}(\epsilon m).$$

We shall choose ϵ depending on R to make the error term best possible.]
   Note that $\varphi_R * \psi_\epsilon(x) = \varphi_R(x)$ if $\operatorname{dist}(x, S_R) > \epsilon$, where S_R is the circle of radius R. Therefore we get an esimate

$$|\text{left} - \text{handside} - N(R)| \ll \epsilon R.$$

Splitting off the term with $m = 0$ on the right-hand side, we find by Theorem 3.4:

$$\sum_{m\neq 0} R^2\hat{\varphi}(Rm)\hat{\psi}(\epsilon m) \ll R^{2-3/2} \sum_{m\neq 0} |m|^{-3/2}\hat{\psi}(\epsilon m).$$

But we can compare this last sum with the integral

$$\int_1^\infty r^{-3/2}\hat{\psi}(\epsilon r)r\,dr = O(\epsilon^{-1/2}).$$

Therefore we find

$$N(R) = \pi R^2 + O(\epsilon R) + O(R^{1/2}\epsilon^{-1/2}).$$

We choose $\epsilon = R^{-1/3}$ to make the error term $O(R^{2/3})$ as desired.

Solution. The proof is almost complete. We verify that $\{\psi_\epsilon\}$ is a Dirac family for $\epsilon \to 0$. We work with two variables x_1 and x_2 because $x \in \mathbf{R}^2$. Then changing variables $x = \epsilon y$ we get

$$\int \psi_\epsilon(x) = \iint \epsilon^{-2} \psi\left(\frac{x}{\epsilon}\right) dx_1 dx_2 = \iint \psi(y) dy_1 dy_2 = 1.$$

Finally the third property is verified because changing variables as before we get

$$\int_{|x|>\delta} \psi_\epsilon(x) dx = \int_{|x|>\delta/\epsilon} \psi(\alpha) d\alpha$$

and the last integral is 0 when ϵ is small because ψ has compact support.

Note that the area of the unit disc is π so changing variables $x = Ry$ we get

$$\hat{\varphi}_R(0) = \int \varphi_R(x) dx = R^2 \int \varphi(y) dy = \pi R^2.$$

Furthermore,

$$\hat{\psi}_\epsilon(0) = \int \psi_\epsilon(x) dx = 1.$$

Generalizing the formula of Exercise 1, §1, we find $\hat{\psi}_\epsilon(m) = \psi(\epsilon m)$ and $\hat{\varphi}_R(m) = R^2 \hat{\varphi}(Rm)$.

Finally, note that $|\text{left} - \text{handside} - N(R)|$ is the number of the lattice points N_A in the annulus $A = \{x : R - \epsilon \le |x| \le R + \epsilon\}$. But considering the area of the annulus $A' = \{x : R - \epsilon - \sqrt{2} \le |x| \le R + \epsilon + \sqrt{2}\}$ ($\sqrt{2}$ is the diameter of one of the squares in $\mathbf{Z}^2$) we find that $N_A = O(\epsilon R)$.

XV
Functions on n-Space

XV.1   Partial Derivatives

In the exercises, assume that all repeated partial derivatives exist and are continuous as needed.

Exercise XV.1.1 *Let f, g be two functions of two variables with continuous partial derivatives of order ≤ 2 in an open set U. Assume that*

$$\frac{\partial f}{\partial x} = -\frac{\partial g}{\partial y} \quad and \quad \frac{\partial f}{\partial y} = \frac{\partial g}{\partial x}.$$

Show that

$$\frac{\partial^2 f}{\partial x^2} + \frac{\partial^2 f}{\partial y^2} = 0.$$

Solution. We simply have

$$\frac{\partial^2 f}{\partial x^2} + \frac{\partial^2 f}{\partial y^2} = -\frac{\partial^2 g}{\partial x \partial y} + \frac{\partial^2 g}{\partial y \partial x} = 0.$$

Exercise XV.1.2 *Let f be a function of three variables, defined for $X \neq O$ by $f(X) = 1/|X|$. Show that*

$$\frac{\partial^2 f}{\partial x^2} + \frac{\partial^2 f}{\partial y^2} + \frac{\partial^2 f}{\partial z^2} = 0$$

if the three variables are (x, y, z). (The norm is the euclidean norm.)

Solution. We compute $\partial f^2/\partial x^2$. First we have

$$\frac{\partial f}{\partial x} = -\frac{x}{|X|^3}$$

and therefore

$$\frac{\partial^2 f}{\partial x^2} = -\frac{|X|^3 - 3x^2|X|}{|X|^6} = \frac{2x^2 - y^2 - z^2}{|X|^5}.$$

By symmetry we conclude that

$$\frac{\partial^2 f}{\partial x^2} + \frac{\partial^2 f}{\partial y^2} + \frac{\partial^2 f}{\partial z^2} = 0.$$

Exercise XV.1.3 *Let $f(x, y) = \arctan(y/x)$ for $x > 0$. Show that*

$$\frac{\partial^2 f}{\partial x^2} + \frac{\partial^2 f}{\partial y^2} = 0.$$

Solution. If $f(x, y) = \arctan(y/x)$, then

$$\frac{\partial f}{\partial x} = \frac{1}{1 + (y/x)^2} \frac{-y}{x^2} = \frac{-y}{x^2 + y^2}$$

and therefore

$$\frac{\partial^2 f}{\partial x^2} = \frac{2xy}{(x^2 + y^2)^2},$$

and similarly we obtain

$$\frac{\partial^2 f}{\partial y^2} = \frac{-2xy}{(x^2 + y^2)^2}.$$

Conclude.

Exercise XV.1.4 *Let θ be a fixed number, and let*

$$x = u \cos \theta - v \sin \theta, \quad y = u \sin \theta + v \cos \theta.$$

Let f be a function of two variables, and let $f(x, y) = g(u, v)$. Show that

$$\left(\frac{\partial g}{\partial u}\right)^2 + \left(\frac{\partial g}{\partial v}\right)^2 = \left(\frac{\partial f}{\partial x}\right)^2 + \left(\frac{\partial f}{\partial y}\right)^2.$$

Solution. We simply have

$$\frac{\partial g}{\partial u} = \frac{\partial f}{\partial x} \cos \theta + \frac{\partial f}{\partial y} \sin \theta$$

and

$$\frac{\partial g}{\partial v} = -\frac{\partial f}{\partial x} \sin \theta + \frac{\partial f}{\partial y} \cos \theta.$$

Adding the squares of the above expressions and using the fact that $\cos^2 \theta + \sin^2 \theta = 1$ we get the desired identity.

Exercise XV.1.5 *Assume that f is a function satisfying*

$$f(tx, ty) = t^m f(x, y)$$

for all numbers x, y, and t. Show that

$$x^2 \frac{\partial^2 f}{\partial x^2} + 2xy \frac{\partial^2 f}{\partial x \partial y} + y^2 \frac{\partial^2 f}{\partial y^2} = m(m-1)f(x, y).$$

[Hint: Differentiate twice with respect to t. Then put $t = 1$.]

Solution. Differentiating once with respect to t we get

$$\frac{\partial f}{\partial x} x + \frac{\partial f}{\partial y} y = mt^{m-1} f(x, y),$$

and differentiating again with respect to t we find

$$x \left(\frac{\partial^2 f}{\partial x^2} x + \frac{\partial^2 f}{\partial y \partial x} y \right) + y \left(\frac{\partial^2 f}{\partial x \partial y} x + \frac{\partial^2 f}{\partial y^2} y \right) = (m-1)mt^{m-2} f(x, y).$$

Collect terms and put $t = 1$.

Exercise XV.1.6 *Let $x = r \cos \theta$ and $y = r \sin \theta$. Let $f(x, y) = g(r, \theta)$. Show that*

$$\frac{\partial f}{\partial x} = \cos \theta \frac{\partial g}{\partial r} - \frac{\sin \theta}{r} \frac{\partial g}{\partial \theta},$$
$$\frac{\partial f}{\partial y} = \sin \theta \frac{\partial g}{\partial r} + \frac{\cos \theta}{r} \frac{\partial g}{\partial \theta}.$$

[Hint: Solve the simultaneous system of linear equations (*) and (**) given in the example of the text.]

Solution. If we form $r \sin \theta (*) + \cos \theta (**)$ we get

$$r \frac{\partial f}{\partial y} \sin^2 \theta + r \frac{\partial f}{\partial y} \cos^2 \theta = r \frac{\partial g}{\partial r} \sin \theta + \frac{\partial g}{\partial \theta} \cos \theta.$$

Dividing by r yields the desired formula for $\partial f / \partial y$. To find the formula for $\partial f / \partial x$, form $r \cos \theta (*) - \sin \theta (**)$ and divide by r.

Exercise XV.1.7 *Let $x = r \cos \theta$ and $y = r \sin \theta$. Let $f(x, y) = g(r, \theta)$. Show that*

$$\frac{\partial^2 g}{\partial r^2} + \frac{1}{r} \frac{\partial g}{\partial r} + \frac{1}{r^2} \frac{\partial^2 g}{\partial \theta^2} = \frac{\partial^2 f}{\partial x^2} + \frac{\partial^2 f}{\partial y^2}.$$

This exercise gives the polar coordinate form of the Laplace operator, and we can write symbolically:

$$\left(\frac{\partial}{\partial x} \right)^2 + \left(\frac{\partial}{\partial y} \right)^2 = \left(\frac{\partial}{\partial r} \right)^2 + \frac{1}{r} \frac{\partial}{\partial r} + \frac{1}{r^2} \left(\frac{\partial}{\partial \theta} \right)^2.$$

[Hint for the proof: Start with (*) and (**) and take the further derivatives as needed. Then take the sum. Lots of things will cancel out leaving you with $D_1^2 f + D_2^2 f$.]

Solution. See Exercise 5, §3, of Chapter XII.

Exercise XV.1.8 *With the same notation as in the preceding exercise, show that*

$$\left(\frac{\partial g}{\partial r}\right)^2 + \frac{1}{r^2}\left(\frac{\partial g}{\partial \theta}\right)^2 = \left(\frac{\partial f}{\partial x}\right)^2 + \left(\frac{\partial f}{\partial y}\right)^2.$$

Solution. Using the formulas in the solution of Exercise 5, §3, of Chapter XII, we get after cancellations

$$\left(\frac{\partial g}{\partial x}\right)^2 + \frac{1}{r^2}\left(\frac{\partial g}{\partial \theta}\right)^2 = \left(\frac{\partial f}{\partial x}\right)^2 (\cos^2 \theta + \sin^2 \theta) + \left(\frac{\partial f}{\partial y}\right)^2 (\cos^2 \theta + \sin^2 \theta).$$

Exercise XV.1.9 *In $\mathbf{R}^2$, suppose that $f(x,y) = g(r)$ where $r = \sqrt{x^2 + y^2}$. Show that*

$$\frac{\partial^2 f}{\partial x^2} + \frac{\partial^2 f}{\partial y^2} = \frac{d^2 g}{dr^2} + \frac{1}{r}\frac{dg}{dr}.$$

Solution. See Exercise 10.

Exercise XV.1.10 *(a) In $\mathbf{R}^3$, suppose that $f(x,y,z) = g(r)$ where $r = \sqrt{x^2 + y^2 + z^2}$. Show that*

$$\frac{\partial^2 f}{\partial x^2} + \frac{\partial^2 f}{\partial y^2} + \frac{\partial^2 f}{\partial z^2} = \frac{d^2 g}{dr^2} + \frac{2}{r}\frac{dg}{dr}.$$

(b) Assume that f is harmonic except possibly at the origin on $\mathbf{R}^n$, and that there is a C^2 function g such that $f(X) = g(r)$ where $r = \sqrt{X \cdot X}$. Let $n \geq 3$. Show that there exist constants C, K such that $g(r) = Kr^{2-n} + C$. What if $n = 2$?

Solution. (a) We prove the general formula in $\mathbf{R}^n$. If $f(x_1, \ldots, x_n) = g(r)$ where $r = (x_1^2 + \cdots + x_n^2)^{1/2}$, then

$$\frac{\partial^2 f}{\partial x_1^2} + \cdots + \frac{\partial^2 f}{\partial x_n^2} = \frac{\partial^2 g}{\partial r^2} + \frac{n-1}{r}\frac{\partial g}{\partial r}.$$

Indeed,

$$\frac{\partial f}{\partial x_j} = \frac{\partial g}{\partial r}\frac{x_j}{r},$$

hence

$$\frac{\partial^2 f}{\partial x_j^2} = \frac{\partial^2 g}{\partial r^2}\left(\frac{x_j}{r}\right)^2 + \frac{\partial g}{\partial r}\left(\frac{\sum_{k \neq j} x_k^2}{r^3}\right).$$

Summing over j we get the desired formula.

(b) For $r > 0$ we have

$$\frac{d}{dr}\left(\frac{g'(r)}{(2-n)r^{1-n}}\right) = \frac{g''(r)(2-n)r^{1-n} - g'(r)(1-n)(2-n)r^{-n}}{[(2-n)r^{1-n}]^2},$$

but f is harmonic, so this last expression is 0 because by (a) we know that

$$g''(r) = g'(r)(1-n)/r.$$

So there exists a constant K such that $g'(r) = K(2-n)r^{1-n}$. Integrating once we see that there exists a constant C such that $g(r) = Kr^{2-n} + C$.

The primitive of $1/x$ is $\log x$, so for $n = 2$ we might expect $\log\sqrt{x_1^2 + x_2^2}$ to be harmonic on $\mathbf{R}^2 - \{0\}$. Indeed, if $f(x_1, x_2) = \log\sqrt{x_1^2 + x_2^2}$, then a straightforward computation shows that

$$\frac{\partial^2 f}{\partial x^2} = \frac{x_2^2 - x_1^2}{(x_1^2 + x_2^2)^2} \quad\text{and}\quad \frac{\partial^2 f}{\partial y^2} = \frac{x_1^2 - x_2^2}{(x_1^2 + x_2^2)^2}.$$

Exercise XV.1.11 *Let $r = \sqrt{x^2 + y^2}$ and let r, θ be the polar coordinates in the plane. Using the formula for the Laplace operator in Exercise 7 verify that the following functions are harmonic:*
(a) $r^n \cos n\theta = g(r, \theta)$.   (b) $r^n \sin n\theta = g(r, \theta)$.
As usual, n denotes a positive integer. So you are supposed to prove that the expression

$$\frac{\partial^2 g}{\partial r^2} + \frac{1}{r}\frac{\partial g}{\partial r} + \frac{1}{r^2}\frac{\partial^2 g}{\partial \theta^2}$$

is equal to 0 for the above functions g.

Solution. (a) We have

$$\frac{\partial g}{\partial r} = nr^{n-1}\cos n\theta$$

and

$$\frac{\partial^2 g}{\partial r^2} = n(n-1)r^{n-2}\cos n\theta$$

and

$$\frac{\partial^2 g}{\partial \theta^2} = -n^2 r^n \cos n\theta,$$

so a simple manipulation gives

$$\frac{\partial^2 g}{\partial r^2} + \frac{1}{r}\frac{\partial g}{\partial r} + \frac{1}{r^2}\frac{\partial^2 g}{\partial \theta^2} = 0.$$

(b) In this case, we have

$$\frac{\partial g}{\partial r} = nr^{n-1}\sin n\theta$$

and

$$\frac{\partial^2 g}{\partial r^2} = n(n-1)r^{n-2} \sin n\theta$$

and

$$\frac{\partial^2 g}{\partial \theta^2} = -n^2 r^n \sin n\theta,$$

so a simple manipulation gives

$$\frac{\partial^2 g}{\partial r^2} + \frac{1}{r}\frac{\partial g}{\partial r} + \frac{1}{r^2}\frac{\partial^2 g}{\partial \theta^2} = 0.$$

Exercise XV.1.12 *For $x \in \mathbf{R}^n$, let $x^2 = x_1^2 + \cdots + x_n^2$. For t real > 0, let*

$$f(x,t) = t^{-n/2} e^{-x^2/4t}.$$

If Δ is the Laplace operator, $\Delta = \sum \partial^2/\partial x_i^2$, show that $\Delta f = \partial f/\partial t$. A function satisfying this differential equation is said to be a solution of the **heat equation.**

Solution. Differentiating once yields

$$\frac{\partial f}{\partial x_j} = t^{-n/2}\left(\frac{-2x_j}{4t}\right)e^{-x^2/4t},$$

and differentiating once more we get

$$\frac{\partial^2 f}{\partial x_j^2} = -\frac{t^{-n/2-1}}{2}\left[e^{-x^2/4t} - \frac{2x_j^2}{4t}e^{-x^2/4t}\right],$$

hence

$$\Delta f = -\frac{t^{-n/2-1}}{2}e^{-x^2/4t}\left(n - \frac{x^2}{2t}\right).$$

Differentiating once with respect to t yields

$$\frac{\partial f}{\partial t} = -\frac{n}{2}t^{-n/2-1}e^{-x^2/4t} + t^{-n/2}\frac{x^2}{4t^2}e^{-x^2/4t} = -\frac{t^{-n/2-1}}{2}e^{-x^2/4t}\left(n - \frac{x^2}{2t}\right),$$

whence $\Delta f = \partial f/\partial t$.

Exercise XV.1.13 *This exercise gives an example of a function whose repeated partials exist but such that $D_1 D_2 f \neq D_2 D_1 f$. Let*

$$f(x,y) = \begin{cases} xy\frac{x^2-y^2}{x^2+y^2} & \text{if } (x,y) \neq (0,0), \\ 0 & \text{if } (x,y) = (0,0). \end{cases}$$

Prove:
(a) The partial derivatives $\partial^2 f/\partial x\partial y$ and $\partial^2 f/\partial y\partial x$ exist for all (x,y) and are continuous except at $(0,0)$.
(b) $D_1 D_2 f(0,0) \neq D_2 D_1 f(0,0)$.

Solution. We have

$$\frac{\partial f}{\partial x} = y\frac{x^4 - y^4 + 4x^2y^2}{(x^2 + y^2)^2}$$

and

$$\frac{\partial f}{\partial y} = -x\frac{y^4 - x^4 + 4x^2y^2}{(x^2 + y^2)^2}.$$

Both these partials are equal to 0 at $(0,0)$. This follows from a direct computation of the Newton quotient. For $(0,0)$ we see that

$$\frac{\partial f/\partial x(0,h) - \partial f/\partial x(0,0)}{h} = -1 = \frac{\partial^2 f}{\partial y\partial x}(0,0),$$

and

$$\frac{\partial f/\partial y(h,0) - \partial f/\partial y(0,0)}{h} = 1 = \frac{\partial^2 f}{\partial x\partial y}(0,0).$$

For $(x, y) \neq (0,0)$ we have

$$\frac{\partial^2 f}{\partial y\partial x} = \frac{(x^4 - 5y^4 + 12x^2y^2)(x^2 + y^2)^2 - (yx^4 - y^5 + 4x^2y^3)(4y(x^2 + y^2))}{(x^2 + y^2)^2}.$$

Then letting $y = 0$ we find that

$$\lim_{(x,0)\to(0,0)} \frac{\partial^2 f}{\partial y\partial x} = 1.$$

By symmetry, we conclude that both $\partial^2/\partial y\partial x$ and $\partial^2/\partial x\partial y$ exist and are continuous except at $(0,0)$.

Green's Functions

Exercise XV.1.14 *Let (a, b) be an open interval, which may be (a, ∞). Let*

$$M_y = -\left(\frac{d}{dy}\right)^2 + p(y),$$

where p is an infinitely differentiable function. We view M_y as a differential operator. If f is a function of the variable y, then we use the notation

$$M_y f(y) = -f''(y) + p(y)f(y).$$

A Green's function for the differential operator M is a suitably smooth function $g(y, y')$ defined for y, y' in (a, b) such that

$$M_y \int_a^b g(y, y')f(y')dy' = f(y)$$

for all infinitely differentiable functions f on (a, b) with compact support (meaning f is 0 outside a closed interval contained in (a, b)). Now let

$g(y, y')$ be any continuous function satisfying the following additional conditions:

GF 1. g is infinitely differentiable in each variable except on the diagonal, that is when $y = y'$.
GF 2. If $y \neq y'$, then $M_y g(y, y') = 0$.

Prove:
Let g be a function satisfying **GF 1** and **GF 2**. Then g is a Green's function for the operator M if and only if g also satisfies the jump condition.

GF 3. $D_1 g(y, y+) - D_1 g(y, y-) = 1$.

As usual, one defines

$$D_1 g(y, y+) = \lim_{\substack{y' \to y \\ y' > y}} D_1 g(y, y'),$$

and similarly for $y-$ instead of $y+$, we take the limit with $y' < y$. [Hint: Write the integral

$$\int_a^b = \int_a^y + \int_y^b .\bigg]$$

Solution. The chain rule shows that under some suitable assumptions we have

$$\frac{d}{dx} \int_a^{\alpha(x)} f(x, t) dt = f(x, \alpha(x)) \alpha'(x) + \int_a^{\alpha(x)} \frac{\partial}{\partial x} f(x, t) dt.$$

Splitting the integral as hinted, and after some cancellations we see that

$$\frac{d^2}{dy^2} \int_a^b g(y, y') f(y') dy' = f(y)(D_1 g(y, y-) - D_1 g(y, y+))$$
$$+ \int_a^y \frac{d^2}{dy^2} g(y, y') f(y') dy' + \int_y^b \frac{d^2}{dy^2} g(y, y') f(y') dy'.$$

Condition **GF 2** implies that the sum of the last two integrals is equal to

$$p(y) \int_a^b g(y, y') f(y') dy'.$$

So we see that g is a Green's function if and only if

$$D_1 g(y, y-) - D_1 g(y, y+) = -1,$$

as was to be shown.

Exercise XV.1.15 *Assume now that the differential equation $f'' - pf = 0$ has two linearly independent solutions J and K. (See Chapter XIX, §3, Exercise 2.) Let $W = JK' - J'K$.*

(a) Show that W is constant $\neq 0$.

(b) Show that there exists a unique Green's function of the form

$$g(y, y') = \begin{cases} A(y')J(y) & \text{if } y' < y, \\ B(y')K(y) & \text{if } y' > y, \end{cases}$$

and that the functions A, B necessarily have the values $A = K/W$, $B = J/W$.

Solution. The quantity W is constant because

$$W' = J'K' + JK'' - J''K - J'K' = JpK - JpK = 0.$$

Suppose W is 0. If J is never 0, then $(K/J)' = 0$ which is impossible because the solutions are linearly independent. If for some t_0, $J(t_0) = 0$, then $J'(t_0)K(t_0) = 0$. If $J'(t_0) = 0$, then J is identically 0 and if $J'(t_0) \neq 0$, then $K(t) = (K'(t_0)/J'(t_0))J(t)$ by the uniqueness theorem. This is a contradiction.

(b) First we show that if $A = K/W$ and $B = J/W$, then the function $g(y, y')$ is a Green's function. The continuity condition is satisfied and we have $g(y, y) = JK/W$. Since J and K are solutions of the differential equation $f'' - pf = 0$ we see at once that if $y \neq y'$, then $M_y g(y, y') = 0$. Finally, we verify the jump condition,

$$D_1 g(y, y+) - D_1 g(y, y-) = \frac{J(y)K'(y) - K(y)J'(y)}{W} = 1.$$

So $g(y, y')$ is a Green's function. This solution is unique. Indeed, the continuity and jump conditions imply

$$\begin{cases} A(y)J(y) - B(y)K(y) = 0, \\ A(y)J'(y) - B(y)K'(y) = 1, \end{cases}$$

but

$$\begin{vmatrix} J(y) & K(y) \\ J'(y) & K'(y) \end{vmatrix} = W \neq 0.$$

Conclude.

Exercise XV.1.16 On the interval $(-\infty, \infty)$ let $M_y = -(d/dy)^2 + c^2$ where c is a positive number, so take $p = c > 0$ constant. Show that e^{cy} and e^{-cy} are two linearly independent solutions and write down explicitly the Green's function for M_y.

Solution. We have

$$-\frac{d^2}{dy^2}e^{cy} + c^2 e^{cy} = -c^2 e^{cy} + c^2 e^{cy} = 0,$$

and

$$-\frac{d^2}{dy^2}e^{-cy} + c^2 e^{-cy} = -c^2 e^{-cy} + c^2 e^{-cy} = 0.$$

Clearly, both solutions are linearly independent. Let $J = e^{-cy}$ and $K = e^{cy}$, then

$$W = ce^{-cy}e^{cy} + ce^{-cy}e^{cy} = 2c,$$

so the expression of the Green's function is

$$\begin{cases} e^{cy'}(e^{-cy}/2c) & \text{if } y' < y, \\ e^{-cy'}(e^{cy}/2c) & \text{if } y' > y. \end{cases}$$

Exercise XV.1.17 *On the interval* $(0, \infty)$ *let*

$$M_y = -\left(\frac{d}{dy}\right)^2 - \frac{s(1-s)}{y^2},$$

where s is some fixed complex number. For $s \neq \frac{1}{2}$, show that y^{1-s} and y^s are two linearly independent solutions and write down explicitly the Green's function for the operator.

Solution. Both functions y^{1-s} and y^s are solutions of the differential equation because

$$-\frac{d^2}{dy^2}y^s = -s(s-1)y^{s-2} = s(1-s)\frac{y^s}{y^2},$$

and

$$-\frac{d^2}{dy^2}y^{1-s} = s(1-s)y^{-1-s} = s(1-s)\frac{y^{1-s}}{y^2}.$$

Clearly, both solutions are linearly independent. Let $J(y) = y^s$ and $K(y) = y^{1-s}$. Then

$$W = y^s(1-s)y^{-s} + sy^{s-1}y^{1-s} = 1,$$

so the expression of the Green's function is

$$\begin{cases} (y')^{1-s}\, y^s & \text{if } y' < y, \\ (y')^s\, y^{1-s} & \text{if } y' > y. \end{cases}$$

XV.2   Differentiability and the Chain Rule

Exercise XV.2.1 *Show that any two points on the sphere of radius 1 (or any radius) in n-space centered at the origin can be joined by a differentiable curve. If the points are not antipodal, divide the straight line between them by its length at each point. Or use another method: taking the plane containing the two points, and using two perpendicular vectors of lengths 1 in this plane, say A, B, consider the unit circle*

$$\alpha(t) = (\cos t)A + (\sin t)B.$$

Solution. Method 1. If A and B are the two points, the equation of the line segment joining them is $L(t) = A + t(B - A)$ where $t \in [0, 1]$. If A and B are not antipodal, then $L(t) \neq O$ for all t so $|L(t)| \neq 0$ for all t. If R is the radius of the sphere, the curve

$$\alpha(t) = R\frac{L(t)}{|L(t)|},$$

is a solution to our problem because α is differentiable, $|\alpha| = R$, $\alpha(0) = A$, and $\alpha(1) = B$.

Method 2. We can always choose A to be the vector determined by one of the points. Note that

$$|\alpha|^2 = |A| \cos^2 t + |B| \sin^2 t = R^2.$$

There exist numbers a and b such that the second point equals $aA + bB$. Taking the square of the norm we see that $a^2 + b^2 = 1$ thus there exists a number t_0 such that the second point equals $\alpha(t_0)$.

Exercise XV.2.2 *Let f be a differentiable function on $\mathbf{R}^n$, and assume that there is a differentiable function h such that*

$$(\text{grad } f)(X) = h(X)X.$$

Show that f is constant on the sphere of radius r centered at the origin in $\mathbf{R}^n$. [Hint: Use Exercise 1.]

Solution. Let A and B be two points on the sphere and let α be as in Exercise 1. Then

$$(f \circ \alpha)'(t) = \text{grad } f(\alpha(t)) \cdot \alpha'(t) = h(\alpha(t))\alpha(t) \cdot \alpha'(t).$$

Since $\alpha \cdot \alpha$ is constant we see that

$$0 = \frac{d}{dt}[\alpha(t) \cdot \alpha(t)] = 2\alpha(t) \cdot \alpha'(t),$$

and therefore $(f \circ \alpha)'(t) = 0$. Hence $f(A) = f(B)$.

Exercise XV.2.3 *Prove the converse of Exercise 2, which is the last statement preceding the exercises, namely if $f(X) = g(r)$, then $\text{grad } f(X) = g'(r)X/r$.*

Solution. The chain rule implies

$$\frac{\partial f}{\partial x_i} = g'(r)\frac{x_i}{\sqrt{x_1^2 + \cdots + x_n^2}},$$

so we immediately get that $\text{grad } f(X) = g'(r)X/r$.

Exercise XV.2.4 *Let f be a differentiable function on $\mathbf{R}^n$ and assume that there is a positive integer m such that $f(tX) = t^m f(X)$ for all numbers $t \neq 0$ and all points X in $\mathbf{R}^n$. Prove **Euler's relation**:*

$$x_1 \frac{\partial f}{\partial x_1} + \cdots + x_n \frac{\partial f}{\partial x_n} = mf(X).$$

Solution. We differentiate with respect to t the identity $f(tX) = t^m f(X)$. The left-hand side becomes grad $f(tX) \cdot X$ and the right-hand side becomes $mt^{m-1} f(X)$. Put $t = 1$ and conclude.

Exercise XV.2.5 *Let f be a differentiable function defined on all of space. Assume that*

$$f(tP) = tf(P)$$

for all numbers t and all points P. Show that

$$f(P) = \operatorname{grad} f(O) \cdot P.$$

Solution. Differentiate $f(tP) = tf(P)$ with respect to t. You get grad $f(tP) \cdot P = f(P)$. Put $t = 0$.

Exercise XV.2.6 *Find the equation of the tangent plane to each of the following surfaces at the specified point.*
(a) $x^2 + y^2 + z^2 = 49$ at $(6, 2, 3)$.
(b) $x^2 + xy^2 + y^3 + z + 1 = 0$ at $(2, -3, 4)$.
(c) $x^2 y^2 + xz - 2y^3 = 10$ at $(2, 1, 4)$.
(d) $\sin xy + \sin yz + \sin xz = 1$ at $(1, \pi/2, 0)$.

Solution. (a) $6x + 2y + 3z = 49$.
(b) $13x + 15y + z = -15$.
(c) $4x + y + z = 13$.
(d) $z = 0$.

Exercise XV.2.7 *Find the directional derivative of the following functions at the specified points in the specified directions.*
(a) $\log(x^2 + y^2)^{1/2}$ at $(1, 1)$, direction $(2, 1)$.
(b) $xy + yz + xz$ at $(-1, 1, 7)$, direction $(3, 4, -12)$.

Solution. (a) $3/(2\sqrt{5})$.
(b) $48/13$.

Exercise XV.2.8 *Let $f(x, y, z) = (x + y)^2 + (y + z)^2 + (z + x)^2$. What is the direction of greatest increase of the function at the point $(2, -1, 2)$? What is the directional derivative of f in this direction at that point?*

Solution. The gradient is simply

$$(4x + 2y + 2z, \, 4y + 2x + 2z, \, 4z + 2x + 2y),$$

so the direction of greatest increase of f at the given point is $(10, 4, 10)$ or any scalar multiple of this vector. The directional derivative is $6\sqrt{6}$.

Exercise XV.2.9 *Let f be a differentiable function defined on an open set U. Suppose that P is a point of U such that f(P) is a maximum, that is suppose we have*

$$f(P) \geq f(X) \quad \text{for all } X \text{ in } U.$$

Show that grad $f(P) = O$.

Solution. Let $g(t) = f(P + tH)$. Then $f(P) \geq g(t)$ for all small t and $g(0) = f(P)$ so

$$0 = g'(0) = \text{grad } f(P) \cdot H.$$

This equality holds for every vector H, $H \neq O$ so grad $f(P) = 0$, as was to be shown.

Exercise XV.2.10 *Let f be a function on an open set U in 3-space. Let g be another function, and let S be the surface consisting of all points X such that*

$$g(X) = 0 \quad \text{but} \quad \text{grad } g(X) \neq O.$$

Suppose that P is a point of the surface S such that f(P) is a maximum for f on S, that is

$$f(P) \geq f(X) \quad \text{for all } X \text{ on } S.$$

Prove that there is a number λ such that

$$\text{grad } f(P) = \lambda \text{grad } g(P).$$

Solution. Let $\alpha(t)$ be a C^1 curve such that $\alpha(t) \in S$ for all t, and $\alpha(t_0) = P$ for some number t_0. Then $f(\alpha(t))$ has a maximum at t_0 so

$$0 = \frac{d}{dt} f(\alpha(t))\Big|_{t_0} = \text{grad } f(P) \cdot \alpha'(t_0).$$

Hence grad $f(P)$ is perpendicular to the level hypersurface at P. Conclude.

Exercise XV.2.11 *Let $f: \mathbf{R}^2 \to \mathbf{R}$ be the function such that $f(0,0) = 0$ and*

$$f(x,y) = \frac{x^3}{x^2 + y^2} \quad \text{if } (x,y) \neq (0,0).$$

Show that f is not differentiable at $(0,0)$. However, show that for any differentiable curve $\varphi: J \to \mathbf{R}^2$ passing through the origin, $f \circ \varphi$ is differentiable.

Solution. Since $f(x,0)/x = 1$ we have

$$\frac{\partial f}{\partial x}(0,0) = 1.$$

Moreover, if $A = \text{grad } f(O)$ and $h = (h_1, h_2)$, then

$$f(O + h) - f(O) - A \cdot h = \frac{h_1^3}{h_1^2 + h_2^2} - h_1 = \frac{-h_1 h_2^2}{h_1^2 + h_2^2} = \psi(h).$$

If $h_1 = h_2$, then $|\psi(h)|/|h| = 1/(2\sqrt{2})$, whence ψ is not $o(h)$ and therefore f is not differentiable at $(0,0)$.

Let φ be a differentiable curve passing through the origin. Write $\varphi(t) = (\varphi_1(t), \varphi_2(t))$ and assume without loss of generality that $\varphi(0) = O$. The Newton quotient at 0 is

$$\frac{f(\varphi(h)) - f(\varphi(0))}{h} = \frac{\varphi_1(h)^3}{h(\varphi_1(h)^2 + \varphi_2(h)^2)} = \frac{\frac{\varphi_1(h)^3}{h^3}}{\left(\frac{\varphi_1(h)^2}{h^2} + \frac{\varphi_2(h)^2}{h^2} \right)}.$$

If $\varphi_1'(0)$ and $\varphi_2'(0)$ are not both zero, then we see that the Newton quotient tends to

$$\frac{[\varphi_1'(0)]^3}{[\varphi_1'(0)]^2 + [\varphi_2'(0)]^2}$$

as h tends to 0. If both $\varphi_1'(0)$ and $\varphi_2'(0)$ are zero, then the inequality

$$\left| \frac{f(\varphi(h)) - f(\varphi(0))}{h} \right| = \left| \frac{\varphi_1(h)}{h} \frac{\frac{\varphi_1(h)^2}{h^2}}{\left(\frac{\varphi_1(h)^2}{h^2} + \frac{\varphi_2(h)^2}{h^2} \right)} \right| \leq \frac{\varphi_1(h)}{h}$$

implies that the Newton quotient tends to 0 as h tends to 0. Therefore $f \circ \varphi$ is differentiable.

XV.3   Potential Functions

Exercise XV.3.1 *Let $X = (x_1, \ldots, x_n)$ denote a vector in $\mathbf{R}^n$. Let $|X|$ denote the euclidean norm. Find a potential function for the vector field F defined for all $X \neq O$ by the formula*

$$F(X) = r^k X,$$

where $r = |X|$. (Treat separately the cases $k = -2$, and $k \neq -2$.)

Solution. If $k \neq -2$, let

$$\varphi(X) = \frac{1}{k+2} |X|^{k+2}.$$

Then

$$\frac{\partial \varphi}{\partial x_i} = x_i |X|^k.$$

If $k = -2$, let $\varphi(X) = \log |X|$. Then

$$\frac{\partial \varphi}{\partial x_i} = \frac{x_i}{|X|^2}.$$

Exercise XV.3.2 *Again, let $r = |X|$. Let g be a differentiable function of one variable. Show that the vector field defined by*

$$F(X) = \frac{g'(r)}{r} X$$

on the open set of all $X \neq O$ has a potential function, and determine this potential function.

Solution. Let $\varphi = g(r)$. See Exercise 3, §2.

Exercise XV.3.3 *Let*

$$G(x, y) = \left(\frac{-y}{x^2 + y^2}, \frac{x}{x^2 + y^2} \right).$$

This vector field is defined on the plane $\mathbf{R}^2$ from which the origin has been deleted.
(a) For this vector field $G = (f, g)$ show that $D_2 f = D_1 g$.
(b) Why does this vector field have a potential function on every rectangle not containing the origin?
(c) Verify that the function $\psi(x, y) = - \arctan x/y$ is a potential function for G on any rectangle not intersecting the line $y = 0$.
(d) Verify that the function $\psi(x, y) = \arccos x/r$ is a potential function for this vector field in the upper half plane.

Solution. (a) A simple computation gives

$$D_2 f = \frac{y^2 - x^2}{(x^2 + y^2)^2} = D_1 g.$$

(b) Theorem 3.3.
(c) See the intermediate steps of Exercise 3, §1.
(d) In the upper half plane, we have $y > 0$. Differentiating yields

$$\frac{\partial \psi}{\partial x} = \frac{-1}{\sqrt{1 - (x/r)^2}} \frac{r - x^2/r}{r^2} = \frac{-y}{x^2 + y^2},$$

and

$$\frac{\partial \psi}{\partial y} = \frac{-1}{\sqrt{1 - (x/r)^2}} \frac{-xy}{r^3} = \frac{x}{x^2 + y^2}.$$

XV.4 Curve Integrals

Compute the curve integrals of the vector field over the indicated curves.

Exercise XV.4.1 $F(x, y) = (x^2 - 2xy, y^2 - 2xy)$ *along the parabola $y = x^2$ from $(-2, 4)$ to $(1, 1)$.*

Solution. $-369/10$. Parametrize the piece of the parabola by (t, t^2) with $-2 \le t \le 1$.

Exercise XV.4.2 $(x, y, xz - y)$ *over the line segment from* $(0,0,0)$ *to* $(1, 2, 4)$.

Solution. $23/6$. Parametrize the line segment by $(t, 2t, 4t)$.

Exercise XV.4.3 $(x^2 y^2, xy^2)$ *along the closed path formed by parts of the line* $x = 1$ *and the parabola* $y^2 = x$, *counterclockwise.*

Solution. $4/15$. Write the integral as the sum of two integrals. The first over the line and the second over the parabola. Parametrize the line segment by $(1, t)$ with $-1 \le t \le 1$ and parametrize the piece of the parabola by $(t^2, -t)$ with $1 \le t \le 1$.

Exercise XV.4.4 *Let*

$$G(x, y) = \left(\frac{-y}{x^2 + y^2}, \frac{x}{x^2 + y^2} \right).$$

(a) Find the integral of this vector field counterclockwise along the circle $x^2 + y^2 = 2$ *from* $(1, 1)$ *to* $(-\sqrt{2}, 0)$.
(b) Counterclockwise around the whole circle.
(c) Counterclockwise around the circle $x^2 + y^2 = a^2$ *for* $a > 0$.

Solution. If $\alpha(\theta) = (r \cos \theta, r \sin \theta)$, then

$$G(\alpha(\theta)) \cdot \alpha'(\theta) = \frac{r \sin \theta}{r^2} r \sin \theta + \frac{r \cos \theta}{r^2} r \cos \theta = 1.$$

So if we integrate counterclockwise from an angle θ_1 to an angle θ_2 we get

$$\int_C G = \int_{\theta_1}^{\theta_2} d\theta = \theta_2 - \theta_1.$$

(a) $3\pi/4$ because $\pi/4 \le \theta \le \pi$.
(b) 2π because $0 \le \theta \le 2\pi$.
(c) 2π because $0 \le \theta \le 2\pi$.

Exercise XV.4.5 *Let* $r = (x^2 + y^2)^{1/2}$ *and* $F(X) = r^{-1} X$ *for* $X = (x, y)$. *Find the integral of* F *over the circle or radius 2, centered at the origin, taken in the counterclockwise direction.*

Solution. 0. Special case of Exercise 6.

Exercise XV.4.6 *Let* C *be a circle of radius 20 with center at the origin. Let* $F(X)$ *be a vector field on* $\mathbf{R}^2$ *such that* $F(X)$ *has the same direction as* X *(that is there exists a differentiable function* $g(X)$ *such that* $f(X) = g(X)X$, *and* $g(X) > 0$ *for all* X). *What is the integral of* F *around* C, *taken counterclockwise?*

Solution. Parametrize the circle by $C(\theta) = (r\cos\theta, r\sin\theta)$ with $r = 20$ and $0 \le \theta \le 2\pi$. Then

$$\int_C F = \int_0^{2\pi} F(C(\theta)) \cdot C'(\theta)d\theta = \int_0^{2\pi} g(C(\theta))C(\theta) \cdot C'(\theta)d\theta = 0$$

because $C(\theta) \cdot C'(\theta) = 0$. This follows at once from differentiating the identity $C(\theta) \cdot C(\theta) = r^2 = \text{constant}$.

Exercise XV.4.7 *Let P, Q be points in 3-spaces. Show that the integral of the vector field given by*

$$F(x, y, z) = (z^2, 2y, 2xz)$$

from P to Q is independent of the curve selected between P and Q.

Solution. The function $z^2 x + y^2$ is a potential function of the vector field. Conclude.

Exercise XV.4.8 *Let $F(x, y) = (x/r^3, y/r^3)$ where $r = (x^2 + y^2)^{1/2}$. Find the integral of F along the curve*

$$\alpha(t) = (e^t \cos t, e^t \sin t)$$

from the point $(1, 0)$ to the point $(e^{2\pi}, 0)$.

Solution. The function $-1/r$ is a potential function of the vector field, so

$$\int_C F = -\frac{1}{r}\Big|_{(1,0)}^{(e^{2\pi},0)} = 1 - e^{-2\pi}.$$

Exercise XV.4.9 *Let $F(x, y) = (x^2 y, xy^2)$.*
(a) Does this vector field admit a potential function?
(b) Compute the integral of this vector field from $(0, 0)$ to the point

$$P = (1/\sqrt{2}, 1/\sqrt{2})$$

along the line segment from $(0, 0)$ to P.
(c) Compute the integral of this vector field from $(0, 0)$ to P along the path which consists of the segment from $(0, 0)$ to $(1, 0)$, and the arc of circle from $(1, 0)$ to P. Compare with the value found in (b).

Solution. (a) No, because $D_1 g \ne D_2 f$.
(b) Parametrize the line segment by $L(t) = \frac{1}{\sqrt{2}}(t, t)$ with $0 \le t \le 1$. Then

$$\int_L F = \int_0^1 F(L(t)) \cdot L'(t)dt = \frac{1}{4}\int_0^1 2t^3 dt = \frac{1}{8}.$$

(c) Parametrize the line segment by $L_1(t) = (t, 0)$ with $0 \leq t \leq 1$. Then $F(L_1(t)) = (0, 0)$, so

$$\int_{L_1} F = 0.$$

Parametrize the arc by $C(t) = (\cos t, \sin t)$ with $0 \leq t \leq \pi/4$. Then

$$F(C(T)) \cdot C'(t) = -\cos^2 t \sin^2 t + \cos^2 t \sin^2 t = 0,$$

hence

$$\int_C F = 0$$

so the answer is 0.

Exercise XV.4.10 *Let*

$$F(x, y) = \left(\frac{x \cos r}{r}, \frac{y \cos r}{r} \right),$$

where $r = \sqrt{x^2 + y^2}$. Find the value of the integral of this vector field:
(a) Counterclockwise along the circle of radius 1, from $(1, 0)$ to $(0, 1)$.
(b) Counterclockwise around the entire circle.
(c) Does this vector field admit a potential function? Why?

Solution. (a) 0.
(b) 0.
(c) The function $g(x, y) = \sin \sqrt{x^2 + y^2}$ is a potential function for the vector field F, because grad $g(x, y) = F(x, y)$.

Exercise XV.4.11 *Let*

$$F(x, y) = \left(\frac{x e^r}{r}, \frac{y e^r}{r} \right).$$

Find the value of the integral of this vector field:
(a) Counterclockwise along the circle of radius 1 centered at the origin.
(b) Counterclockwise along the circle of radius 5 centered at the point $(14, -17)$.
(c) Does this vector field admit a potential function? Why?

Solution. (a) 0.
(b) 0.
(c) The function $g(x, y) = e^r$ is a potential function for the vector field F because grad $g(x, y) = F(x, y)$.

Exercise XV.4.12 *Let*

$$G(x, y) = \left(\frac{-y}{x^2 + y^2}, \frac{x}{x^2 + y^2} \right).$$

(a) Find the integral of G along the line $x + y = 1$ from $(0, 1)$ to $(1, 0)$.
(b) From the point $(2, 0)$ to the point $(-1, \sqrt{3})$ along the path shown on the figure:

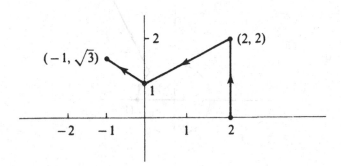

Solution. (a) $-\pi/2$.

(b) $2\pi/3$.

In Exercise 3, §3, we found a potential function for G in the upper half plane, namely $g(x, y) = \arccos(x/r)$. In (a) we have

$$\arccos\left(\frac{1}{\sqrt{1+0}}\right) - \arccos\left(\frac{0}{\sqrt{1+0}}\right) = -\frac{\pi}{2}$$

and in (b) we have

$$\arccos\left(\frac{-1}{\sqrt{1+3}}\right) - \arccos\left(\frac{2}{\sqrt{4}}\right) = \frac{2\pi}{3}.$$

Exercise XV.4.13 *Let F be a smooth vector field on $\mathbf{R}^2$ from which the origin has been deleted, so F is not defined at the origin. Let $F = (f, g)$. Assume that $D_2 f = D_1 g$ and let*

$$k = \frac{1}{2\pi} \int_C F,$$

where C is the circle of radius 1 centered at the origin. Let G be the vector field

$$G(x, y) = \left(\frac{-y}{x^2 + y^2}, \frac{x}{x^2 + y^2}\right).$$

Show that there exists a function φ defined on $\mathbf{R}^2$ from which the origin has been deleted such that

$$F = \operatorname{grad} \varphi + kG.$$

[*Hint: Follow the same method as in the proof of Theorem 4.2 in the text, but define $\varphi(P)$ by integrating F from the point $(1,0)$ to P as shown on the figure.*]

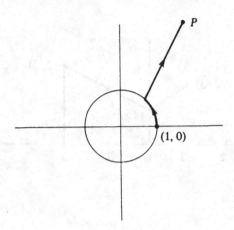

Solution. Suppose $\int_C F = 0$. Then F has a potential function. Indeed, define $\varphi(P)$ to be the integral of F along the path shown on the figure:

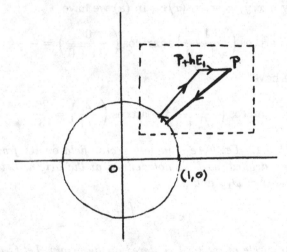

By assumption, φ is well defined. We want to show that $D_1\varphi(P) = f(P)$, so we proceed as in the text, we form the Newton quotient, namely

$$\frac{\varphi(P + he_1) - \varphi(P)}{h}.$$

But if we add the line segment from P to $P+he_1$ we see after a cancellation that we formed a closed path (in bold) and that

$$\varphi(P + he_1) - \varphi(P) - \int_{[P,P+he_1]} F = \int_{\text{closed path}} F.$$

For all small h we can inscribe the closed paths in a rectangle which does not intersect the origin. The assumptions imply that F has a potential function in the rectangle, thus

$$\int_{\text{closed path}} F = 0.$$

This implies

$$\frac{\varphi(P + he_1) - \varphi(P)}{h} = \frac{1}{h} \int_{[P,P+he_1]} F.$$

From Theorem 4.2 we know that the right-hand side tends to $f(P)$ as $h \to 0$, thus $D_1\varphi(P) = f(P)$.

If $\int_C F \neq 0$, then we have

$$\int_C F - kG = \int_C F - k \int_C G = 2\pi k - 2\pi k = 0.$$

Hence there exists a function φ such that $F - kG = \text{grad } \varphi$.

XV.5   Taylor's Formula

Exercise XV.5.1 *Let f be a differentiable function defined for all of $\mathbf{R}^n$. Assume that $f(O) = 0$ and that $f(tX) = tf(X)$ for all numbers t and vectors $X = (x_1, \ldots, x_n)$. Show that for all $X \in \mathbf{R}^n$ we have $f(X) = \text{grad } f(O) \cdot X$.*

Solution. See Exercise 5, Chapter XV, §2.

Exercise XV.5.2 *Let f be a function with continuous partial derivatives of order ≤ 2, that is of class C^2 on $\mathbf{R}^n$. Assume that $f(O) = 0$ and $f(tX) = t^2 f(X)$ for all numbers t and all vectors X. Show that for all X we have*

$$f(X) = \frac{(X \cdot \nabla)^2 f(O)}{2}.$$

Solution. Differentiate both sides of $f(tX) = t^2 f(X)$ with respect to t. For the left-hand side, use Theorem 5.1, so that

$$(X \cdot \nabla)^2 f(tX) = 2f(X).$$

Put $t = 0$.

Exercise XV.5.3 *Let f be a function defined on an open ball centered at the origin in $\mathbf{R}^n$ and assume that f is of class C^∞. Show that one can write*

$$f(X) = f(O) + g_1(X)x_1 + \cdots + g_n(X)x_n,$$

where $g_1, \ldots, g_n$ are functions of class C^∞. [Hint: Use the fact that

$$f(X) - f(O) = \int_0^1 \frac{d}{dt} f(tX) dt.\bigg]$$

Solution. The chain rule implies that

$$\frac{d}{dt} f(tX) = x_1 \frac{\partial f}{\partial x_1}(tX) + \cdots + x_n \frac{\partial f}{\partial x_n}(tX).$$

Then one can take

$$g_j(X) = \int_0^1 \frac{\partial f}{\partial x_j}(tX) dt.$$

Exercise XV.5.4 *Let f be a C^∞ function defined on an open ball centered at the origin in $\mathbf{R}^n$. Show that one can write*

$$f(X) = f(O) + \operatorname{grad} f(O) \cdot X + \sum_{i,j} g_{ij}(X) x_i x_j,$$

where g_{ij} are C^∞ functions. [Hint: Assume first that $f(O) = 0$ and $\operatorname{grad} f(O) = O$. In Exercises 3 and 4, use an integral form for the remainder.]

Solution. We use the integral expression twice. First we have

$$f(X) - f(O) = \int_0^1 \frac{d}{dt} f(tX) dt = \int_0^1 \sum_i D_i f(tX) x_i dt$$

$$= \int_0^1 \sum_i (D_i f(tX) - D_i f(O)) x_i dt + \operatorname{grad} f(O) \cdot X.$$

However,

$$D_i f(tX) - D_i f(O) = \int_0^1 \frac{d}{du} D_i f(tuX) du = \int_0^1 \sum_j D_j D_i f(tuX) x_j du,$$

so by plugging this expression in the above formula for $f(X) - f(O)$ and taking the sum signs out, we find the desired result.

Exercise XV.5.5 *Generalize Exercise 4 near an arbitrary point $A = (a_1, \ldots, a_n)$, expressing*

$$f(X) = f(A) + \sum_{i=1}^n D_i f(A)(x_i - a_i) + \sum_{i,j} h_{ij}(X)(x_i - a_i)(x_j - a_j).$$

Solution. Use the same method as in Exercise 4, or combine the result of Exercise 4 together with a change of variable. The proof goes as follows. Consider the function g defined by $g(Y) = f(Y + A)$, where $Y = X - A$. By Exercise 4 we have the expression

$$g(Y) - g(O) = \operatorname{grad} g(O) \cdot Y + \sum_{i,j} g_{ij}(Y) y_i y_j,$$

so using the expression for g, and the change of variable, we get

$$f(X - A + A) - f(A) = \operatorname{grad} f(A) \cdot (X - A) + \sum_{i,j} g_{ij}(X - A)(x_i - a_i)(x_j - a_j),$$

as was to be shown.

Exercise XV.5.6 *Let F_∞ be the set of all C^∞ functions defined on an open ball centered at the origin in $\mathbf{R}^n$. By a* **derivation** *D of F_∞ into itself, one means a map $D : F_\infty \to F_\infty$ satisfying the rules*

$$D(f + g) = Df + Dg, \quad D(cf) = cDf,$$
$$D(fg) = fD(g) + D(f)g,$$

for C^∞ functions f, g and constant c. Let $\lambda_1, \ldots, \lambda_n$ be the coordinate functions, that is $\lambda_i(X) = x_i$ for $i = 1, \ldots, n$. Let D be a derivation as above, and let $\psi_i = D(\lambda_i)$. Show that for any C^∞ function f on the ball, we have

$$D(f) = \sum_{i=1}^{n} \psi_i D_i f,$$

where $D_i f$ is the i-th partial derivative of f. [Hint: Show first that $D(1) = 0$ and $D(c) = 0$ for every constant c. Then use the representation of Exercise 5.]

Solution. We first show that $D(1) = 0$. This follows from the fact that $D(1) = D(1 \cdot 1)$ which implies

$$D(1) = 1D(1) + D(1)1 = 2D(1),$$

hence $D(1) = 0$ as was to be shown. We also have $D(c) = 0$ for all constants c because $D(c) = cD(1)$. Let A be a point. Now using the expression of Exercise 5 and the properties of the derivation we find

$$D(f) = \sum_{i=1}^{n} D_i f(A) D\lambda_i + \sum_{i,j} D\left[h_{ij}(\lambda_i - a_i)(\lambda_j - a_j)\right].$$

We now show that the last sum is zero when we put $X = A$. Using the properties of the derivation, we see that $D\left[h_{ij}(\lambda_i - a_i)(\lambda_j - a_j)\right]$ is equal to

$$D(h_{ij})(\lambda_i - a_i)(\lambda_j - a_j) + h_{ij}\left[(\lambda_i - a_i)D(\lambda_j) + D(\lambda_i)(\lambda_j - a_j)\right],$$

and evaluating this expression at A we find 0, so

$$D(f) = \sum_{i=1}^{n} D_i f(A) D\lambda_i(A).$$

This last expression holds for all points $A \in \mathbf{R}^n$ so $D(f) = \sum_{i=1}^{n} \psi_i D_i f$, as was to be shown.

Exercise XV.5.7 *Let $f(X)$ and $g(X)$ be polynomials in n variables $(x_1, \ldots, x_n)$ of degrees $\leq s - 1$. Assume that there is a number $a > 0$ and a constant C such that*

$$|f(X) - g(X)| \leq C|X|^s$$

for all X such that $|X| \leq a$. Show that $f = g$. In particular, the polynomial of Taylor's formula is uniquely determined.

Solution. It is sufficient to prove that if P is a polynomial of degrees $\leq s - 1$, and

$$|P(X)| \leq C|X|^s,$$

then $P = 0$. Set $x_1 = x_2 = \cdots = x_n$ and use Exercise 3, §3, of Chapter V to conclude.

Exercise XV.5.8 *Let U be open in $\mathbf{R}^n$ and let $f: U \to \mathbf{R}$ be a function of class C^p. Let $g: \mathbf{R} \to \mathbf{R}$ be a function of class C^p. Prove by induction that $g \circ f$ is of class C^p. Furthermore, assume that at a certain point $P \in U$ all partial derivatives*

$$D_{i_1} \cdots D_{i_r} f(P) = 0$$

for all choices of $i_1, \ldots, i_r$ and $r \leq k$. In other words, assume that all partials of f up to order k vanish at P. Prove that the same is true for $g \circ f$. [Hint: Induction.]

Solution. For each $0 \leq k \leq p$ the function $g^{(k)}$ exists and is continuous. We have

$$D_k(g \circ f)(X) = g'(f(X)) D_k(f(X)),$$

so $g \circ f$ is of class C^1. Suppose $g \circ f$ is of class C^r. Any expression of the form $D_1^{i_1} \cdots D_n^{i_n}(g \circ f)$ where $i_1 + \cdots + i_n = r$ can be written as a linear combination of terms of the form

$$g^{(k)}(f) \left[D_1^{a_1} \cdots D_n^{a_n} f \right] \left[(D_1 f)^{b_1} \cdots (D_n f)^{b_n} \right], \tag{XV.1}$$

where $k \leq r$, $a_1 + \cdots + a_n \leq r$, and $0 \leq b_1 + \cdots + b_n \leq r$. This is proved by induction. Applying D_j to the expression (XV.1) we get terms of the form given in (XV.1) but now the sum of the powers is bounded by $r + 1$.

Under the assumptions we see that we can differentiate one more time with respect to any variable and that the result is continuous for $g \circ f$ is of class C^{r+1}. By induction, we conclude that $g \circ f$ is of class C^p.

Furthermore, we see from this analysis that if all partial derivatives of f up to order k vanish at P, then all the partial derivatives of $g \circ f$ also vanish at P.

XV.6   Maxima and the Derivative

Exercise XV.6.1 *Find the maximum of $6x^2 + 17y^4$ on the subset of $\mathbf{R}^2$ consisting of those points (x, y) such that*

$$(x - 1)^3 - y^2 = 0.$$

Solution. Given any large x, there exists $y \geq 0$ such that $(x - 1)^3 - y^2 = 0$ and y is also large, so $6x^2 + 17y^4$ does not attain a maximum on the given set.

Exercise XV.6.2 *Find the maximum of $x^2 + xy + y^2 + yz + z^2$ on the sphere of radius 1 centered at the origin.*

Solution. Since $g(x, y, z) = x^2 + y^2 + z^2 = 1$, we can choose to maximize the function

$$f(x, y, z) = 1 + xy + yz.$$

We have grad $g = 2(x, y, z)$ which is never 0 when $g = 1$. So at an extremum point on the sphere there exists a number λ such that λgrad $g = $ grad f, thus

$$y = 2\lambda x, \quad x + z = 2\lambda y, \quad y = 2\lambda z \qquad \text{(XV.2)}$$

Suppose $\lambda = 0$. Then $y = 0$ and $x = -z$. From the equation $g = 1$ we conclude that $x = \pm 1/\sqrt{2}$, so we consider the two points

$$P_1 = (1/\sqrt{2}, 0, -1/\sqrt{2}) \quad \text{and} \quad P_2 = (-1/\sqrt{2}, 0, 1/\sqrt{2}).$$

Since $y = 0$ we have

$$f(P_1) = f(P_2) = 1.$$

Suppose $\lambda \neq 0$. Then from the first and third equation of (XV.2) we see that $x = z$. The second equation implies $2x = \lambda y = \lambda^2 x$ so $x(\lambda^2 - 2) = 0$. We cannot have $x = 0$ because $(0, 0, 0)$ is not a point on the sphere, so we must have $\lambda = \pm\sqrt{2}$. If $\lambda = \sqrt{2}$, substituting in the equation $g = 1$ we find the two points

$$Q_1 = (1/2, -\sqrt{2}/2, 1/2) \quad \text{and} \quad Q_2 = (-1/2, \sqrt{2}/2, -1/2).$$

Evaluate f at each of these two points to find that the maximum of f on the sphere is $1 + 1/\sqrt{2}$.

Exercise XV.6.3 *Let f be a differentiable function on an open set U in $\mathbf{R}^n$, and suppose that P is a minimum for f on U, that is $f(P) \leq f(X)$ for all X in U. Show that all partial derivatives $D_i f(P) = 0$.*

Solution. See Exercise 9, Chapter XV, §2, or consider functions of one variable and use the fact that if a function of one variable attains an extremum at an interior point, then the derivative vanishes.

Exercise XV.6.4 *Let A, B, C be three distinct points in $\mathbf{R}^n$. Let*

$$f(X) = (X - A)^2 + (X - B)^2 + (X - C)^2.$$

Find the point where f reaches its minimum and find the minimum value.

Solution. If $|X|$ is large, we see that $f(X)$ is large, so the function f reaches a minimum in some closed ball of large radius centered at the origin. The minimum is global and is not on the boundary of the ball, hence is a critical point (a critical point is a point where all the partial derivatives of f are zero, see Exercise 3). We use the notation

$$A = (a_1, a_2, a_3), \quad B = (b_1, b_2, b_3), \quad \text{and} \quad C = (c_1, c_2, c_3).$$

We have for $i = 1, 2, 3$

$$\frac{\partial f}{\partial x_i} = 2\left((x_i - a_1) + (x_i - b_1) + (x_i - c_1)\right) = 0$$

if and only if

$$x_i = \frac{a_1 + b_1 + c_1}{3}.$$

So the unique critical point of f is

$$M = \frac{1}{3}(A + B + C).$$

We then find the value of f at this point

$$f(M) = \frac{2}{3}(A^2 + B^2 + C^2 - AB - AC - BC).$$

Exercise XV.6.5 *Find the maximum of the function $f(x, y, z) = xyz$ subject to the constraints $x \geq 0, y \geq 0, z \geq 0$ and $xy + yz + xz = 2$.*

Solution. Let $g(x, y, z) = xy + yz + xz$. If one of the coordinates x, y, or z is 0, then $f(x, y, z) = 0$. Suppose that $x > 0$, $y > 0$, and $z > 0$. Then

$$\text{grad } g = (y + z, x + z, y + x) \quad \text{and} \quad \text{grad } f = (yz, xz, yx).$$

So there is a number λ such that

$$yz = \lambda(y+z), \quad xz = \lambda(x+z), \quad yx = \lambda(y+x). \qquad \text{(XV.3)}$$

By assumption we cannot have $\lambda = 0$ so we can take the ratio of the first and second equation in (XV.3), which gives

$$\frac{y}{x} = \frac{y+z}{x+z},$$

so $x = y$. Repeating the same argument shows that $x = y = z$. So f attains a maximum of $(2/3)^{3/2}$ at $\sqrt{2/3}(1,1,1)$.

Exercise XV.6.6 *Find the shortest distance from a point on the ellipse* $x^2 + 4y^2 = 4$ *to the line* $x + y = 4$.

Solution. The problem has a solution, as was shown in Exercise 5 of Chapter VIII, §2. Let $f(x_1, y_1) = x_1^2 + 4y_1^2, g(x_2, y_2) = x_2 + y_2$, and $h(x_1, y_1, x_2, y_2) = (x_1 - x_2)^2 + (y_1 - y_2)^2$. Our goal is to minimize h under the constraints $f = 4$ and $g = 4$. We can eliminate the second constraint by substituting $y_2 = 4 - x_2$. So we must minimize

$$h(x_1, y_1, x_2) = (x_1 - x_2)^2 + (y_1 - 4 + x_2)^2$$

subject to $f(x_1, y_1, x_2) = x_1^2 + 4y_1^2 = 4$. Setting

$$\text{grad } h = \lambda f$$

we get the system

$$\begin{cases} 2(x_1 - x_2) & = 2\lambda x_1, \\ 2(y_1 - 4 + x_2) & = 8\lambda y_1, \\ -2(x_1 - x_2) + 2(y_1 - 4 + x_2) & = 0, \end{cases}$$

this system is equivalent to

$$\begin{cases} x_1 - x_2 & = \lambda x_1, \\ y_1 - 4 + x_2 & = 4\lambda y_1, \\ x_1 - x_2 & = y_1 - 4 + x_2. \end{cases}$$

We find that $x_1 = 4y_1$ so plugging this in the equation of the ellipse we find that we have two possibilities,

$$A = (4/\sqrt{5}, 1/\sqrt{5}) \quad \text{and} \quad B = (-4/\sqrt{5}, -1/\sqrt{5}).$$

We can now find the various possibilities for x_2, and then plug in to find which combination gives us the smallest value of h. These simple computations show that the minimum distance between the ellipse and the line is

$$\frac{4 - \sqrt{5}}{2}.$$

Exercise XV.6.7 *Let S be the set of points $(x_1, \ldots, x_n)$ in $\mathbf{R}^n$ such that*

$$\sum x_i = 1 \quad and \quad x_i > 0 \quad for\ all\ i.$$

Show that the maximum of $g(x) = x_1 \cdots x_n$ occurs at $(1/n, \ldots, 1/n)$ and that

$$g(x) \leq n^{-n} \quad for\ all\ x \in S.$$

[Hint: Consider $\log g$.] Use the result to prove that the geometric mean of n positive numbers is less than or equal to the arithmetic mean.

Solution. Let $f(x) = \sum_i x_i$. Then S union its boundary is a compact set, and therefore g attains its maximum and minimum on this set. The minimum of f, namely 0, is attained on the boundary where one of the coordinates is 0. So at the maximum, which is an interior point, we have

$$\operatorname{grad} g = \lambda \operatorname{grad} f.$$

This equality yields the system of n linear equations

$$\begin{cases} x_2 x_3 \cdots x_n & = \lambda, \\ x_1 x_3 \cdots x_n & = \lambda, \\ \quad\vdots \\ x_1 x_2 \cdots x_{n-1} & = \lambda. \end{cases}$$

Multiplying the k-th equation by x_k we see that $\lambda x_1 = \cdots = \lambda x_n$. But $x_k > 0$ so $\lambda \neq 0$ hence $x_1 = \cdots = x_n$. From the constraint $f = 1$ we conclude that g attains its maximum $1/n^n$ at $(1/n, \ldots, 1/n)$. So $g(x) \leq n^{-n}$ for all $x \in S$.

Suppose we are given n positive numbers $a_1, \ldots, a_n$. Let

$$x_i = \frac{a_i}{\sum_k a_k}.$$

Then $\sum_i x_i = 1$. The inequality $g(x) \leq n^{-n}$ implies that

$$(a_1 \cdots a_n)^{1/n} \leq \frac{a_1 + \cdots + a_n}{n}.$$

Exercise XV.6.8 *Find the point nearest the origin on the intersection of the two surfaces*

$$x^2 - xy + y^2 - z^2 = 1 \quad and \quad x^2 + y^2 = 1.$$

Solution. The square of the distance function is

$$f(x, y, z) = x^2 + y^2 + z^2.$$

But $x^2 + y^2 = 1$, so the first constraint becomes $z^2 = -xy$, and we are reduced to the problem of minimizing the function $f(x, y) = 1 - xy$ subject to the constraint $g(x, y) = x^2 + y^2 - 1 = 0$ and $xy \leq 0$. If $x = 0$, then $y = \pm 1$ and similarly, if $y = 0$, then $x = \pm 1$. In the other cases, we have

$$\text{grad } f = \lambda \text{grad } g$$

which implies

$$-y = 2\lambda x \quad \text{and} \quad -x = 2\lambda y.$$

The number λ cannot be 0 because both x and y cannot be 0 at the same time. Taking the ratio of both equations, we find that $x^2 = y^2$, thus $x = \pm 1/\sqrt{2}$ and $y = \pm 1/\sqrt{2}$. By direct evaluation we find that the minimum of f is 1.

Exercise XV.6.9 *Find the maximum and minimum of the function* $f(x, y, z) = xyz$:
(a) *on the ball* $x^2 + y^2 + z^2 \leq 1$; *and*
(b) *on the plane triangle* $x + y + z = 4$, $x \geq 1$, $y \geq 1$, $z \geq 1$.

Solution. (a) Since grad $f = (yz, xz, xy)$, we see that f attains its maximum on the boundary of the ball. Let $g(x, y, z) = x^2 + y^2 + z^2$, then on the sphere, grad $g \neq 0$ so we solve grad $f = \lambda$grad g. We obtain the following three equations

$$yz = 2\lambda x, \quad xz = 2\lambda y, \quad xy = 2\lambda z.$$

If $\lambda = 0$, then at least one coordinate is 0, hence f equals 0.
If $\lambda \neq 0$ and none of the coordinates is 0, then we can take the ratio of two equations and we get $x^2 = y^2 = z^2$ and we find that the maximum of f is $(1/\sqrt{3})^3$ and the minimum is $-(1/\sqrt{3})^3$.
(b) On the boundary, we have $x = 1$ or $y = 1$ or $z = 1$. Suppose that $z = 1$, then we want to investigate xy subject to $x + y = 3$ and $x \geq 1, y \geq 1$. Write $y = 3 - x$, so that the study is reduced to the polynomials $3x - x^2$ for $x \geq 1$ and $x \leq 2$. This polynomial reaches a maximum of $9/4$ when $x = 3/2$ and it reaches a minimum of 2 when $x = 1$ or $x = 2$. For the other sides of the triangle we simply permute the coordinates to see that the maximum and minimum remain the same.
Now let $g(x, y, z) = x + y + z$. Then grad $g \neq 0$ so we solve

$$\text{grad } f = \lambda \text{grad } g.$$

We otain

$$yz = \lambda, \quad xz = \lambda, \quad xy = \lambda.$$

Since $xyz \neq 0$ we must have $\lambda \neq 0$, so we can take the ratio of two equations. We obtain $x = y = z$, so $x = y = z = 4/3$. Then $f(4/3, 4/3, 4/3) = (4/3)^3$.
We conclude that f attains a maximum of $(4/3)^3$ and a minimum of 2 in the given region.

Exercise XV.6.10 *Find the maxima and minima of the function*

$$(ax^2 + by^2)e^{-x^2-y^2}$$

if a, b are numbers with $0 < a < b$.

Solution. Let $f(x,y) = (ax^2 + by^2)e^{-x^2-y^2}$. Then for all x, y we have $f(x,y) \geq 0$ and $f(0,0) = 0$ so 0 is the minimum value of f. If we set $\partial f/\partial x = \partial f/\partial y = 0$, then we are reduced to the following two equations:

$$x(a - ax^2 - by^2) = 0 \quad \text{and} \quad y(b - ax^2 - by^2) = 0.$$

We cannot have both x and y nonzero, for otherwise $a = b$. If $x = 0$, then from the second equation we get $y = \pm 1$. If $y = 0$ see that $x = \pm 1$. Since $a < b$ we conclude that the maximum of f is be^{-1}.

Exercise XV.6.11 *Let A, B, C denote the intercepts which the tangent plane at (x, y, z)*

$$(x > 0, y > 0, z > 0)$$

on the ellipsoid

$$\frac{x^2}{a^2} + \frac{y^2}{b^2} + \frac{z^2}{c^2} = 1$$

makes on the coordinate axes. Find the point on the ellipsoid such that the following functions are a minimum:
(a) $A + B + C$.
(b) $\sqrt{A^2 + B^2 + C^2}$.

Solution. (a) The gradient is a vector orthogonal to the surface, so an equation of the plane passing through (x, y, z) on the ellipse and tangent to the ellipse is given by

$$\frac{x}{a^2}X + \frac{y}{b^2}Y + \frac{z}{c^2}Z - 1 = 0,$$

so $A = a^2/x$, $B = b^2/y$, and $C = c^2/z$. Let

$$g(x,y,z) = \frac{x^2}{a^2} + \frac{y^2}{b^2} + \frac{z^2}{c^2},$$

and $f_1(x,y,z) = A + B + C = a^2/x + b^2/y + c^2/z$. Then from the equation $\lambda \operatorname{grad} f_1 = \operatorname{grad} g$ we get the following:

$$-\lambda a^4 = 2x^3, \quad -\lambda b^4 = 2y^3, \quad \text{and} \quad -\lambda c^4 = 2z^3.$$

Writting z and y as functions of x and substituting in the equation of the ellipse we get

$$x^2 \left(\frac{a^{-2/3} + b^{-2/3} + c^{-2/3}}{a^{4/3}} \right) = 1.$$

By symmetry, it is easy to find similar expressions for y and z. Finally we find that the minimum value of f_1 is

$$(a^{4/3} + b^{4/3} + c^{4/3})(a^{-2/3} + b^{-2/3} + c^{-2/3})^{1/2}.$$

(b) Let $f_2(x, y, z) = A^2 + B^2 + C^2 = a^4/x^2 + b^4/y^2 + c^4/z^2$. Then the equation grad $f_2 = \lambda$grad g yields

$$-a^6 = \lambda x^4, \quad -b^6 = \lambda y^4, \quad \text{and} \quad -c^6 = \lambda z^4. \tag{XV.4}$$

Taking the square root and substituting in the equation of the ellipse we get

$$\sqrt{\frac{-1}{\lambda}}(a + b + c) = 1$$

hence $-\lambda = (a + b + c)^2$. From the equations in (XV.4) we find that the minimum value of f_2 is $(a + b + c)^2$.

Exercise XV.6.12 *Find the maximum of the expression*

$$\frac{x^2 + 6xy + 3y^2}{x^2 - xy + y^2}.$$

Because there are only two variables, the following method will work: Let $y = tx$, and reduce the question to the single variable t.

Solution. When $x = 0$ and $y \neq 0$, the fraction equals 3. Suppose $x \neq 0$ and let $t = y/x$, so that the fraction becomes

$$f(t) = \frac{1 + 6t + 3t^2}{1 - t + t^2}.$$

Setting the derivative of f equal to 0 and disregarding the denominator, we find

$$-9t^2 + 4t + 7 = 0.$$

An analysis of the graph of f shows that the maximum of f happens at the largest of the roots of the above equation, namely at

$$t_0 = \frac{2 + \sqrt{67}}{9}.$$

Then we find that

$$f(t_0) = \frac{14 + 2\sqrt{67}}{3}.$$

Exercise 12 can be generalized to more variables, in which case the above method has to be replaced by a different conceptual approach, as follows.

Exercise XV.6.13 *Let A be a symmetric $n \times n$ matrix. Denote column vectors in $\mathbf{R}^n$ by X, Y, etc. Let $X \in \mathbf{R}^n$, let $f(X) = \langle AX, X \rangle$, so f is a quadratic form. Prove that the maximum of f on the sphere of radius 1 is the largest eigenvalue of A.*

Remark. If you know some linear algebra, you should know that the roots of the characteristic polynomial of A are precisely the eigenvalues of A.
Solution. We know that f attains its maximum at an eigenvector, say X. Let λ be the corresponding eigenvalue. Then

$$f(X) = {}^t X A X = {}^t X \lambda X = \lambda.$$

Conclude.

Exercise XV.6.14 *Let C be a symmetric $n \times n$ matrix, and assume that $X \mapsto \langle CX, X \rangle$ defines a symmetric positive definite scalar product on $\mathbf{R}^n$. Such a matrix is called* **positive definite**. *From linear algebra, prove that there exists a symmetric positive definite matrix B such that for all $X \in \mathbf{R}^n$ we have*

$$\langle CX, X \rangle = \langle BX, BX \rangle = \|BX\|^2.$$

Thus B is a square root of C, denoted by $C^{1/2}$. [Hint: The vector space $V = \mathbf{R}^n$ has a basis consisting of eigenvectors of C, so one can define the square root of C by the linear map operating diagonally by the square roots of the eigenvalues of C.]

Solution. The Spectral Theorem for symmetric operators guarantees the existence of a basis of eigenvectors of C. In this basis, the matrix is diagonal with eigenvalues on the diagonal. These eigenvalues are positive because C is positive definite. Indeed, if X is a non-zero eigenvector with eigenvalue λ, then

$$0 < \langle AX, X \rangle = \lambda \langle X, X \rangle$$

so $0 < \lambda$. Hence taking the square roots of the eigenvalues we see that there exists a matrix whose square is C.

Exercise XV.6.15 *Let A, C be symmetric $n \times n$ matrices, and assume that C is positive definite. Let $Q_A(X) = \langle AX, X \rangle$ and $Q_C(X) = \langle CX, X \rangle = \langle BX, BX \rangle$ with $B = C^{1/2}$. Let*

$$f(X) = Q_A(X)/Q_C(X) \quad for \ X \neq O.$$

Show that the maximum of f (for $X \neq O$) is the maximal eigenvalue of $B^{-1}AB^{-1}$. [Hint: Change variables, write $X = BY$.]

Solution. Changing variables $Y = BX$, and using the fact that B is symmetric we get

$$\frac{Q_A(X)}{Q_C(X)} = \frac{\langle AB^{-1}Y, B^{-1}Y \rangle}{\langle Y, Y \rangle} = \frac{\langle B^{-1}AB^{-1}Y, Y \rangle}{\langle Y, Y \rangle}$$
$$= \langle B^{-1}AB^{-1}(Y/|Y|), Y/|Y| \rangle.$$

But $Y/|Y|$ has norm 1, hence we want to maximize the form $Q_{B^{-1}AB^{-1}}$ on the unit sphere. Exercise 12 concludes the proof.

Exercise XV.6.16 *Let a, b, c, e, f, g be real numbers. Show that the maximum value of the expression*

$$\frac{ax^2 + 2bxy + cy^2}{ex^2 + 2fxy + gy^2} \quad (eg - f^2 > 0)$$

is equal to the greater of the roots of the equation

$$(ac - b^2) - T(ag - 2bf + ec) + T^2(eg - f^2) = 0.$$

Solution. Let

$$A = \begin{pmatrix} a & b \\ b & c \end{pmatrix} \quad \text{and} \quad C = \begin{pmatrix} e & f \\ f & g \end{pmatrix}.$$

Then the problem is to maximize

$$\frac{Q_A(X)}{Q_C(X)}.$$

By multiplying all numbers by -1 if necessary, we may assume that C is positive definite, because $\det C = eg - f^2 > 0$. Let B be a square root of C. Then by the previous exercises, the maximum of the above quotient is equal to the greater of the roots of the equation

$$\det(B^{-1}AB^{-1} - tI) = 0.$$

However,

$$B^{-1}AB^{-1} - tI = B^{-1}AB^{-1} - tB^{-1}CB^{-1} = B^{-1}(A - tC)B^{-1},$$

so the solution to the problem is equal to the greater of the roots of $\det(A - tC)$, which when we expand is the equation given in the exercise.

$$\frac{\partial \lambda(X)}{\psi_e(X)} = \frac{(4B^2 + E^2(B + E))}{\psi_e(X)} = \frac{D_e(\lambda + e)D}{\psi_e(X)}$$

$$= \psi_e(X) = \ln(\psi(X) - \psi_e(X))$$

that $\psi(X)$ [illegible] is the derivative to form since the form $D_e(\lambda + e)$ on the final sphere. Therefore D concludes the proof.

Exercise. X V.8.16 *Let* a, *b*, *c* *be two numbers. Show that the work* [illegible] *that of*A*b*. *comparing.*

$$\frac{dX}{\sqrt{2(\lambda + e)\psi}} = \sqrt{\frac{\frac{X^2 + eX}{\sqrt{\psi}}}{\frac{X^2 + eX}{\psi}}}$$ (by [illegible]) (by [illegible] [illegible])

is equal to the product of [formula] of the data by

$$(a + b)_e e^{\lambda(e)e} \frac{1}{\psi(X)} = \sqrt{e} + [illegible]^e + e^{e e + [illegible]} \frac{1}{\psi(X)} = \frac{1}{\psi}$$

Point in the e.

$$\psi = \left(\frac{e^e}{\psi}\right) - \partial_e \ln e - \left(\frac{e^e}{e\psi}\right) + \partial_e \psi$$

That the function is to derive e.

$$\frac{\frac{X}{e^e}}{\psi(X)}$$

By multiplying each of these by e_e all together, we now assume that C is [illegible] positive definite, omitting $dcD = \psi + \frac{\lambda^2 \psi D}{D_e}$. Take the positive square root of [illegible] on IV the previous exercise. The maximum of the above quotient is consistent in a point in the zones of the equation:

$$\lambda e(e^e = D_e^e \sqrt{e} + [illegible] e$$

so:

$$e^e = e(\frac{\psi^e}{e^e} + e)e^e + e + e_{ee}\frac{eX}{\psi} + e^e\psi^e\frac{e}{\psi} = D_e^e(e)D_e^e$$

is the solution, which is obtained by squaring the roots of the equation defined ψ_e, when we exponent is the solution given above in the exercise.

XVI
The Winding Number and Global Potential Functions

XVI.2 The Winding Number and Homology

Exercise XVI.2.1 *In Theorem 2.7, let γ_i be a small circle centered at P_i. Determine the value*

$$\int_{\gamma_i} G_{P_j}.$$

Solution. If $i \neq j$, then consider an open disc V such that $\gamma_i \subset V$ and $P_j \notin V$. Then by the integrability theorem we see that

$$\int_{\gamma_i} G_{P_j} = 0.$$

If $i = j$, then translating to the origin and using Exercise 4, §4, of Chapter XV we find that

$$\int_{\gamma_i} G_{P_j} = \int_{\gamma} G = 2\pi.$$

One can also do the computation with the parametrization

$$\gamma_i(\theta) = (x_i + r\cos\theta, y_i + r\sin\theta),$$

so that

$$\int_{\gamma_i} G_{P_i} = \int_{\gamma_i} \frac{-(y - y_i)}{(x - x_i)^2 + (y - y_i)^2} dx + \frac{(x - x_i)}{(x - x_i)^2 + (y - y_i)^2} dy$$

$$= \int_0^{2\pi} \frac{(-r\sin\theta)(-r\sin\theta)}{r^2} + \frac{(r\cos\theta)(r\cos\theta)}{r^2} d\theta = 2\pi.$$

Exercise XVI.2.2 *Give a complete proof of Theorem 2.7, using Theorem 2.6. [Hint: Let*

$$a_i = \frac{1}{2\pi} \int_{\gamma_i} F = \operatorname{res}_{P_i}(F).]$$

Solution. Define a_i as in the hint. The vector field $F - \sum a_i G_i$ is locally integrable on U^*. Moreover, U^* is connected because U is connected and if a path passes through a point P_j, then this path can be modified in a small disc around P_j so that P_j does not belong to the new path.

Let γ be a closed path in U^*. Then Theorem 2.6 implies

$$\int_\gamma F = \sum_i m_i \int_{\gamma_i} F \quad \text{and} \quad \int_\gamma G_{P_i} = 2\pi m_i,$$

so

$$\int_\gamma \left(F - \sum a_i G_i \right) = 0.$$

Theorem 4.2 of Chapter XV concludes the exercise.

XVI.5   The Homotopy Form of the Integrability Theorem

Exercise XVI.5.1 *Let A be a closed annulus bounded by two circles $|X| = r_1$ and $|X| = r_2$ with $0 < r_1 < r_2$. Let F be a locally integrable vector field on an open set containing the annulus. Let γ_1 and γ_2 be the two circles, oriented counterclockwise. Show that*

$$\int_{\gamma_1} F = \int_{\gamma_2} F.$$

Solution. We must exhibit a homotopy of closed paths between the two circles, so we must deform continuously one circle onto the other. This can be done as the figure shows, namely by considering a family of circles centered at the origin and whose radius increases from r_1 to r_2.

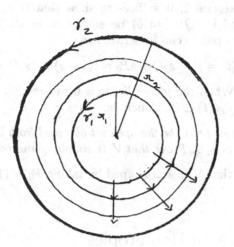

We contend that
$$\psi\colon [0,1] \times [0,1] \ \to \ \mathbf{R}^2,$$
$$(t,s) \ \mapsto \ (1-s)\gamma_1(t) + s\gamma_2(t),$$

does exactly what we want. Indeed, we have $\psi(t,0) = \gamma_1(t)$ and $\psi(t,1) = \gamma_2(t)$. Moreover, ψ is continuous and if we parametrize the circles by $\gamma_i(t) = (r_i \cos t, r_i \sin t)$ we see at once that ψ is a circle of radius $(1-s)r_1 + sr_2$.

Hence $\gamma_1 \approx \gamma_2$ and therefore

$$\int_{\gamma_1} F = \int_{\gamma_2} F.$$

Exercise XVI.5.2 *A set S is called* **star-shaped** *if there exists a point P_0 in S such that the line segment between P_0 and any point P in S is contained in S. Prove that a star-shaped set is simply connected, that is, every closed path is homotopic to a point.*

Solution. Clearly, S is connected because given two points A, B in S we can connect A to P_0 and then P_0 to B by line segments.

Given any continuous closed path γ parametrized on $[a,b]$ we must show that it is homotopic to a point. Let

$$\psi\colon [a,b] \times [0,1] \ \to \ \mathbf{R}^2,$$
$$(t,s) \ \mapsto \ (1-s)\gamma(t) + sP_0.$$

Then ψ is a homotopy of closed paths deforming γ in P_0. Indeed, γ is continuous, $\psi(t,o) = \gamma(t)$, $\psi(t,1) = P_0$, and for each s, the path $\psi(s)$ is closed because γ is closed, thus $\gamma \approx P_0$.

Exercise XVI.5.3 *Let U be the open set obtained from $\mathbf{R}^2$ by deleting the set of real numbers ≥ 0. Prove that U is simply connected.*

Solution. By Exercise 2, it suffices to show that U is star-shaped. Let $P_0 = (-1, 0)$, and let $Q = (a, b)$ be any point in U. The line segment between these two points can be written as

$$L_{Q,P_0}(t) = (1 - s)Q + sP_0 = ((1 - s)a - s, (1 - s)b),$$

where $s \in [0, 1]$. When the y coordinate is 0 we see that the x coordinate is negative, so $L_{Q,P_0}(t) \subset U$. Conclude.

Exercise XVI.5.4 *Let V be the open set obtained from $\mathbf{R}^2$ by deleting the set of real numbers ≤ 0. Prove that V is simply connected.*

Solution. Show that V is star-shaped by taking $P_0 = (1, 0)$. See the previous exercise.

XVI.6   More on Homotopies

In this section, all the homotopies are defined on the unit square

$$S_1 = [0, 1] \times [0, 1].$$

Exercise XVI.6.1 (Proposition 6.1) *Let $P, Q \in S$. If $\alpha, \beta, \gamma \in \mathrm{Path}(P, Q)$ and $\alpha \approx \beta$, $\beta \approx \gamma$, then $\alpha \approx \gamma$. If $\alpha \approx \beta$, then $\beta \approx \alpha$.*

Solution. Let $\alpha \approx_{h_1} \beta$ and $\beta \approx_{h_2} \gamma$. Then define h on S_1 by

$$h(t, u) = \begin{cases} h_1(t, u) & \text{if } 0 \leq u \leq 1/2, \\ h_2(t, 2u - 1) & \text{if } 1/2 \leq u \leq 1. \end{cases}$$

We see at once that $\alpha \approx_h \gamma$.

If $\alpha \approx_h \beta$, then $\beta \approx_{\tilde{h}_1} \alpha$ where $\tilde{h}(t, u) = h(t, 1 - u)$. This proves the second assertion.

Exercise XVI.6.2 (Proposition 6.2) *If $\alpha \approx \alpha_1$ and $\beta \approx \beta_1$, then $\alpha \# \beta \approx \alpha_1 \# \beta_1$.*
[Hint: Let $h(t, u) = h_1(2t, u)$ for $0 \leq t \leq 1/2$; $h(t, u) = h_2(2t - 1, u)$ for $1/2 \leq t \leq 1$.]

Solution. The idea behind the homotopy given in the hint, is that for each u we consider the union of the paths given by h_1 and h_2.

Exercise XVI.6.3 (Proposition 6.6) *Let $P \in S$ and let $\gamma \in \mathrm{Path}(P, P)$ be a closed curve in S. Suppose that γ is homotopic to a point Q in S, by a homotopy which does not necessarily leave the point P fixed. Then γ is also homotopic to P itself, by a homotopy which leaves P fixed.*
[Hint: Use Proposition 6.5 when γ_1 has the constant value Q. Then $\alpha \# \gamma_1 \# \alpha^-$ simply consists of first going along α, and then retracting your steps backward. You can then use Proposition 5.3(c).]

Solution. Proposition 6.5 implies that $\gamma \approx \alpha \# \gamma_1 \# \alpha^-$ where the homotopy leaves P fixed. But γ_1 has the constant value Q, so we have $\gamma = \alpha \# \alpha^-$ and since $\alpha \# \alpha^- \approx P$ by a homotopy which leaves P fixed, the proposition follows.

XVII
Derivatives in Vector Spaces

XVII.1   The Space of Continuous Linear Maps

Exercise XVII.1.1 *Let E be a vector space and let $v_1, \ldots, v_n \in E$. Assume that every element of E has a unique expression as a linear combination $x_1 v_1 + \cdots + x_n v_n$ with $x_i \in \mathbf{R}$. That is, given $v \in E$, there exist unique numbers $x_i \in \mathbf{R}$ such that*

$$v = x_1 v_1 + \cdots + x_n v_n.$$

Show that any linear map $\lambda \colon E \to F$ into a normed vector space is continuous.

Solution. Define the sup norm on E by $|v| = \max_i |x_i|$. Any norm in E is equivalent to the sup norm, because E is finite dimensional. The continuity of λ follows from

$$|\lambda(v)| = \left| \sum_{i=1}^n x_i \lambda(v_i) \right| \leq \sum_{i=1}^n |x_i| |\lambda(v_i)| \leq C|v|,$$

where $C = \sum_{i=1}^n |\lambda(v_i)|$.

Exercise XVII.1.2 *Let $\mathrm{Mat}_{m,n}$ be the vector space of all $m \times n$ matrices with components in $\mathbf{R}$. Show that $\mathrm{Mat}_{m,n}$ has elements e_{ij} ($i = 1, \ldots, m$ and $j = 1, \ldots, n$) such that every element A of $\mathrm{Mat}_{m,n}$ can be written in the form*

$$A = \sum_{i=1}^{m} \sum_{j=1}^{n} a_{ij} e_{ij},$$

with number a_{ij} uniquely determined by A.

Solution. Let e_{ij} be the matrix with 1 in the ij-entry and 0 in all the other entries. Then the set $\{e_{ij}\}$ does the job.

Exercise XVII.1.3 *Let E, F be normed vector spaces. Show that the association*

$$L(E, F) \times E \to F$$

given by

$$(\lambda, y) \mapsto \lambda(y)$$

is a product in the sense of Chapter VII, §1.

Solution. By definition we have $(\lambda_1 + \lambda_2)(v) = \lambda_1(v) + \lambda_2(v)$ and $(c\lambda)(v) = c\lambda(v)$. The map λ is linear, so the first two conditions of a product are verified. We must now verify the norm condition. By definition we see that

$$|\lambda(v)| \le |\lambda||v|$$

thereby proving that the given association is a product.

Exercise XVII.1.4 *Let E, F, G be normed vector spaces. A map*

$$\lambda \colon E \times F \to G$$

is said to be **bilinear** *if it satisfies the conditions*

$$\lambda(v, w_1 + w_2) = \lambda(v, w_1) + \lambda(v, w_2),$$
$$\lambda(v_1 + v_2, w) = \lambda(v_1, w) + \lambda(v_2, w),$$
$$\lambda(cv, w) = c\lambda(v, w) = \lambda(v, cw),$$

for all $v, v_i \in E$, $w, w_i \in F$, and $c \in \mathbf{R}$.
(a) Show that a bilinear map is continuous if and only if there exists $C > 0$ such that for all $(v, w) \in E \times F$ we have

$$|\lambda(v, w)| \le C|v||w|.$$

(b) Let $v \in E$ be fixed. Show that if λ is continuous, then the map $\lambda_v \colon F \to G$ given by $w \mapsto \lambda(v, w)$ is a continuous linear map.

Solution. (a) Suppose that λ is continuous. Then λ is continuous at O so there exists $\delta > 0$ such that if $|v| \le \delta$ and $|w| \le \delta$, then $|\lambda(v, w)| \le 1$. For arbitrary v and w, both non-zero, we can write

$$\lambda(v, w) = \frac{|v||w|}{\delta^2} \lambda\left(\frac{\delta v}{|v|}, \frac{\delta w}{|w|}\right).$$

Let $C = 1/\delta^2$ and conclude.

Conversely, suppose that $|\lambda(v, w)| \le C|v||w|$ for some number $C > 0$. We want to prove the continuity of λ at the point $(v_0, w_0) \in E \times F$. We have

$$\lambda(v, w) - \lambda(v_0, w_0) = \lambda(v, w - w_0) + \lambda(v - v_0, w_0)$$

so

$$|\lambda(v, w) - \lambda(v_0, w_0)| \le C|v||w - w_0| + C|v - v_0||w_0|.$$

But $|v| \le |v - v_0| + |v_0|$, so we see that λ is continuous.

(b) We simply have

$$|\lambda(v, w) - \lambda(v, w_0)| \le C|v||w - w_0|$$

so λ_v is continuous.

XVII.2   The Derivative as a Linear Map

Exercise XVII.2.1 *Find explicitly the Jacobian matrix of the polar coordinate map*

$$x = r\cos\theta \quad and \quad y = r\sin\theta.$$

Compute the determinant of this 2×2 matrix. The determinant of the matrix

$$\begin{pmatrix} a & b \\ c & d \end{pmatrix}$$

is by definition $ad - bc$.

Solution. Taking partial derivatives we find

$$J = \begin{pmatrix} \cos\theta & -r\sin\theta \\ \sin\theta & r\cos\theta \end{pmatrix}.$$

Therefore

$$\det(J) = r\cos^2\theta + r\sin^2\theta = r.$$

Exercise XVII.2.2 *Find the Jacobian matrix of the map $(u, v) = F(x, y)$ where*

$$u = e^x \cos y, \quad v = e^x \sin y.$$

Compute the determinant of this 2×2 matrix. The determinant of the matrix

$$\begin{pmatrix} a & b \\ c & d \end{pmatrix}$$

is by definition $ad - bc$.

Solution. Taking partial derivatives we find

$$J = \begin{pmatrix} e^x \cos y & -e^x \sin y \\ e^x \sin y & e^x \cos y \end{pmatrix}.$$

Therefore

$$\det(J) = e^{2x} \cos^2 \theta + e^{2x} \sin^2 \theta = e^{2x}.$$

Exercise XVII.2.3 *Let* $\lambda : \mathbf{R}^n \to \mathbf{R}^m$ *be a linear map. Show that* λ *is differentiable at every point, and that* $\lambda'(x) = \lambda$ *for all* $x \in \mathbf{R}^n$.

Solution. We have

$$\lambda(x + h) = \lambda(x) + \lambda(h) + |h|\psi(h),$$

where ψ is identically 0. The uniqueness of the derivative implies $D\lambda(x) = \lambda$ for all $x \in \mathbf{R}^n$.

XVII.3   Properties of the Derivative

Exercise XVII.3.1 *Let* U *be open in* E. *Assume that any two points of* U *can be connected by a continuous curve. Show that any two points can be connected by a piecewise differentiable curve.*

Solution. Let A and B be two points in U and let $\alpha : [a, b] \to U$ be a continuous curve such that $\alpha(0) = A$ and $\alpha(1) = B$. If $U = E$ we can just join A and B by a straight line. If $U \neq E$, then we proceed as follows. The image of α is compact, and $E - U$ is non-empty and closed, so by Exercise 5, §2, of Chapter VIII, the distance d between $\alpha([a, b])$ and $E - U$ exists and $d > 0$. Select r such that $0 < r < d/4$. By the uniform continuity of α, we can find a partition $\{a = a_0 < a_1 < \cdots < a_n = b\}$ of $[a, b]$ such that

$$\alpha(a_{i+1}) \in B_{r/2}(\alpha(a_i)).$$

Then we can connect the two consecutive points $\alpha(a_i)$ and $\alpha(a_{i+1})$ by a straight line segment which is entirely contained in U because of the convexity of the ball. Considering the union of all these line segments, we obtain a piecewise differentiable curve connecting A and B:

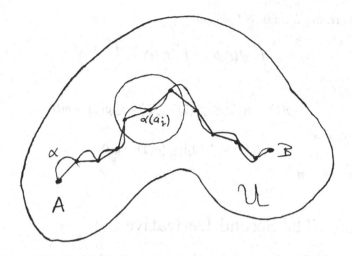

Exercise XVII.3.2 *Let* $f : U \to F$ *be a differentiable map such that* $f'(x) = 0$ *for all* $x \in U$. *Assume that any two points of* U *can be connected by a piecewise differentiable curve. Show that* f *is constant on* U.

Solution. Let A and B be two points in U and let $\alpha = \{\alpha_1, \ldots, \alpha_t\}$ be a piecewise differentiable curve connecting A and B. By the chain rule we have

$$(f \circ \alpha_i)'(t) = f'(\alpha_i(t)) \circ \alpha_i'(t) = 0,$$

hence $f \circ \alpha_i$ is constant. By induction, we find that if P_i is the end point of α_i, we get

$$f(A) = f(P_1) = f(P_2) = \cdots = f(P_r) = f(B)$$

and therefore we conclude that f is constant.

XVII.4   Mean Value Theorem

Exercise XVII.4.1 *Let* $f : [0, 1] \to \mathbf{R}^n$ *and* $g : [0, 1] \to \mathbf{R}$ *have continuous derivatives. Suppose* $|f'(t)| \leq g'(t)$ *for all* t. *Prove that* $|f(1) - f(0)| \leq |g(1) - g(0)|$.

Solution. We write $f(t) = (f_1(t), \ldots, f_n(t))$, where $f_i : [0, 1] \to \mathbf{R}$. Then

$$|f(1) - f(0)| = |f_m(1) - f_m(0)|$$

for some m, with $1 \leq m \leq n$. This follows at once from the definition of the sup norm. We have $|f_k'(t)| \leq g'(t)$ for each k where $1 \leq k \leq n$, so

$$-g'(t) \leq f_k'(t) \leq g'(t),$$

and integrating from 0 to 1 we get

$$-\int_0^1 g'(t)dt \le \int_0^1 f_k'(t) \le \int_0^1 g'(t).$$

Therefore

$$-(g(1) - g(0)) \le f_k(1) - f_k(0) \le g(1) - g(0)$$

which implies

$$|f_k(1) - f_k(0)| \le |g(1) - g(0)|,$$

for all $1 \le k \le n$.

XVII.5   The Second Derivative

Exercise XVII.5.1 *Let $E_1, \ldots, E_n, F$ be normed vector spaces and let*

$$\lambda : E_1 \times \cdots \times E_n \to F$$

be a multilinear map. Show that λ is continuous if and only if there exists a number $C > 0$ such that for all $v_i \in E_i$ we have

$$|\lambda(v_1, \ldots, v_n)| \le C|v_1||v_2| \cdots |v_n|.$$

Solution. If λ is continuous, it is continuous at the origin, so there exists $\delta > 0$ such that the inequalities $|v_1| \le \delta, \ldots, |v_n| \le \delta$ imply

$$|\lambda(v_1, \ldots, v_n)| \le 1.$$

Then for arbitrary $v_1, \ldots, v_n$, we have

$$|\lambda(v_1, \ldots, v_n)| = \frac{|v_1| \cdots |v_n|}{\delta^n} \lambda\left(\frac{\delta v_1}{|v_1|}, \ldots, \frac{\delta v_n}{|v_n|}\right).$$

Let $C = 1/\delta^n$ and conclude.

Conversely, suppose there exists a constant C such that

$$|\lambda(v_1, \ldots, v_n)| \le C|v_1| \cdots |v_n|.$$

Let $(a_1, a_2, \ldots, a_n) \in E_1 \times \cdots \times E_n$. Write

$$\lambda(v_1, \ldots, v_n) - \lambda(a_1, \ldots, a_n) = \sum_{i=1}^n h_i,$$

where

$$h_i = \lambda(a_1, \ldots, a_{i-1}, v_i, \ldots, v_n) - \lambda(a_1, \ldots, a_i, v_{i+1}, \ldots, v_n).$$

But
$$h_i = \lambda(a_1, \ldots, a_{i-1}, v_i - a_i, v_{i+1} \ldots v_n)$$

so

$$|\lambda(v_1, \ldots, v_n) - \lambda(a_1, \ldots, a_n)| \leq C \sum_{i=1}^{n} |a_1| \cdots |a_{i-1}| \cdot |v_i - a_i| \cdot |v_{i+1}| \cdots |v_n|.$$

Let $(v_1, \ldots, v_n)$ be in some open ball centered at $(a_1, \ldots, a_n)$. Then there exists a number $B > 0$ such that $|a_i|, |v_i| \leq B$ for all i so

$$|\lambda(v_1, \ldots, v_n) - \lambda(a_1, \ldots, a_n)| \leq CB^{n-1} \sum_{i=1}^{n} |v_i - a_i| \leq nCB^{n-1}|v - a|,$$

hence λ is continuous.

Exercise XVII.5.2 *Denote the space of continuous multilinear maps as above by $L(E_1, \ldots, E_n; F)$. If λ is in this space, define $|\lambda|$ to be the greatest lower bound of all numbers $C > 0$ such that*

$$|\lambda(v_1, \ldots, v_n)| \leq C|v_1||v_2| \cdots |v_n|$$

for all $v_i \in E_i$. Show that this defines a norm.

Solution. Clearly, $|\lambda| \geq 0$ and $|\lambda| = 0$ whenever $\lambda = 0$. Conversely, suppose $|\lambda| = 0$. Fix $(v_1, \ldots, v_n)$ and note that the inequality

$$|\lambda(v_1, \ldots, v_n)| \leq \epsilon |v_1||v_2| \cdots |v_n|$$

holds for every $\epsilon > 0$. Letting $\epsilon \to 0$ we see that $\lambda = 0$.

Clearly, $|c\lambda| = |c||\lambda|$ because

$$|c\lambda(v_1, \ldots, v_n)| = |c||\lambda(v_1, \ldots, v_n)|.$$

Finally, the triangle inequality follows because

$$|\lambda_1(v_1, \ldots, v_n) + \lambda_2(v_1, \ldots, v_n)| \leq |\lambda_1(v_1, \ldots, v_n)| + |\lambda_2(v_1, \ldots, v_n)|.$$

Exercise XVII.5.3 *Consider the case of bilinear maps. We denote by $L^2(E, F)$ the space of continuous bilinear maps of $E \times E \to F$. If $\lambda \in L(E, L(E, F))$, denote by f_λ the bilinear map such that $f_\lambda(v, w) = \lambda(v)(w)$. Show that $|\lambda| = |f_\lambda|$.*

Solution. Since
$$|\lambda(v)(w)| \leq |\lambda(v)||w| \leq |\lambda||v||w|,$$

we conclude that $|f_\lambda| \leq |\lambda|$. Now suppose $C = |f_\lambda| < |\lambda|$. By definition, there exists a vector v_0 of norm 1 such that

$$C < |\lambda(v_0)| \le |\lambda|.$$

But, by definition of C, for all vectors w of norm 1 we have

$$|\lambda(v_0)(w)| \le C|v_0||w| = C,$$

and therefore, for all vectors $u \ne O$, we have

$$\left|\lambda(v_0)\left(\frac{u}{|u|}\right)\right| \le C$$

which implies that

$$|\lambda(v_0)(u)| \le C|u|$$

hence $|\lambda(v_0)| \le C$. This contradiction implies that $|f_\lambda| = |\lambda|$.

Exercise XVII.5.4 *Let E, F, G be normed vector spaces. Show that the composition of mappings*

$$L(E, F) \times L(F, G) \to L(E, G)$$

given by $(\lambda, \omega) \mapsto \omega \circ \lambda$ is continuous and bilinear. Show that the constant C of Exercise 1 is equal to 1.

Solution. The map ω is linear, so

$$\omega \circ (\lambda_1 + \lambda_2) = \omega \circ \lambda_1 + \omega \circ \lambda_2 \quad \text{and} \quad \omega \circ (c\lambda) = c\omega \circ \lambda.$$

Similarly,

$$(\omega_1 + \omega_2) \circ \lambda = \omega_1 \circ \lambda + \omega_2 \circ \lambda \quad \text{and} \quad (c\omega) \circ \lambda = c\omega \circ \lambda,$$

so the composition of mappings $(\lambda, \omega) \mapsto \omega \circ \lambda$ is bilinear. Furthermore, this composition is continuous because

$$|\omega \circ \lambda(v)| \le |\omega||\lambda(v)| \le |\omega||\lambda||v|,$$

whence $|\omega \circ \lambda| \le |\omega||\lambda|$.

Exercise XVII.5.5 *Let f be a function of class C^2 on some open ball U in $\mathbf{R}^n$ centered at A. Show that*

$$f(X) = f(A) + Df(A) \cdot (X - A) + g(X)(X - A, X - A),$$

where $g : U \to L^2(\mathbf{R}^n, \mathbf{R})$ is a continuous map of U into the space of bilinear maps of $\mathbf{R}^n$ into $\mathbf{R}$. Show that one can select $g(X)$ to be symmetric for each $X \in U$.

Solution. We can write

$$f(X) - f(A) = \int_0^1 \frac{d}{dt} f(A + t(X - A)) dt.$$

Integrating by parts, we see that the integral equals

$$= \left[-(1-t)\frac{d}{dt} f(A + t(X - A)) \right]_0^1 + \int_0^1 (1-t)\frac{d^2}{dt^2} f(A + t(X - A)) dt$$

$$= Df(A) \cdot (X - A) + \int_0^1 (1-t)\frac{d^2}{dt^2} f(A + t(X - A)) dt.$$

But

$$\frac{d^2}{dt^2} f(A + t(X - A)) = \frac{d}{dt} \sum_i D_i f(A + t(X - A))(x_i - a_i)$$

$$= \sum_j \sum_i D_j D_i f(A + t(X - A))(x_i - a_i)(x_j - a_j),$$

so we see that we can write

$$\int_0^1 (1-t)\frac{d^2}{dt^2} f(A + t(X - A)) dt = g(X)(X - A, X - A),$$

where $g(X)$ is the bilinear map whose matrix is

$$\left(\int_0^1 (1-t) D_j D_i f(A + t(X - A)) dt \right)_{1 \le i,j \le n}.$$

Then g is symmetric because $D_j D_i = D_i D_j$.

XVII.6   Higher Derivatives and Taylor's Formula

Exercise XVII.6.1 *Let U be open in E and V open in F. Let*

$$f : U \to V \quad and \quad g : V \to G$$

be of class C^p. Let $x_0 \in U$. Assume that $D^k f(x_0) = 0$ for all $k = 0, \ldots, p$. Show that $D^k(g \circ f)(x_0) = 0$ for $0 \le k \le p$. [Hint: Induction.] Also, prove that if $D^k(g(f(x_0)) = 0$ for $0 \le k \le p$, then $(D^k g \circ f))(x_0) = 0$ for $0 \le k \le p$.

Solution. By the chain rule, we know that

$$D(f \circ g)(x) = Dg(f(x)) \circ Df(x),$$

hence $D(f \circ g)(x_0)$. The point now, is that composition of linear maps acts as a product, so differentiating one more time we find

$$D^2(g \circ f)(x) = D(Dg(f(x))) \circ Df(x) + Dg(f(x)) \circ D^2 f(x)$$
$$= D^2 g(f(x)) \circ Df(x) \circ Df(x) + Dg(f(x)) \circ D^2 f(x).$$

By induction we see that $D^k(g \circ f)(x)$ is a linear combination of terms of the form

$$D^m g(f(x)) \cdots D^n f(x)$$

with $1 \leq m, n \leq k$. Conclude.

XVIII
Inverse Mapping Theorem

XVIII.1 The Shrinking Lemma

Exercise XVIII.1.1 (Tate) *Let E, F be complete normed vector spaces. Let $f : E \to F$ be a map having the following property. There exists a number $C > 0$ such that for all $x, y \in E$ we have*

$$|f(x + y) - f(x) - f(y)| \leq C.$$

(a) Show that there exists a unique additive map $g \colon E \to F$ such that $g - f$ is bounded for the sup norm. [Hint: Show that the limit

$$g(x) = \lim_{n \to \infty} \frac{f(2^n x)}{2^n}$$

exists and satisfies $g(x + y) = g(x) + g(y)$.]
(b) If f is continuous, prove that g is continuous and linear.

Solution. (a) Let

$$f_n(x) = \frac{f(2^n x)}{2^n}.$$

We then have

$$|f_{n+1}(x) - f_n(x)| \leq \frac{C}{2^{n+1}}$$

because

$$|f(2^{n+1}x) - f(2^n x) - f(2^n x)| \leq C,$$

and therefore

$$|f_{n+m}(x) - f_n(x)| \leq |f_{n+m}(x) - f_{n+m-1}(x)| + |f_{n+m-1}(x) - f_{n+m-2}(x)|$$
$$+ \cdots + |f_{n+1}(x) - f_n(x)|$$
$$\leq \frac{C}{2^{n+1}} \left(\frac{1}{2^{m-1}} + \cdots + \frac{1}{2} + 1 \right)$$
$$\leq \frac{C}{2^n},$$

thus $\{f_n\}$ converges uniformly to a map $g = \lim_{n \to \infty} f_n$. Since

$$|f_n(x+y) - f_n(x) - f_n(y)| \leq \frac{C}{2^n}$$

in the limit we get $g(x + y) = g(x) + g(y)$, so g is additive. Clearly, $f - g$ is bounded because $f_0 = f$ and therefore

$$|f_m(x) - f(x)| \leq C$$

so it suffices to let $m \to \infty$. If there exists two additive functions g_1 and g_2 such that $f - g_1$ and $f - g_2$ are bounded, then $g_1 - g_2$ is additive and bounded. But if h is a bounded additive function, then

$$h(nx) = nh(x),$$

so h must be identically zero. This proves the uniqueness of g.

(b) If f is continuous, then each f_n is continuous, so the uniform limit g is also continuous. If p and q are integers, with $q \neq 0$, we see that additivity implies

$$qg\left(\frac{p}{q}x\right) = g(px) = pg(x)$$

so for all rational numbers r we have $g(rx) = rg(x)$. The continuity of g implies that the above relation also holds for all real numbers r. This proves that g is linear.

Exercise XVIII.1.2 *Generalize Exercise 1 to the bilinear case. In other words, let $f: E \times F \to G$ be a map and assume that there is a constant C such that*

$$|f(x_1 + x_2, y) - f(x_1, y) - f(x_2, y)| \leq C,$$
$$|f(x, y_1 + y_2) - f(x, y_1) - f(x, y_2)| \leq C,$$

for all $x, x_1, x_2 \in E$ and $y, y_1, y_2 \in F$. Show that there exists a unique biadditive map $g: E \times F \to G$ such that $f - g$ is bounded for the sup norm. If f is continuous, then g is continuous and bilinear.

Solution. Let

$$f_n(x, y) = \frac{f(2^n x, 2^n y)}{2^{2n}}.$$

Then
$$f_{n+1}(x,y) - f_n(x,y) = \varphi_n(x,y) + \psi_n(x,y),$$

where
$$\varphi_n(x,y) = \frac{f(2^{n+1}x, 2^{n+1}y)}{2^{2n+2}} - \frac{f(2^n x, 2^{n+1}y)}{2^{2n+1}}$$

and
$$\psi_n(x,y) = \frac{f(2^n x, 2^{n+1}y)}{2^{2n+1}} - \frac{f(2^n x, 2^n y)}{2^{2n}}.$$

Using the inequalities on f and arguing as in Exercise 1, we find that
$$|\varphi_n(x,y)| \leq \frac{C}{2^{2n+2}} \quad \text{and} \quad |\psi_n(x,y)| \leq \frac{C}{2^{2n+1}}.$$

So the triangle inequality implies
$$|f_{n+1}(x,y) - f_n(x,y)| \leq \frac{C}{2^{2n+2}} + \frac{C}{2^{2n+1}} \leq \frac{C}{2^n}.$$

The same argument as in Exercise 1 shows that the sequence $\{f_n\}$ converges uniformly to a map g and that g is additive in each variable, $f - g$ is bounded, and g is unique. If f is continuous, then so is each f_n and therefore g is continuous and bilinear.

Exercise XVIII.1.3 *Prove the following statement. Let $\overline{B}_r$ be the closed ball of radius r centered at 0 in E. Let $f : \overline{B}_r \to E$ be a map such that:*
(a) $|f(x) - f(y)| \leq b|x - y|$ with $0 < b < 1$.
(b) $|f(0)| \leq r(1 - b)$.
Show that there exists a unique point $x \in \overline{B}_r$ such that $f(x) = x$.

Solution. It is sufficient to show that the image of f is contained in $\overline{B}_r$, because then we can apply the shrinking lemma. The following inequalities give us what we want
$$|f(x)| \leq |f(x) - f(0)| + |f(0)| \leq br + (1 - b)r = r.$$

Exercise XVIII.1.4 *Notation as in Exercise 3, let g be another map of $\overline{B}_r$ into E and let $c > 0$ be such that $|g(x) - f(x)| \leq c$ for all x. Assume that g has a fixed point x_2, and let x_1 be the fixed point of f. Show that $|x_2 - x_1| \leq c/(1 - b)$.*

Solution. This result follows at once from
$$|x_2 - x_1| = |g(x_2) - f(x_1)| \leq |g(x_2) - f(x_2)| + |f(x_2) - f(x_1)|$$
$$\leq c + b|x_2 - x_1|.$$

Exercise XVIII.1.5 *Let K be a continuous function of two variables, defined for (x,y) in the square $a \leq x \leq b$ and $a \leq y \leq b$. Assume that*

$\|K\| \leq C$ *for some constant* $C > 0$. *Let* f *be a continuous function on* $[a, b]$ *and let* r *be a real number satisfying the inequality*

$$|r| < \frac{1}{C(b-a)}.$$

Show that there is one and only one function g *continuous on* $[a, b]$ *such that*

$$f(x) = g(x) + r \int_a^b K(t, x) g(t) dt.$$

Solution. Let $C^0[a, b]$ be the vector space of all continuous functions on $[a, b]$ equipped with the sup norm. Consider the map $\psi \colon C^0[a, b] \to C^0[a, b]$ defined by

$$g \mapsto f - r \int_a^b K g \, dt,$$

where

$$f - r \int_a^b K g \, dt \colon x \mapsto f(x) - r \int_a^b K(t, x) g(t) dt.$$

There exists a number k such that $0 < k < 1$ and $|r| \leq k/C(b-a)$. Then we have

$$\|\psi(g_1) - \psi(g_2)\| \leq k \|g_1 - g_2\|,$$

so by the shrinking lemma, there exists a unique function $g \in C^0[a, b]$ such that $\psi(g) = g$.

Exercise XVIII.1.6 (Newton's Method) *This method serves the same purpose as the shrinking lemma but sometimes is more efficient and converges more rapidly. It is used to find zeros of mappings.*

Let B_r *be a ball of radius* r *centered at a point* $x_0 \in E$. *Let* $f \colon B_r \to E$ *be a* C^2 *mapping, and assume that* f'' *is bounded by some number* $C \geq 1$ *on* B_r. *Assume that* $f'(x)$ *is invertible for all* $x \in B_r$ *and that* $|f'(x)^{-1}| \leq C$ *for all* $x \in B_r$. *Show that there exists a number* δ *depending only on* C *and* r *such that if* $|f(x_0)| \leq \delta$, *then the sequence defined by*

$$x_{n+1} = x_n - f'(x_n)^{-1} f(x_n)$$

lies in B_r *and converges to an element* x *such that* $f(x) = 0$. *[Hint: Show inductively that*

$$|x_{n+1} - x_n| \leq C |f(x_n)|,$$

$$|f(x_{n+1})| \leq |x_{n+1} - x_n|^2 \frac{C}{2},$$

and hence that

$$|f(x_n)| \leq \frac{(C^3 \delta)^{2^n}}{2^{2^n - 1} C^3},$$

$$|x_{n+1} - x_n| \leq \frac{(C^3 \delta)^{2n}}{2^{2^n - 1} C^2}.]$$

Solution. Select $\delta > 0$ such that

$$CC^{3(1+2+\cdots+2^n)}\delta^{2^n} < r/2^{n+2}$$

for all $n \geq 0$. Such a choice of δ is possible because

$$CC^{3(1+2+\cdots+2^n)} = CC^{3(2^{n+1}-1)} \leq C^{3(2^{n+1})}.$$

Now suppose that $|f(x_0)| < \delta$. We prove inductively, that

$$\begin{cases} |x_{n+1} - x_n| \leq CC^{3(1+2+\cdots+2^n)}\delta^{2^n} < r/2^{n+2}, \\ |f(x_{n+1})| \leq C^{3(1+2+\cdots+2^{n+1})}\delta^{2^{n+1}}. \end{cases}$$

We check the base step $n = 0$. We have

$$|x_1 - x_0| = |f'(x_0)^{-1}f(x_0)| \leq C|f(x_0)| \leq CC^3\delta.$$

But our choice of δ gives $CC^3\delta < r/2^2$, so $x_1 \in B_r$. Note that

$$f'(x_0)x_1 = f'(x_0)x_0 - f'(x_0)f'(x_0)^{-1}f(x_0) = f'(x_0)x_0 - f(x_0),$$

hence $f'(x_0)(x_1 - x_0) = -f(x_0)$. Now we use Taylor's formula given in §6 of Chapter XVII. We let $x = x_0$ and $y = x_1 - x_0$ and we use R_2. The preceding computations, the estimate of the remainder, and the bound for f'' on B_r imply

$$|f(x_1)| \leq C|x_1 - x_0|^2 \leq CC^2C^6\delta^2 = C^{3(1+2)}\delta^2,$$

and this proves the base step of the induction. Assume that the formulas are true for all integers $\leq n$. Then the recurrence formula for x_{n+2}, the induction assumption, and our choice of δ gives

$$|x_{n+2} - x_{n+1}| \leq C|f(x_{n+1})| \leq CC^{3(1+2+\cdots+2^{n+1})}\delta^{2^{n+1}} < \frac{r}{2^{n+3}}.$$

So we see that $x_{n+2} \in B_r$ because

$$|x_{n+2} - x_0| \leq |x_{n+2} - x_{n+1}| + |x_{n+1} - x_n| + \cdots + |x_1 - x_0|$$
$$\leq r\left(\frac{1}{2^{n+3}} + \cdots + \frac{1}{2^2}\right) \leq \frac{r}{2}.$$

From the recurrence formula for x_{n+2} we immediately get

$$f'(x_{n+1})(x_{n+2} - x_{n+1}) = -f(x_{n+1}),$$

so appyling Taylor's formula, with the bound on f'', we find that

$$|f(x_{n+2})| \leq C|x_{n+2} - x_{n+1}|^2,$$

and since

$$C|x_{n+2} - x_{n+1}|^2 \le CC^2 C^{3(2+\cdots+2^{n+2})}\delta^{2^{n+2}} = C^{3(1+2+\cdots+2^{n+2})}\delta^{2^{n+2}}.$$

This completes the proof by induction.

The sequence $\{x_n\}$ is Cauchy because

$$|x_{n+k} - x_n| \le r\left(\frac{1}{2^{n+k+1}} + \cdots + \frac{1}{2^{n+2}}\right) \le \frac{r}{2^{n+1}},$$

hence $\{x_n\}$ has a limit, say x which is in B_r because the above argument shows that in fact $\{x_i\}_{i=0}^{\infty}$ is contained in $B_{r/2}$, the ball of radius $r/2$ centered at x_0. The second formula we proved by induction implies

$$|f(x_{n+1})| \le \frac{Cr}{2^{n+3}},$$

so by continuity we conclude that $f(x) = 0$.

Exercise XVIII.1.7 *Apply Newton's method to prove the following statement. Assume that $f: U \to E$ is of class C^2 and that for some point $x_0 \in U$ we have $f(x_0) = 0$ and $f'(x_0)$ is invertible. Show that given y sufficiently close to 0, there exists x close to x_0 such that $f(x) = y$. [Hint: Consider the map $g(x) = f(x) - y$.]*

Solution. Applying Theorem 2.1 of this chapter and using the fact that f is of class C^2, we see that there exists an open ball B centered at x_0 and a constant $C \ge 1$ such that for all $x \in B$, the linear map $f'(x)$ is invertible and we have the bounds on f'' and $|f'(x)^{-1}|$ as in the previous exercise. Since

$$|g(x_0)| = |y|$$

we see that if y is sufficiently close to 0, we can apply Newton's method. Conclude.

Exercise XVIII.1.8 *The following is a reformulation due to Tate of a theorem of Michael Shub.*
(a) Let n be a positive integer, and let $f: \mathbf{R} \to \mathbf{R}$ be a differentiable function such that $f'(x) \ge r > 0$ for all x. Assume that $f(x+1) = f(x) + n$. Show that there exists a strictly increasing continuous map $\alpha: \mathbf{R} \to \mathbf{R}$ satisfying

$$\alpha(x+1) = \alpha(x) + 1$$

such that

$$f(\alpha(x)) = \alpha(nx).$$

[Hint: Follow Tate's proof. Show that f is continuous, strictly increasing, and let g be its inverse function. You want to solve $\alpha(x) = g(\alpha(nx))$. Let M be the set of all continuous functions which are increasing (not necessarily strictly) and satisfying $\alpha(x+1) = \alpha(x) + 1$. On M, define the norm

$$\|\alpha\| = \sup_{0 \le x \le 1} |\alpha(x)|.$$

Let $T: M \to M$ be the map such that

$$(T\alpha)(x) = g(\alpha(nx))).$$

Show that T maps M into M and is a shrinking map. Show that M is complete, and that a fixed point for T solves the problem.] Since one can write

$$nx = \alpha^{-1}(f(\alpha(x))$$

one says that the map $x \mapsto nx$ is conjugate to f. Interpreting this on the circle, one gets the statement originally due to Shub that a differentaible function on the circle, with positive derivative, is conjugate to the n-th power for some n.

(b) Show that the differentiability condition can be replaced by the weaker condition: There exist numbers r_1, r_2 with $1 < r_1 < r_2$ such that for all $x \ge 0$ we have

$$r_1 s \le f(x+s) - f(x) \le r_2 s.$$

Further problems involving similar ideas, and combined with another technique will be found at the end of the next section. It is also recommended that the first theorem on differential equations be considered simultaneously with these problems.

Solution. We prove (b). By assumption, f is continuous, strictly increasing, and therefore f has an inverse g.

We contend that T maps M into M. Clearly, $T\alpha$ is continuous and increasing because g and α are continuous and increasing. By induction, we find

$$\alpha(n(x+1)) = \alpha(nx) + n$$

so

$$f[(T\alpha)(x+1)] = \alpha(nx) + n,$$

and

$$f[(T\alpha)(x) + 1] = f[(T\alpha)(x)] + n = \alpha(nx) + n.$$

The function f is injective so $(T\alpha)(x+1) = (T\alpha)(x) + 1$ which proves our contention. The map T is a shrinking map because the condition on f implies

$$r_1(g(x) - g(y)) \le f(g(x)) - f(g(y)) \le r_2(g(x) - g(y)),$$

thus

$$\frac{x-y}{r_2} \le g(x) - g(y) \le \frac{x-y}{r_1},$$

so there exists a constant $0 < K < 1$ such that

$$|g(x) - g(y)| \leq K|x - y|.$$

Finally we show that M is complete. Let $\{\alpha_n\}$ be a Cauchy sequence in M. By induction, we see that $\alpha(x + j) = \alpha(x) + j$ for all integers j, so if $\|\alpha_n - \alpha_m\| < \epsilon$, then $|\alpha_n(x) - \alpha_m(x)| < \epsilon$ for all x. Use an argument as in Theorem 3.1 of Chapter VII, and the fact that the limit of a uniformly convergent sequence of continuous functions is continuous, to show that there exists a continuous function α such that $\alpha_n \to \alpha$ as $n \to \infty$. Since $\alpha_n(x + 1) = \alpha_n(x) + 1$ in the limit we have $\alpha(x + 1) = \alpha(x) + 1$, and α is increasing, whence M is complete.

The shrinking lemma implies that there exists a map α_0 such that $T\alpha_0 = \alpha_0$ or equivalently

$$g(\alpha_0(nx)) = \alpha_0(x),$$

thus $\alpha_0(nx) = f(\alpha_0(x))$.

XVIII.2 Inverse Mappings, Linear Case

Exercise XVIII.2.1 *Let E be the space of $n \times n$ matrices with the usual norm $|A|$ such that*

$$|AB| \leq |A||B|.$$

Everything that follows would also apply to an arbitrary complete normed vector space with an associative product $E \times E \to E$ into itself, and an element I which acts like a multiplicative identity, such that $|I| = 1$.
(a) Show that the series

$$\exp(A) = \sum_{n=0}^{\infty} \frac{A^n}{n!}$$

converges absolutely, and that $|\exp(A) - I| < 1$ if $|A|$ is sufficiently small.
(b) Show that the series

$$\log(I + B) = \frac{B}{1} - \frac{B^2}{2} + \cdots + (-1)^{n+1}\frac{B^n}{n} + \cdots$$

converges absolutely if $|B| < 1$ and that in that case

$$|\log(I + B)| \leq |B|/(1 - |B|).$$

If $|I - C| < 1$, show that the series

$$\log C = (C - I) - \frac{(C - I)^2}{2} + \cdots + (-1)^{n+1}\frac{(C - I)^n}{n} + \cdots$$

converges absolutely.

*(c) If $|A|$ is sufficiently small show that $\log \exp(A) = A$ and if $|C - I| < 1$
show that $\exp \log(C) = C$. [Hint: Approximate $\exp$ and $\log$ by the polynomials of the usual Taylor series, estimating the error terms.]*
(d) Show that if A, B commute, that is $AB = BA$, then

$$\exp(A + B) = \exp A \exp B.$$

State and prove the similar theorem for the $\log$.
*(e) Let C be a matrix sufficiently close to I. Show that given an integer
$m > 0$, there exists a matrix X such that $X^m = C$, and that one can choose
X so that $XC = CX$.*

Solution. (a) Since $|AB| \leq |A||B|$ we have $|A^n| \leq |A|^n$. The ratio test
implies at once the absolute convergence of the series. One could also compare the series with the series of the exponential function defined for real
numbers and use the comparison test.

We have

$$| \exp(A) - I| \leq \sum_{n \geq 1} \frac{|A|^n}{n!} = e^{|A|} - 1,$$

but for x sufficiently small and positive, we know that $e^x - 1$ is < 1 so
$|\exp(A) - I| < 1$ if $|A|$ is sufficiently small. We could also estimate the
series on the right using the fact that for all $n \geq 2$ we have $n! \geq 2^n$.
(b) The estimate

$$\left| (-1)^{n+1} \frac{B^n}{n} \right| \leq \frac{|B|^n}{n} \leq |B|^n,$$

and the fact that $|B| < 1$ imply the absolute convergence of the series. If
$|B| < 1$ we have

$$| \log(I + B)| \leq |B| + \cdots + \frac{|B|^n}{n} + \cdots$$

$$\leq |B| + \cdots + |B|^n + \cdots = \frac{|B|}{1 - |B|},$$

as was to be shown. To show the absolute convergence of the second series,
let $C - I = B$ and argue as we just did.
(c) We prove that $\log \exp(A) = A$ when A is sufficiently small. In fact,
suppose that $|A| < \log 2$. Then $|\exp(A) - I| \leq e^{|A|} - 1 < 2 - 1 = 1$ so the
expression $\log \exp(A)$ makes sense. It is sufficient to show that for large r
and s, the expression

$$\sum_{n=1}^{r} \frac{(-1)^{n-1}}{n} \left(\sum_{j=1}^{s} \frac{1}{j!} A^j \right)^n - A \qquad \text{(XVIII.1)}$$

has small absolute value. Suppose $r, s \geq t$ where t is a large positive integer.
We view the term on the left as a polynomial in A. Writing $\log e^x$ as a power

series (here x is a real number) and using the fact that $\log e^x - x = 0$ we see that the coefficients of all the terms of degree $\leq t$ are 0. Choose $0 < a < 1$ such that $|A| < a < \log 2$. Then there is a constant C depending only on a such that the absolute value of the coefficient of the term A^p in (XVIII.1) is $\leq C/a^p$. Putting absolute values in (XVIII.1) and using the triangle inequality we see that the desired expression is

$$\leq \sum_{p=t+1}^{rs} C\frac{|A|^p}{a^p} = C \sum_{p=t+1}^{rs} \left(\frac{|A|}{a}\right)^p \leq \frac{Ca}{a - |A|} \left(\frac{|A|}{a}\right)^{t+1}$$

whenever $r, s \geq t$. The above expression on the right, tends to 0 as $t \to \infty$, thereby concluding the proof of the first formula. The second formula $\exp \log C = C$ is proved using the same argument.

(d) The series $\sum A^n/n!$ and $\sum B^n/n!$ converge absolutely and the general term of the product of these two series is given by

$$\sum_{k=0}^{n} \frac{A^k}{k!} \frac{B^{n-k}}{(n-k)!}.$$

Since A and B commute, the binomial formula implies that

$$\sum_{k=0}^{n} \frac{A^k}{k!} \frac{B^{n-k}}{(n-k)!} = \frac{(A+B)^n}{n!},$$

and therefore $\exp(A + B) = \exp(A)\exp(B)$.

(e) If we take

$$X = \exp\left(\frac{1}{m}\log C\right),$$

then one sees at once that $X^m = \exp(\log C) = C$. Moreover

$$
\begin{aligned}
XC &= \exp\left(\frac{1}{m}\log C\right)\exp(\log C) \\
&= \exp\left(\frac{1}{m}\log C + \log C\right) \\
&= \exp\left(\log C + \frac{1}{m}\log C\right) \\
&= \exp(\log C)\exp\left(\frac{1}{m}\log C\right) = CX.
\end{aligned}
$$

Exercise XVIII.2.2 *Let U be the open ball of radius 1 centered at I. Show that the map* $\log: U \to E$ *is differentiable.*

Solution. The difficulty is that multiplication of matrices is not commutative, so we cannot simply differentiate the series term by term. Let $L_{A,n+1}$ be the linear map defined by

$$L_{A,n+1}(X) = A^n X + A^{n-1} X A + \cdots + A X A^{n-1} + X A^n.$$

We define a linear map L_C by

$$L_C = \sum_{n=1}^{\infty} (-1)^{n+1} \frac{L_{C-I,n}}{n}.$$

This series converges absolutely by comparison to the geometric series

$$\left| (-1)^{n+1} \frac{L_{C-I,n}}{n} \right| \le |C - I|^{n-1}.$$

We contend that $\log' C = L_C$. With the notation $C - I = C_I$, and the fact that we can write

$$(C_I + h)^n - C_I^n = L_{C-I,n}(h) + P_n(C_I, h)$$

(note that the binomial formula does not apply because the multiplication is not commutative) we get

$$\log(C_I + h) - \log C_I - L_C(h) = \sum_{n=1}^{\infty} (-1)^{n+1} \frac{P_n(C_I, h)}{n}.$$

Since the series on the left is absolutely convergent, so is the series on the right. If $|h| < \delta$ and $|C_I| < 1$, then we have the estimate

$$\left| \frac{P_n(C_I, h)}{n} \right| \le \frac{1}{n} \sum_{k=2}^{n} \binom{n}{k} |h|^k |C_I|^{n-k} \le \frac{|h|^2}{n} \sum_{k=2}^{n} \binom{n}{k} |\delta|^{k-2} = |h|^2 Q_n(\delta).$$

Conclude.

Exercise XVIII.2.3 *Let V be the open ball of radius 1 centered at 0. Show that the map* $\exp: V \to E$ *is differentiable.*

Solution. Let $L_{A,n+1}$ be the linear map defined by

$$L_{A,n+1}(X) = A^n X + A^{n-1} X A + \cdots + A X A^{n-1} + X A^n.$$

We define a linear map L_A by

$$L_A = \sum_{n=1}^{\infty} \frac{L_{A,n}}{n!}.$$

This series converges absolutely because of the estimate

$$\frac{|L_{A,n}|}{n!} \le \frac{n |A|^{n-1}}{n!} = \frac{|A|^{n-1}}{(n-1)!}$$

and the comparison test. We contend that $\exp' A = L_A$. We have

$$\exp(A+h) - \exp A = \sum_{n=1}^{\infty} \frac{(A+h)^n - A^n}{n!},$$

and we can write $(A+h)^n - A^n = L_{A,n}(h) + P_n(A,h)$ so that

$$\exp(A+h) - \exp A - L_A(h) = \sum_{n=1}^{\infty} \frac{P_n(A,h)}{n!}.$$

Since the left-hand side converges absolutely, so does the right-hand side. If $|A| < 1$ and $|h| < \delta$, then we have the estimate

$$\left| \frac{P_n(A,h)}{n!} \right| \leq \frac{1}{n!} \sum_{k=2}^{n} \binom{n}{k} |h|^k |A|^{n-k} \leq \frac{|h|^2}{n!} \sum_{k=2}^{n} \binom{n}{k} |\delta|^{k-2} = |h|^2 Q_n(\delta),$$

hence

$$\left| \sum_{n=1}^{\infty} \frac{P_n(A,h)}{n!} \right| \leq |h|^2 \sum_{n=1}^{\infty} Q_n(\delta).$$

Conclude. Note that we have never used the fact that $A \in V$ because the exponential is differentiable on all of E.

Exercise XVIII.2.4 *Let K be a continuous function of two variables, defined for (x,y) in the square $a \leq x \leq b$ and $a \leq y \leq b$. Assume that $\|K\| \leq C$ for some constant $C > 0$. Let f be a continuous function on $[a,b]$ and let r be a real number satisfying the inequality*

$$|r| < \frac{1}{C(b-a)}.$$

Show that there is one and only one function g continuous on $[a,b]$ such that

$$f(x) = g(x) + r \int_a^b K(t,x)g(t)dt.$$

(This exercise was also given in the preceding section. Solve it here by using Theorem 2.1.)

Solution. Consider the linear transformation $L \colon C^0[a,b] \to C^0[a,b]$ defined by

$$(Lg)(x) = -r \int_a^b K(t,x)g(t)dt.$$

Then $|L| < 1$ because there exists a number k with $0 < k < 1$ such that $|r| \leq k/C(b-a)$ hence

$$|(Lg)(x)| \leq k \sup_{t \in [a,b]} |g(t)|,$$

which means that $\|Lg\| \leq k\|g\|$, where $\| \cdot \|$ is the sup norm. Theorem 2.1 implies that $I - L$ is invertible. Conclude.

Exercise XVIII.2.5 *Exercises 5 and 6 develop a special case of a theorem of Anosov, by a proof due to Moser.*

*First we make some definitions. Let $A \colon \mathbf{R}^2 \to \mathbf{R}^2$ be a linear map. We say that A is **hyperbolic** if there exist numbers $b > 1$, $c < 1$, and two linearly independent vectors v, w in $\mathbf{R}^2$ such that $Av = bv$ and $Aw = cw$. As an example, show that the matrix (linear map)*

$$A = \begin{pmatrix} 2 & 1 \\ 3 & 2 \end{pmatrix}$$

has this property.

Next we introduce the C^1 norm. If f is a C^1 map, such that both f and f' are bounded, we define the C^1 norm to be

$$\|f\|_1 = \max(\|f\|, \|f'\|),$$

where $\| \cdot \|$ is the usual sup norm. In this case, we also say that f is C^1-bounded.

The theorem we are after runs as follows:

Theorem. *Let $A \colon \mathbf{R}^2 \to \mathbf{R}^2$ be a hyperbolic linear map. There exists δ having the following property. If $f \colon \mathbf{R}^2 \to \mathbf{R}^2$ is a C^1 map such that*

$$\|f - A\|_1 \leq \delta,$$

then there exists a continuous bounded map $h \colon \mathbf{R}^2 \to \mathbf{R}^2$ satisfying the equation

$$f \circ h = h \circ A.$$

First prove a lemma.

Lemma. *Let M be the vector space of continuous bounded maps of $\mathbf{R}^2$ into $\mathbf{R}^2$. Let $T \colon M \to M$ be the map defined by $Tp = p - A^{-1} \circ p \circ A$. Then T is a continuous linear map, and is invertible.*

To prove the lemma, write

$$p(x) = p^+(x)v + p^-(x)w$$

where p^+ and p^- are functions, and note that symbolically,

$$Tp^+ = p^+ - b^{-1}p^+ \circ A,$$

that is $Tp^+ = (I - S)p^+$ where $\|S\| < 1$. So find an inverse for T on p^+. Analogously, show that $Tp^- = (I - S_0^{-1})p^-$ where $\|S_0\| < 1$, so that $S_0 T = S_0 - I$ is invertible on p^-. Hence T can be inverted componentwise, as it were.

To prove the theorem, write $f = A + g$ where g is C^1-small. We want to solve for $h = I + p$ with $p \in M$, satisfying $f \circ h = h \circ A$. Show that this is equivalent to solving

$$Tp = -A^{-1} \circ g \circ h,$$

or equivalently

$$p = -T^{-1}(A^{-1} \circ g \circ (I + p)).$$

This is then a fixed point condition for the map $R: M \to M$ given by

$$R(p) = -T^{-1}(A^{-1} \circ g \circ (I + p)).$$

Show that R is a shrinking map to conclude the proof.

Solution. The numbers b and c are eigenvalues, and v, w are eigenvectors, so we must solve

$$\begin{vmatrix} 2 - \lambda & 1 \\ 3 & 2 - \lambda \end{vmatrix} = 0.$$

The solutions of this equation are $b = 2 + \sqrt{3}$ and $c = 2 - \sqrt{3}$. If we let $v = (1, \sqrt{3})$ and $w = (1, -\sqrt{3})$, then we have $Av = bv$ and $Aw = cw$ so the linear map A is hyperbolic.

To prove the lemma, we see that we can write

$$\begin{aligned} Tp &= p^+ v + p^- w - A^{-1}(p^+(A)v + p^-(A)w) \\ &= (p^+ - b^{-1}p^+(A))v + (p^- - c^{-1}p^-(A))w. \end{aligned}$$

On the first component we have

$$Tp^+ = p^+ - b^{-1}p^+(A) = (I - S)p^+,$$

where $S: p^+ \mapsto b^{-1}p^+(A)$. We find that $\|S\| < 1$ (for the sup norm on M) because $b > 1$ and

$$|Sp^+| = b^{-1}|p^+(A)| \le b^{-1}|p^+|.$$

Actually, we have equality because A is invertible. Hence $I - S$ is invertible and its inverse is given by the geometric series $I + S + S^2 + \cdots + S^n + \cdots$ (see Theorem 2.1).

On the second component we have

$$Tp^- = p^- - c^{-1}p^-(A) = (I - S_0^{-1})p^-,$$

where $S_0: p^- \mapsto cp^-(A^{-1})$, and hence $S_0^{-1}: p^- \mapsto c^{-1}p^-(A)$. We also have $\|S_0\| < 1$ because $c < 1$, so $I - S_0$ is invertible, whence $S_0^{-1}(I - S_0) = S_0^{-1} - I$ is invertible. This concludes the proof of the lemma.

We see that the equation $f \circ h = h \circ A$ is equivalent to

$$(A + g) \circ h = (I + p) \circ A,$$
$$A \circ p + g \circ h = p \circ A,$$
$$p = A^{-1}(p \circ A - g \circ h),$$
$$p - A^{-1}p \circ A = -A^{-1} \circ g \circ h,$$
$$Tp = -A^{-1} \circ g \circ h.$$

To apply the shrinking lemma which would guarantee the existence of p, we must show that R is a shrinking map. We have

$$R(p)(x) - R(q)(x) = -T^{-1}A^{-1}(g(x + p(x)) - g(x + q(x)))$$

$$(\text{XVIII.2})$$

for all x. Since $\|T^{-1}\|$ depends only on A, we can select $B > 0$ (depending only on A) such that $\|T^{-1}\|\|A^{-1}\| \le B$. Putting absolute values in (XVIII.2) and using the mean value theorem with $\|g\|_1 < \delta$ we get

$$|R(p)(x) - R(q)(x)| \le B\delta|x + p(x) - x - q(x)|$$

for all x, so

$$\|R(p) - R(q)\| \le B\delta\|p - q\|,$$

where $\| \cdot \|$ denotes the sup norm. It is now clear that if we choose δ small enough, then R is a shrinking map and we are done.

Exercise XVIII.2.6 *One can formulate a variant of the preceding exercise (actually the very case dealt with by Anosov-Moser). Assume that the matrix A with respect to the standard basis of $\mathbf{R}^2$ has integer coefficients. A vector $z \in \mathbf{R}^2$ is called an* **integral** *vector if its coordinates are integers. A map $p: \mathbf{R}^2 \to \mathbf{R}^2$ is said to be* **periodic** *if*

$$p(x + z) = p(x)$$

for all $x \in \mathbf{R}^2$ and all integral vectors z. Prove:

Theorem. *Let A be hyperbolic, with integer coefficients. There exists δ having the following property. If g is a C^1, periodic map, and $\|g\|_1 < \delta$, and if*

$$f = A + g,$$

then there exists a periodic continuous map h satisfying the equation

$$f \circ h = h \circ A.$$

Note. With a bounded amount of extra work, one can show that the map h itself is C^0-invertible, and so $f = h \circ A \circ h^{-1}$.

Solution. Let M_{per} denote the vector space of continuous and periodic maps of $\mathbf{R}^2$ into $\mathbf{R}^2$. With the notation of Exercise 5, we may view M_{per} as a subspace of M. This subspace is closed because the uniform limit of continuous periodic functions is continuous and periodic. To use the same proof as in Exercise 5 with M_{per} instead of M, we must show that T maps M_{per} into M_{per}, and that R maps M_{per} into M_{per}. Since A has integral coefficients, a brute force computation shows that if $x \in \mathbf{R}^2$, then

$$A(v + \text{integral vector}) = Av + \text{integral vector},$$

where the integral vector on the right-hand side need not be the same as the integral vector on the left-hand side. It follows that if z is an integral vector, then
$$Tp(x + z) = Tp(x)$$

whenever $p \in M_{\mathrm{per}}$, hence T maps M_{per} into M_{per}. It is clear that R maps M_{per} into M_{per}.

XVIII.3   The Inverse Mapping Theorem

Exercise XVIII.3.1 *Let $f : U \to F$ be of class C^1 on an open set U of E. Suppose that the derivative of f at every point of U is invertible. Show that $f(U)$ is open.*

Solution. Let $y \in f(U)$ and select x such that $f(x) = y$. By the inverse mapping theorem we know that f is locally C^1-invertible at x, so by definition there exists an open set U_1 such that $x \in U_1$ and $f(U_1)$ is open. Since $y \in f(U_1)$ we conclude that $f(U)$ is open.

Exercise XVIII.3.2 *Let $f(x,y) = (e^x + e^y, e^x - e^y)$. By computing Jacobians, show that f is locally invertible around every point of $\mathbf{R}^2$. Does f have a global inverse on $\mathbf{R}^2$ itself?*

Solution. The Jacobian of f at a point (x, y) is

$$J_f(x, y) = \begin{pmatrix} e^x & e^y \\ e^x & -e^y \end{pmatrix},$$

whose determinant is $-2e^x e^y \neq 0$, so f is locally invertible around every point of $\mathbf{R}^2$. Note that f is injective because if $f(x_1, y_1) = f(x_2, y_2)$, then

$$\begin{cases} e^{x_1} + e^{y_1} = e^{x_2} + e^{y_2}, \\ e^{x_1} - e^{y_1} = e^{x_2} - e^{y_2}, \end{cases}$$

so adding the two equations we get $x_1 = x_2$ and subtracting the two equations we see that $y_1 = y_2$. This shows that

$$\overline{f} \colon \mathbf{R}^2 \to f(\mathbf{R}^2)$$

has a set inverse. Note that

$$f(\mathbf{R}^2) = \{(x, y) \in \mathbf{R}^2 : x > y\} = V$$

because $e^x + e^y > e^x - e^y$ and if $(a, b) \in V$, then

$$\overline{f}\left(\log \frac{a+b}{2}, \log \frac{a-b}{2}\right) = (a, b).$$

From this analysis, we also see that the map $g \colon V \to \mathbf{R}^2$ defined by

$$g(x, y) = \left(\log \frac{x+y}{2}, \log \frac{x-y}{2}\right)$$

is a C^1-inverse for $\overline{f}$ because $g(\overline{f}(x, y)) = (x, y)$ and all the partial derivatives of g exist and are continuous on V.

Exercise XVIII.3.3 *Let $f : \mathbf{R}^2 \to \mathbf{R}^2$ be given by $f(x, y) = (e^x \cos y,$ $e^x \sin y)$. Show that $Df(x, y)$ is invertible for all $(x, y) \in \mathbf{R}^2$, that f is locally invertible at every point, but does not have an inverse defined on all of $\mathbf{R}^2$.*

Solution. The Jacobian of f at (x, y) is

$$J_f(x, y) = \begin{pmatrix} e^x \cos y & -e^x \sin y \\ e^x \sin y & e^x \cos y \end{pmatrix}$$

and its determinant is $e^{2x} \neq 0$, thus f is locally invertible around every point of $\mathbf{R}^2$. However, f does not have an inverse defined on all of $\mathbf{R}^2$ because it is not injective on all of $\mathbf{R}^2$. Indeed,

$$f(x, y) = f(x, y + 2\pi).$$

Exercise XVIII.3.4 *Let $f \colon \mathbf{R}^2 \to \mathbf{R}^2$ be given by $f(x, y) = (x^2 - y^2, 2xy)$. Determine the points of $\mathbf{R}^2$ at which f is locally invertible, and determine whether f has an inverse defined on all of $\mathbf{R}^2$.*

Solution. We have

$$J_f(x, y) = \begin{pmatrix} 2x & -2y \\ 2y & 2x \end{pmatrix}$$

so that $\det J_f(x, y) = 4x^2 + 4y^2$. Hence f is locally invertible at every point of $\mathbf{R}^2 - \{O\}$. The map f does not have an inverse defined on all of $\mathbf{R}^2$ because f is not injective on all of $\mathbf{R}^2$. Indeed, $f(1, 1) = f(-1, -1)$. In fact,

$$f(t, t) = f(-t, -t)$$

for all $t \in \mathbf{R}$.

XVIII.5   Product Decompositions

Exercise XVIII.5.1 *Let $f: \mathbf{R}^2 \to \mathbf{R}$ be a function of class C^1. Show that f is not injective, that is there must be points $P, Q \in \mathbf{R}^2$, $P \neq Q$, such that $f(P) = f(Q)$.*

Solution. Assuming f not constant, we can apply Exercise 2. Here we give a proof where we only assume that f is continuous. After translation and multiplication by a real number, we may assume without loss of generality that $f(0,0) = 0$ and $f(1,0) = 1$ (if $f(1,0) = 0$, there is nothing to prove). We look at lines passing through the origin
Case 1. If $f(0,1) = 0$ we are done.
Case 2. If $\alpha = f(0,1) > 0$, select $\beta > 0$ such that $\beta < \alpha$ and $\beta < 1$. The intermediate value theorem guarantees the existence of $0 < x, y < 1$ such that

$$f(x,0) = \beta \quad \text{and} \quad f(0,y) = \beta,$$

so f is not injective.
Case 3. If $\alpha = f(0,1) < 0$, then consider the point $(-1,0)$. If $f(-1,0) = 0$ we are done. If $f(-1,0) > 0$, then arguing as in case 2 above, we see that there exists $0 < x, t < 1$ such that

$$f(-t,0) = f(x,0).$$

If $f(-1,0) < 0$ the same argument shows that there exists $0 < s, y < 1$ such that

$$f(-s,0) = f(0,y),$$

thereby concluding the exercise.

Exercise XVIII.5.2 *Let $f: \mathbf{R}^n \to \mathbf{R}^m$ be a mapping of class C^1 with $m < n$. Assume $f'(x_0)$ is surjective for some x_0. Show that f is not injective. (Actually much more is true, but it's harder to prove.)*

Solution. Let $E = \operatorname{Ker} f'(x_0)$. Then by assumption on the dimensions, $E \neq \emptyset$. Select F such that $E \oplus F = \mathbf{R}^n$. Since we assume $f'(x_0)$ surjective, we know that $D_2 f(x_0): F \to \mathbf{R}^m$ is an isomorphism. By Theorem 5.1 the map $\psi: E \times F \to E \times \mathbf{R}^m$ given by

$$(x,y) \mapsto (x, f(x,y))$$

is locally C^1-invertible at x_0. Let h be its local inverse and let π_2 be the projection of $E \times \mathbf{R}^m$ onto $\mathbf{R}^m$. Then we can select open sets V_1, V_2, and U in E, $\mathbf{R}^m$, and $\mathbf{R}^n$, respectively, such that the following diagram is commutative:

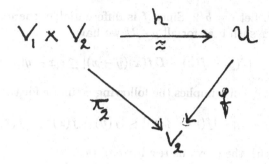

This implies that after a change of charts the map f becomes a projection, and therefore f is not injective.

Exercise XVIII.5.3 *Let $f: \mathbf{R} \to \mathbf{R}$ be a C^1 function such that $f'(x) \neq 0$ for all $x \in \mathbf{R}$. Show that f is a C^1-isomorphism of $\mathbf{R}$ with the image of f.*

Solution. Since the derivative of f is continuous and never 0, it is of constant sign. Assume without loss of generality that $f'(x) > 0$ for all $x \in \mathbf{R}$. Conclude, using Theorem 3.2 of Chapter III.

Exercise XVIII.5.4 *Let U be open in $\mathbf{R}^n$ and let $f: U \to \mathbf{R}^m$ be C^∞ with $f'(x): \mathbf{R}^n \to \mathbf{R}^m$ surjective for all x in U. Prove that $f(U)$ is open.*

Solution. Let $x \in U$. We contend that $f(x)$ has an open neighborhood which is contained on $f(U)$. Let $E = \operatorname{Ker} f'(x)$. Then E is a subspace of $\mathbf{R}^n$ so we can write $\mathbf{R}^n = E \times F$, where $F \approx \mathbf{R}^m$ because we assumed that $f'(x)$ is surjective. Now we can apply Theorem 5.1 of this section. Since ψ is a local isomorphism, it maps open sets onto open sets. Our contention follows from choosing a sufficiently small open neighborhood about x and using the product topology on $E \times F$.

Exercise XVIII.5.5 *Let $f: \mathbf{R}^m \to \mathbf{R}^n$ be a C^1 map. Suppose that $x \in \mathbf{R}^m$ is a point at which $Df(x)$ is injective. Show that there is an open set U containing x such that $f(y) \neq f(x)$ for all $y \in U$.*

Solution. The continuous linear map $Df(x)$ is a bijection from $\mathbf{R}^m$ onto its image. Let L be its inverse. Then L is continuous so there exists a constant $c > 0$ such that $|L(z)| \leq c|z|$, so if we let $z = Df(x)(y)$ we see that for all y we have

$$b|y| \leq |Df(x)y|,$$

where $b = c^{-1}$. Let $\epsilon = b/2$. Since f is differentiable, there exists a neighborhood U of x such that for all $y \in U$ we have

$$|f(y) - f(x) - Df(x)(y - x)| \leq \epsilon |x - y|.$$

The triangle inequality implies the following estimate for the left-hand side

$$|Df(x)(y - x)| - |f(y) - f(x)| \leq |f(y) - f(x) - Df(x)(y - x)|,$$

so combined with the previous result we obtain

$$b|x - y| - \epsilon|x - y| \leq |f(y) - f(x)|.$$

Since $\epsilon = b/2$ we have

$$\frac{b}{2}|y - x| \leq |f(y) - f(x)|$$

for all $y \in U$. This proves that $f(y) \neq f(x)$ whenever $y \neq x$ and $y \in U$.

Exercise XVIII.5.6 *Let $[a, b]$ be a closed interval J and let $f : J \to \mathbf{R}^2$ be a map of class C^1. Show that the image $f(J)$ has* **measure 0** *in $\mathbf{R}^2$. By this we mean that given ϵ, there exists a sequence of squares $\{S_1, S_2, \ldots\}$ in $\mathbf{R}^2$ such that the area of the square S_n is equal to some number K_n and we have*

$$f(J) \subset \bigcup S_n \quad and \quad \sum K_n < \epsilon.$$

Genrealize this to a map $f : J \to \mathbf{R}^3$, in which case measure 0 is defined by using cubes instead of squares.

Solution. See Exercise 7.

Exercise XVIII.5.7 *Let U be open in $\mathbf{R}^2$ and let $f : U \to \mathbf{R}^3$ be a map of class C^1. Let A be a compact subset of U. Show that $f(A)$ has measure 0 in $\mathbf{R}^3$. (Can you generalize this, to maps of $\mathbf{R}^m$ into $\mathbf{R}^n$ when $n > m$?)*

Solution. Propositions 2.1 and 2.2 in §2 of Chapter XX are also valid when replacing "negligible set" by "set of measure 0". We see at once that combined with the remark after Proposition 2.2 of Chapter XX, these results imply Exercises 6 and 7.

Exercise XVIII.5.8 *Let U be open in $\mathbf{R}^n$ and let $f : U \to \mathbf{R}^m$ be a C^1 map. Assume that $m \leq n$ and let $a \in U$. Assume that $f(a) = 0$, and that the rank of the matrix $(D_j f_i(a))$ is m, if $(f_1, \ldots, f_m)$ are the coordinate functions of f. Show that there exists an open subset U_1 of U containing a and a C^1-isomorphism $\varphi : V_1 \to U_1$ (where V_1 is open in $\mathbf{R}^n$) such that*

$$f(\varphi(x_1, \ldots, x_n)) = (x_{n-m+1}, \ldots, x_n).$$

Solution. We use the notation $x = (x_1, \ldots, x_n)$. Since the rank of the matrix $(D_j f_i(a))$ is m, we may assume that the matrix $(D_j f_i(a))$ ($i = 1, \ldots, m$) and ($j = n - m + 1, \ldots, n$) is invertible. By Theorem 5.1, the map $\psi : U \to \mathbf{R}^{n-m} \times \mathbf{R}^m$ given by

$$(x_1, \ldots, x_n) \mapsto (x_1, \ldots, x_{n-m}, f(x))$$

is a local C^1-isomorphism at a. Let U_1 be an open neighborhood of a where ψ has an inverse which we denote by $\varphi \colon V_1 \to U_1$, where V_1 is open in $\mathbf{R}^n$. If we write $\varphi = (\varphi_1, \ldots, \varphi_n)$, then

$$x = \psi(\varphi(x)) = (\varphi_1, \ldots, \varphi_{n-m}, f(\varphi(x))),$$

so that

$$(x_{n-m+1}, \ldots, x_n) = f(\varphi(x)),$$

thereby concluding the exercise.

Exercise XVIII.5.9 *Let $f : \mathbf{R} \times \mathbf{R} \to \mathbf{R}$ be a C^1 function such that $D_2 f(a, b) \neq 0$, and let g solve the implicit function theorem, so that $f(x, g(x)) = 0$ and $g(a) = b$. Show that*

$$g'(x) = -\frac{D_1 f(x, g(x))}{D_2 f(x, g(x))}.$$

Solution. If $\psi : x \mapsto f(x, g(x))$, then we can write $\psi = f \circ h$ where $h \colon x \mapsto (x, g(x))$. The chain rule implies that

$$\psi'(x) = D_1 f(x, g(x)) + g'(x) D_2 f(x, g(x)).$$

But ψ is identically 0, so the desired formula drops out.

Exercise XVIII.5.10 *Generalize Exercise 9, and show that in Theorem 5.4, the derivative of g is given by*

$$g'(x) = -(D_2 f(x, g(x)))^{-1} \circ D_1 f(x, g(x)).$$

Solution. If $\psi \colon x \mapsto f(x, g(x))$ and $h \colon x \mapsto (x, g(x))$, then $\psi = f \circ h$. The chain rule implies that

$$D_1 f(x, g(x)) + D_2 f(x, g(x)) \circ g'(x) = 0,$$

and therefore

$$g'(x) = -(D_2 f(x, g(x)))^{-1} \circ D_1 f(x, g(x)).$$

Exercise XVIII.5.11 *Let $f : \mathbf{R} \to \mathbf{R}$ be of class C^1 and such that $|f'(x)| \leq c < 1$ for all x. Define*

$$g : \mathbf{R}^2 \to \mathbf{R}^2$$

by

$$g(x, y) = (x + f(y), y + f(x)).$$

Show that the image of g is all of $\mathbf{R}^2$.

Solution. Let v be a point of $\mathbf{R}^2$. Consider the map $T_v \colon \mathbf{R}^2 \to \mathbf{R}^2$ defined by

$$(x, y) \mapsto -g(x, y) + (x, y) + v.$$

Using the fact that $|f(z_2) - f(z_1)| \leq c|z_2 - z_1|$ and the sup norm (or the euclidean norm) we get

$$|T_v(x_1, y_1) - T_v(x_2, y_2)| \leq c|(x_1, y_1) - (x_2, y_2)|,$$

which proves that T_v is a shrinking map. The shrinking lemma concludes the exercise.

Exercise XVIII.5.12 *Let $f \colon \mathbf{R}^n \to \mathbf{R}^n$ be a C^1 map, and assume that $|f'(x)| \leq c < 1$ for all $x \in \mathbf{R}^n$. Let $g(x) = x + f(x)$. Show that $g \colon \mathbf{R}^n \to \mathbf{R}^n$ is surjective.*

Solution. Let $v \in \mathbf{R}^n$. Consider the map $T_v \colon \mathbf{R}^n \to \mathbf{R}^n$ defined by

$$(x, y) \mapsto -g(x) + x + v.$$

Then we see that

$$|T_v(x_1) - T_v(x_2)| = |f(x_2) - f(x_1)| \leq c|x_2 - x_1|,$$

so T_v is a shrinking map. The shrinking lemma concludes the exercise.

Exercise XVIII.5.13 *Let $\lambda \colon E \to \mathbf{R}$ be a continuous linear map. Let F be its kernel, that is the set of all $w \in E$ such that $\lambda(w) = 0$. Assume that $F \neq E$ and let $v_0 \in E, v_0 \notin F$. Let F_1 be the subspace of E generated by v_0. Show that E is a direct sum $F \oplus F_1$ (in particular, prove that the map*

$$(w, t) \mapsto w + tv_0$$

is an invertible linear map from $F \times \mathbf{R}$ onto E).

Solution. Let v be an element of E, and let $t = \lambda(v)/\lambda(v_0)$. Let $w = v - tv_0$. Then

$$\lambda(w) = \lambda(v) - \frac{\lambda(v)}{\lambda(v_0)}\lambda(v_0) = 0$$

and therefore w belongs to the kernel of λ. So the continuous linear map $\alpha \colon F \times \mathbf{R} \to E$ given by

$$(w, t) \mapsto w + tv_0$$

is surjective. This map is also injective because if we assume that

$$w + tv_0 = w' + t'v_0,$$

where $w, w' \in F$ and $t, t' \in \mathbf{R}$, then

$$\lambda(w + tv_0) = \lambda(t'v_0)$$

which implies that $t = t'$ and therefore $w = w'$. Conclude.

Exercise XVIII.5.14 *Let $f(x, y) = (x \cos y, x \sin y)$. Show that the determinant of the Jacobian of f in the rectangle $1 < x < 2$ and $0 < y < 7$ is positive. Describe the image of the rectangle under f.*

Solution. The Jacobian of f at the point (x, y) is given by

$$J_f(x, y) = \begin{pmatrix} \cos y & -x \sin y \\ \sin y & x \cos y \end{pmatrix}$$

thus

$$\det J_f(x, y) = x,$$

and we see that the determinant of the Jacobian of f is indeed positive in the given rectangle. Since f describes the change from rectangular coordinates to polar coordinates, we see that the image of the rectangle under f is the annulus A defined by

$$A = \{(x, y) \in \mathbf{R}^2 : 1 < |(x, y)| < 2\},$$

where the norm is the euclidean norm

Exercise XVIII.5.15 *Let S be a submanifold of E, and let $P \in S$. If*

$$\psi_1 : U_1 \cap S \to V_1 \quad and \quad \psi_2 : U_2 \cap S \to V_2$$

are two charts for S at P (where U_1, U_2 are open in $\mathbf{R}^3$), show that there exists a local isomorphism between V_1 at $\psi_1(P)$ and V_2 at $\psi_2(P)$, mapping $\psi_1(P)$ on $\psi_2(P)$.

Solution. We can write

$$\psi_1(S \cap U_1 \cap U_2) = V_1' \quad and \quad \psi_2(S \cap U_1 \cap U_2) = V_2',$$

where V_1' is open in V_1 and V_2' is open in V_2. The map given by the composition $\psi_2 \circ \psi_1^{-1} : V_1' \to V_2'$ gives a local isomorphism between V_1 at $\psi_1(P)$ and V_2 at $\psi_2(P)$. The picture that describes the situation is the following:

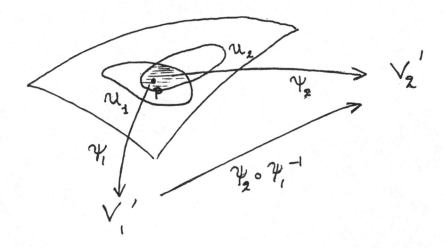

Exercise XVIII.5.16 *Let $\psi_1: U_1 \cap S \to V_1$ be a chart for S at P and let $g_1: V_1 \to U_1 \cap S$ be its inverse mapping. Suppose V_1 is open in F_1, and let $x_1 \in F_1$ be the point such that*

$$g_1(x_1) = P.$$

Show that the image of $g_1'(x_1): F_1 \to E$ is independent of the chart for S at P. (It is called the subspace of E which is parallel to the tangent space of S at P.)

Solution. Let $\psi_2: U_2 \cap S \to V_2$ be another chart for S at P, and let $g_2: V_2 \to U_2 \cap S$ be its inverse mapping. As in Exercise 15, let

$$\psi_1(S \cap U_1 \cap U_2) = V_1' \quad \text{and} \quad \psi_2(S \cap U_1 \cap U_2) = V_2'$$

and let $h: V_1' \to V_2'$ be the local C^1-isomorphism defined by

$$h = \psi_2 \circ \psi_1^{-1} = \psi_2 \circ g_1.$$

Then $g_1 = g_2 \circ h$ and by the chain rule we get

$$g_1'(x_1) = g_2'(h(x_1)) \circ h'(x_1) = g_2'(x_2) \circ h'(x_1).$$

Since $h'(x_1): F_1 \to F_2$ is invertible, it follows that the image of $g_1'(x_1)$ is the same as the image of $g_2'(x_2)$.

XIX
Ordinary Differential Equations

XIX.1 Local Existence and Uniqueness

Exercise XIX.1.1 *Let f be a C^1 vector field on an open set U in E. If $f(x_0) = 0$ for some $x_0 \in U$, if $\alpha : J \to U$ is an integral curve for f, and there exists some $t_0 \in J$ such that $\alpha(t_0) = x_0$, show that $\alpha(t) = x_0$ for all $t \in J$. (A point x_0 such that $f(x_0) = 0$ is called a **critical point** of the vector field.)*

Solution. Consider the map $\beta : J \to U$ defined by $\beta(t) = x_0$. Then β is an integral curve for f because for all $t \in J$ we have

$$0 = \beta'(t) = f(\beta(t)).$$

The uniqueness of integral curves with the same initial condition implies that $\alpha(t) = \beta(t)$ for all $t \in J$.

Exercise XIX.1.2 *Let f be a C^1 vector field on an open set U of E. Let $\alpha : J \to U$ be an integral curve for f. Assume that all numbers $t > 0$ are contained in J, and that there is a point P in U such that*

$$\lim_{t \to \infty} \alpha(t) = P.$$

Prove that $f(P) = 0$. (Exercises 1 and 2 have many applications, notably *when $f = \operatorname{grad} g$ for some function g. In this case, we see that P is a* **critical point** *of the vector field.)*

Solution. When t is large and $t' > t$, we have

$$\alpha(t') - \alpha(t) = \int_t^{t'} f(\alpha(u))du.$$

Write $f(\alpha(u)) = f(P) + g(u)$, where $\lim_{u \to \infty} g(u) = 0$. Then integrating and putting absolute values we see that

$$|f(P)||t' - t| \leq |\alpha(t') - \alpha(t)| + |t' - t| \sup_{t \leq u \leq t'} |g(u)|.$$

Divide by $|t' - t|$ and conclude that $|f(P)|$ is arbitrarily small.

Exercise XIX.1.3 *Let U be open in $\mathbf{R}^n$ and let $g : U \to \mathbf{R}$ be a function of class C^2. Let $x_0 \in U$ and assume that x_0 is a critical point of g (that is $g'(x_0) = 0$). Assume also that $D^2 g(x_0)$ is negative definite. By definition, take this to mean that there exists a number $c > 0$ such that for all vectors v we have*

$$D^2 g(x_0)(v, v) \leq -c|v|^2.$$

Prove that if x_1 is a point in the ball $B_r(x_0)$ of radius r, centered at x_0, and if r is sufficiently small, then the intergral curve α of $\operatorname{grad} g$ having x_1 as initial condition is defined for all $t \geq 0$ and

$$\lim_{t \to \infty} \alpha(t) = x_0.$$

[Hint: Let $\psi(t) = (\alpha(t) - x_0) \cdot (\alpha(t) - x_0)$ be the square of the distance from $\alpha(t)$ to x_0. Show that ψ is strictly decreasing, and in fact satisfies

$$\psi'(t) \leq -2c_1 \psi(t),$$

where $c_1 > 0$ is near c, and is chosen so that

$$D^2 g(x)(v, v) \leq -c_1|v|^2$$

for all x in a sufficiently small neighborhood of x_0.
    Divide by $\psi(t)$ and integrate to see that

$$\log \psi(t) - \log \psi(0) \leq -ct.$$

Alternatively, use the mean value theorem on $\psi(t_2) - \psi(t_1)$ to show that this difference has to approach 0 when $t_1 < t_2$ and t_1, t_2 are large.]

Solution. If $\psi(t) = (\alpha(t) - x_0) \cdot (\alpha(t) - x_0)$, then

$$\psi'(t) = 2\alpha'(t) \cdot (\alpha(t) - x_0) = 2Dg(\alpha(t)) \cdot (\alpha(t) - x_0).$$

Using the fact that $Dg(x_0) = 0$ and the usual integral expression we find that

$$Dg(y) = \int_0^1 D^2 g(x_0 + u(y - x_0))(y - x_0)du. \qquad \text{(XIX.1)}$$

Note that we are integrating linear maps. By continuity there exists a constant $K > 0$ and a ball B centered at x_0 such that if $z \in B$, then we have $D^2 g(z)(v, v) \leq -K|v|^2$ for all vectors v. Suppose y is in B. Then applying (XIX.1) to $y - x_0$ and using the above inequality, we find that

$$Dg(y)(y - x_0) \leq \int_0^1 -K|y - x_0|^2 du \leq -K|y - x_0|^2.$$

If $x_1 \in B$, then by continuity of α we see that for all small $t \geq 0$ we have $\alpha(t) \in B$ and therefore the above inequality can be applied with $y = \alpha(t)$, so that for all small $t \geq 0$ we have

$$\frac{1}{2}\psi'(t) = Dg(\alpha(t)) \cdot (\alpha(t) - x_0) \leq -K|\alpha(t) - x_0|^2 = -K\psi(t).$$

Hence for all small $t \geq 0$ we get

$$\psi'(t) \leq -2K\psi(t).$$

Thus ψ is decreasing and therefore $\alpha(t)$ gets closer to x_0 for increasing small values of t, hence for all $t \geq 0$ the inequality $\psi'(t) \leq -2K\psi(t)$ holds.

Now let $2K = c_1$. If $\alpha(t_0) = x_0$ for some positive t_0, then $\alpha(t) = x_0$ for all large t because ψ is decreasing and positive. If $\psi(t)$ is never 0 we can divide by it and then integrate, so that

$$\log \psi(t) - \log \psi(0) \leq -c_1 t,$$

and therefore $\lim_{t \to \infty} \psi(t) = 0$ as was to be shown.

Exercise XIX.1.4 *Let U be open in E and let $f : U \to E$ be a C^1 vector field on U. Let $x_0 \in U$ and assume that $f(x_0) = v \neq 0$. Let α be a local flow for f at x_0. Let F be a subspace of E which is complementary to the one-dimensional space generated by v, that is the map*

$$\mathbf{R} \times F \to E$$

given by $(t, y) \mapsto tv + y$ is an invertible continuous linear map.

(a) If $E = \mathbf{R}^n$ show that such a subspace exists.

(b) Show that the map $\beta : (t, y) \mapsto \alpha(t, x_0 + y)$ is a local C^1-isomorphism at $(0, 0)$. You may assume that $D_2\alpha$ exists and is continuous, and that $D_2\alpha(0, x) = \text{id}$. This will be proved in §4. Compute $D\beta$ in terms of $D_1\alpha$ and $D_2\alpha$.

(c) The map $\sigma : (t, y) \mapsto x_0 + y + tv$ is obviously a C^1-isomorphism, because it is composed of a translation and an invertible linear map. Define locally at x_0 the map φ by $\varphi = \beta \circ \sigma^{-1}$, so that by definition,

$$\varphi(x_0 + y + tv) = \alpha(t, x_0 + y).$$

Using the chain rule, show that for all x near x_0 we have

$$D\varphi(x)v = f(\varphi(x)).$$

If we view φ as a change of chart near x_0, then this result shows that the vector field f when transported by this change of chart becomes a constant vector field with value v. Thus near a point where a vector field does not vanish, we can always change the chart so that the vector field is straightened out. This is illustrated in the following picture:

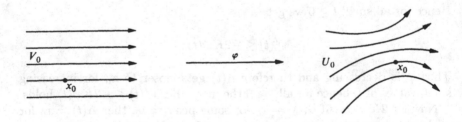

*In this picture we have drawn the flow, which is normalized on the left, the vector field being constant. In general, suppose $\varphi : V_0 \to U_0$ is a C^1-isomorphism. We say that a vector field g on V_0 and f on U_0 correspond to each other under φ, or that f is **transported** to V_0 by φ if we have the relation*

$$f(\varphi(x)) = D\varphi(x)g(x),$$

which can be regarded as coming from the following diagram:

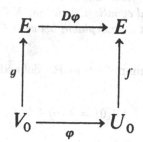

In the special case of our Exercise, g is the constant map such that $g(x) = v$ for all $x \in V_0$.

Solution. (a) If $E = \mathbf{R}^n$, then we can choose F to be the orthogonal complement of the one-dimensional space generated by v which we denote by V. Then

$$\mathbf{R}^n = V \oplus F.$$

One can simply construct a basis $\{v, v_2, \ldots, v_n\}$ of $\mathbf{R}^n$ and take F to be the subspace spanned by $\{v_2, \ldots, v_n\}$.

(b) Consider the maps χ_1 and χ_2 such that $\chi_1(t, y) = t$ and $\chi_2(t, y) = y$. Then $\beta = \alpha(\chi_1, x_0 + \chi_2)$ so that

$$D\beta(t, y) = D_1\alpha(t, x_0 + y)\chi_1 + D_2\alpha(t, x_0 + y)\chi_2.$$

Evaluating at $(t, y) = (0, 0)$ we get

$$D\beta(0, 0) = v\chi_1 + \mathrm{id}\chi_2.$$

So

$$D\beta(0, 0)(t', y') = vt' + y'$$

and therefore $D\beta$ is invertible at $(0, 0)$. The inverse mapping theorem implies that β is a local C^1-isomorphism.

(c) Differentiating with respect to t we see that

$$D\varphi(x)v = \alpha'(t, x_0 + y) = f(\alpha(t, x_0 + y)) = f(\varphi(x)).$$

XIX.3   Linear Differential Equations

Exercise XIX.3.1 *Let $A : J \to \mathrm{Mat}_{n \times n}$ be a continuous map from an open interval J containing 0 into the space of $n \times n$ matrices. Let S be the vector space of solutions of the differential equation*

$$X'(t) = A(t)X(t).$$

Show that the map $X \mapsto X(0)$ is a linear map from S into $\mathbf{R}^n$, whose kernel is $\{O\}$. Show that given any n-tuple $C = (c_1, \ldots, c_n)$ there exists a solution of the differential equation such that $X(0) = C$. Conclude that the map $X \mapsto X(0)$ gives an isomorphism between the space of solutions and $\mathbf{R}^n$.

Solution. Consider the map $\Phi \colon S \to \mathbf{R}^n$ defined by $X \mapsto X(0)$. Then by definition we have

$$\Phi(X + Y) = (X + Y)(0) = X(0) + Y(0) = \Phi(X) + \Phi(Y),$$

and

$$\Phi(cX) = (cX)(0) = cX(0) = c\Phi(X),$$

so Φ is linear. Since the function which is identically 0 satisfies the differential equation, the uniqueness of the solutions implies that the kernel of Φ is $\{O\}$. Moreover, given any n-tuple C, Theorem 3.1 guarantees the existence of a solution verifying $X(0) = C$. Thus the map Φ is bijective and therefore this map gives an isomorphism between the space of solutions and $\mathbf{R}^n$.

Exercise XIX.3.2 *(a) Let $g_0, \ldots, g_{n-1}$ be continuous functions from an open interval J containing 0 into $\mathbf{R}$. Show that the study of the differential equation*

$$D^n y + g_{n-1} D^{n-1} y + \cdots + g_0 y = 0$$

can be reduced to the study of a linear differential equation in n-space. [Hint: Let $x_1 = y, x_2 = y', \ldots, x_n = y^{(n-1)}$.]
(b) Show that the space of solutions of the equation in part (a) has dimension n.

Solution. Let (*) be the equation $D^n y + g_{n-1} D^{n-1} y + \cdots + g_0 y = 0$. We see that with the notation of the hint we have

$$Dx_1 = x_2, \ldots, Dx_{n-1} = x_n$$

and

$$Dx_n = -g_{n-1} x_n - \cdots - g_0 x_1.$$

Consider the matrix

$$A(t) = \begin{pmatrix} 0 & 1 & 0 & 0 & \cdots & 0 \\ 0 & 0 & 1 & 0 & \cdots & 0 \\ \vdots & \vdots & & & & \vdots \\ 0 & 0 & 0 & 0 & \cdots & 1 \\ -g_0(t) & -g_1(t) & \cdots & \cdots & \cdots & -g_{n-1}(t) \end{pmatrix}$$

and let $X(t) = (x_1(t), \ldots, x_n(t))$. If y solves (*), then we see that X solves the equation $X'(t) = A(t)X(t)$. Conversely, if X solves $X'(t) = A(t)X(t)$, then x_1 solves (*), so we have reduced the study of (*) to the study of a linear differential equation in n-space.

(b) Exercise 1 implies that the space of solutions of (*) has dimension n.

Exercise XIX.3.3 *Give an explicit power series solution for the differential equation*

$$\frac{du}{dt} = Au(t),$$

where A is a constant $n \times n$ matrix, and the solution $u(t)$ is in the space of $n \times n$ matrices.

Solution. The power series

$$u(t) = \sum_{n=0}^{\infty} \frac{(tA)^n}{n!}$$

converges absolutely and uniformly on every compact interval because

$$\left| \frac{(tA)^n}{n!} \right| \le \frac{|tA|^n}{n!}.$$

The derived series is given by

$$\sum_{n=0}^{\infty} A\frac{(tA)^n}{n!}$$

which also converges absolutely and uniformly on every compact interval. Therefore $u'(t) = Au(t)$, as was to be shown.

Exercise XIX.3.4 *Let $A\colon J \to L(E, E)$ and let $\psi\colon J \to E$ be continuous. Show that the integral curves of the differential equation*

$$\beta'(t) = A(t)\beta(t) + \psi(t)$$

are defined on all of J.

Solution. We proceed as in Theorem 3.1. Let $f(t, v) = A(t)v + \psi(t)$. Since

$$f(t, v_2) - f(t, v_1) = A(t)(v_2 - v_1),$$

we see that f verifies the Lipschitz condition of Theorem 2.4. Condition (i) is also satisfied. Proceeding as in the proof of Theorem 3.1 we see that $f(t, \beta(t))$ is bounded on $[0, t_0)$. Conclude.

Exercise XIX.3.5 *For each point $(t_0, x_0) \in J \times E$ let $v(t, t_0, x_0)$ be the integral curve of the differential equation*

$$\alpha'(t) = A(t)\alpha(t)$$

satisfying the condition $\alpha(t_0) = x_0$. Prove the following statements:
(a) For each $t \in J$, the map $x \mapsto v(t, s, x)$ is an invertible continuous linear map of E onto itself, denoted by $C(t, s)$.
(b) For fixed s, the map $t \mapsto C(t, s)$ is an integral curve of the differential equation

$$\omega'(t) = A(t) \circ \omega(t)$$

on $L(E, E)$, with initial condition $\omega(s) = \mathrm{id}$.
(c) For $s, t, u \in J$ we have

$$C(s, u) = C(s, t)C(t, u) \quad \text{and} \quad C(s, t) = C(t, s)^{-1}.$$

(d) The map $(s, t) \mapsto C(s, t)$ is continuous.

Solution. (a) Let α and β be solutions of the differential equation with $\alpha(t_0) = x_1$ and $\beta(t_0) = x_2$. Then $\alpha + \beta$ is a solution of the differential equation with $(\alpha + \beta)(t_0) = x_1 + x_2$ so the uniqueness theorem implies that

$$C(t, s)(x_1 + x_2) = C(t, s)(x_1) + C(t, s)(x_2).$$

Similarly, we see that $C(t, s)(ax) = aC(t, s)(x)$. Theorem 2.1 (with $\epsilon = 0$) implies at once the continuity of $C(t, s)$. We contend that the map $C(t, s)$ is bijective. Let $x_0 \in E$ and let ξ be the integral curve such that $\xi(t) = x_0$. Let $y = \xi(s)$. Then $C(t, s)(y) = x_0$ which shows that the map $C(t, s)$ is surjective. Clearly, the uniqueness theorem implies that $C(t, s)(x) = O$ if and only if $x = O$, so $C(t, s)$ is injective, thereby proving our contention.
(b) For every $x \in E$, we have

$$A(t)\omega(t)x = A(t)C(t, s)x = A(t)(v(t, s, x)) = v'(t, s, x)$$

and

$$(\omega(t)x)' = v'(t, s, x).$$

The rule for differentiating a product implies that $(\omega(t)x)' = \omega'(t)x$ so we see that ω is an integral curve for the differential equation $\omega'(t) = A(t) \circ \omega(t)$. Moreover we have $\omega(s) = C(s, s)$ and

$$C(s, s)x = v(s, s, x) = x,$$

thus $\omega(s) = \mathrm{id}$.
(c) We see that $\lambda: s \mapsto C(s, t)C(t, u)x$ solves the differential equation and we have $\lambda(t) = v(t, u, x)$ so the uniqueness theorem implies that

$$C(s, t)C(t, u)x = C(s, u)x$$

as was to be shown. For the second assertion, put $s = u$ in the formula we just proved.
(d) Observe that

$$C(s, t) = C(s, t_0)C(t_0, t) = C(s, t_0)C(t, t_0)^{-1}$$

and use part (a) to conclude.

Exercise XIX.3.6 *Show that the integral curve of the non-homogeneous differential equation*

$$\beta'(t) = A(t)\beta(t) + \psi(t)$$

such that $\beta(t_0) = x_0$ *is given by*

$$\beta(t) = C(t, t_0)x_0 + \int_{t_0}^t C(t, s)\psi(s)ds.$$

Solution. Aspplying the chain rule we get

$$
\begin{aligned}
\beta'(t) &= C'(t, t_0)x_0 + C(t, t)\psi(t) + \int_{t_0}^t C'(t, s)\psi(s)ds \\
&= C'(t, t_0)x_0 + \psi(t) + \int_{t_0}^t C'(t, s)\psi(s)ds,
\end{aligned}
$$

and we also have

$$
\begin{aligned}
A(t)\beta(t) + \psi(t) &= A(t)C(t, t_0)x_0 + \int_{t_0}^t A(t)C(t, s)\psi(s)ds + \psi(t) \\
&= C'(t, t_0)x_0 + \int_{t_0}^t C'(t, s)\psi(s)dx + \psi(t),
\end{aligned}
$$

thus proving that $\beta'(t) = A(t)\beta(t) + \psi(t)$. At $t = t_0$ we have $\beta(t_0) = x_0$ thereby proving that the integral curve of the given non-homogeneous differential equation is given by

$$\beta(t) = C(t, t_0)x_0 + \int_{t_0}^t C(t, s)\psi(s)ds.$$

XX
Multiple Integrals

XX.1   Elementary Multiple Integration

The first set of exercises shows how to generalize the class of integrable functions.

Exercise XX.1.1 *Let A be a subset of $\mathbf{R}^n$ and let $a \in A$. Let f be a bounded function defined on A. For each $r > 0$ define the **oscillation** of f on the ball of radius r centered at a to be*

$$o(f, a, r) = \sup |f(x) - f(y)|$$

the sup *being taken for all $x, y \in B_r(a)$. Define the **oscillation** at a to be*

$$o(f, a) = \lim_{r \to 0} o(f, a, r).$$

Show that this limit exists. Show that f is continuous at a if and only if

$$o(f, a) = 0.$$

Solution. The function f is bounded so $o(f, a, r)$ exists. If $r_1 < r_2$, then

$$o(f, a, r_1) \leq o(f, a, r_2),$$

so $\lim_{r \to 0} o(f, a, r)$ exists.

Suppose f is continuous at a. Then given $\epsilon > 0$ there exists r_0 such that if $|z - a| < r_0$, then $|f(z) - f(a)| < \epsilon/2$. So if $x, y \in B_{r_0}(a)$, then

$$|f(x) - f(y)| \leq |f(x) - f(a)| + |f(a) - f(y)| < \epsilon,$$

hence $o(f, a) = 0$. Conversely, if $o(f, a) = 0$, then given $\epsilon > 0$ there exists an r_0 such that if $r \leq r_0$, then $o(f, a, r) < \epsilon$. Hence if $x \in B_r(a)$, then

$$|f(x) - f(a)| < \epsilon$$

proving that f is continuous.

Exercise XX.1.2 *Let A be a closed set, and let f be a bounded function on A. Given ϵ, show that the subset of elements $x \in A$ such that $o(f, x) \geq \epsilon$ is closed.*

Solution. Assume that $\epsilon > 0$ and let

$$S = \{x \in S : o(f, x) \geq \epsilon\}.$$

Let $\{a_n\}$ be a sequence in S converging to some $a \in A$. We must show that $a \in S$. Suppose not, so that $o(f, a) < \epsilon$, and write $o(f, a) = \epsilon - \delta$, where $\delta > 0$. Select r_0 such that $o(f, a, r_0) < \epsilon - \delta/2$ and choose n_0 such that $a_{n_0} \in B_{r_0/2}(a)$. Then $o(f, a_{n_0}, r_0/2) < \epsilon$, which is a contradiction.

Exercise XX.1.3 *A set A is said to have **measure** 0 if given ϵ, there exists a sequence of rectangles $\{R_1, R_2, \ldots\}$ covering A such that*

$$\sum_{j=1}^{\infty} v(R_j) < \epsilon.$$

Show that a denumerable union of sets of measure 0 has measure 0. Show that a compact set of measure 0 is negligible.

Solution. Let $S_1, S_2, \ldots$ be sets of measure 0. For each i select a covering of S_i by rectangles $\{R_{1i}, R_{2i}, \ldots\}$ such that

$$\sum_{j=1}^{\infty} v(R_{ji}) < \epsilon 2^{-i}.$$

Then $\{R_{ji}\}_{1 \leq i,j < \infty}$ covers $\bigcup S_i$ and

$$\sum_{i=1}^{\infty} \sum_{j=1}^{\infty} v(R_{ji}) \leq \epsilon \sum_{i=1}^{\infty} 2^{-i} = \epsilon,$$

so $\bigcup S_i$ has measure 0.

Now suppose that S is compact and has measure 0. From any covering of S we can extract a finite subcovering, so from the definition of measure 0, we see that there exists a finite number of rectangles whose union covers S and such that the sum of their volumes is $< \epsilon$. This proves that S is negligible whenever S is compact and has measure 0.

Exercise XX.1.4 *Let f be a bounded function on a rectangle R. Let D be the subset of R consisting of points where f is not continuous. If D has measure 0, show that f is integrable on R. [Hint: Given ϵ, consider the set A of points x such that $o(f, x) \geq \epsilon$. Then A has measure 0 and is compact.]*

Solution. Since $A \subset D$, the set A has measure 0, and since A is bounded and closed, it is compact, thus A is negligible. Cover A by a finite number of open rectangles $R_1, \ldots, R_k$ such that $\sum v(R_i) < \epsilon$. Then the set

$$R \cap \left(\mathbf{R}^n - \bigcup_{j=1}^k R_j \right)$$

is closed and bounded, therefore compact, and by definition of the oscillation at each point $x \in C$ we can find a rectangle Q_x centered at x such that for all $y, z \in Q_x$, we have

$$|f(y) - f(z)| < \epsilon.$$

This condition will play the same role as uniform continuity in the proof of Theorem 1.3. From the covering $\{Q_x\}_{x \in C}$ of C we can extract a finite subcovering $\{Q_j\}_{1 \leq j \leq m}$. Then by projecting the sides of each rectangle $R_1, \ldots, R_k$ and $Q_1, \ldots, Q_m$ on each factor we get a partition P of R. Refine P so that if $S_1, \ldots, S_p$ are the rectangles of P which intersect $R_1, \ldots, R_k$, then $\sum v(S_i) < 2\epsilon$. To conclude, estimate $U(P, f) - L(P, f)$ as in the proof of Theorem 1.3.

Exercise XX.1.5 *Prove the converse of Exercise 4, namely: If f is integrable on R, then its set of discontinuities has measure 0. [Hint: Let $A_{1/n}$ be the subset of R consisting of all x such that $o(f, x) \geq 1/n$. Then the set of discontinuities of f is the union of all $A_{1/n}$ for $n = 1, 2, \ldots$ so it suffices to prove that each $A_{1/n}$ has measure 0, or equivalently that $A_{1/n}$ is negligible.]*

Solution. Given $\epsilon > 0$, there exists a partition $P = (S_1, \ldots, S_r)$ of R such that

$$U(P, f) - L(P, f) \leq \frac{\epsilon}{n}.$$

Let $S_1, \ldots, S_k$ be a finite number of rectangles of P which cover $A_{1/n}$ and let $S_1, \ldots, S_m$ be the rectangles whose interior intersects $A_{1/n}$. Then we have

$$\sum_{j=1}^m \left[\sup_{S_j}(f) - \inf_{S_j}(f) \right] v(S_j) < \frac{\epsilon}{n},$$

however $A_{1/n} = \{x \in R : o(f, x) \geq 1/n\}$ so

$$\sum_{j=1}^m v(S_j) < \epsilon.$$

The union of the boundaries of the rectangles $S_{m+1}, \ldots, S_k$ has measure 0, so we conclude that $A_{1/n}$ has measure 0.

Exercise XX.1.6 *Let A be a subset of $\mathbf{R}^n$. Let t be a real number. Show that $\partial(tA) = t\partial(A)$ (where tA is the set of all points tx with $x \in A$).*

Solution. Assume that $t \neq 0$. Suppose that $x \in \partial(tA)$ and write $x' = tx$ for some $x \in A$. By assumption, for every $\epsilon > 0$ the open ball $B_\epsilon(x')$ intersect tA, so there exists an element $y' = ty$ with $y \in A$, and

$$|y' - x'| < \epsilon.$$

The above inequality implies $|x - y| < \epsilon/|t|$, so $B_{\epsilon/|t|}(x)$ intersects A. Similarly, we find see that $B_{\epsilon/|t|}(x)$ intersects the complement of A, thus $\partial(tA) \subset t\partial(A)$. Reversing the argument proves that $t\partial(A) \subset \partial(tA)$.

Exercise XX.1.7 *Let R be a rectangle, and x, y two points of R. Show that the line segment joining x and y is contained in R.*

Solution. Write $x = (x_1, \ldots, x_n)$ and $y = (y_1, \ldots, y_n)$, and suppose that the rectangle R is given by the product $I_1 \times \cdots \times I_n$. To show that the line segment between x and y is contained in R it suffices to show that for each j, the projection of the segment on the j-th factor is contained in the interval I_j. The desired projection is given by $x_j + t(y_j - x_j)$. Since $x_j, y_j \in I_j$, $0 \leq t \leq 1$, and I_j is convex, the line segment between these two points is contained in I_j.

Exercise XX.1.8 *Let A be a subset of $\mathbf{R}^n$ and let A^0 be the interior of A. Let $x \in A^0$ and let y be in the complement of A. Show that the line segment joining x and y intersects the boundary of A. [Hint: The line segment is given by $x + t(y - x)$ with*

$$0 \leq t \leq 1.$$

Consider those values of t such that $[0, t]$ is contained in A^0, and let s be the least upper bound of such values.]

Solution. Let S be the set of $u \in [0, 1]$ such that for all $t \in [0, u]$, $x + t(y - x)$ is contained in A^0. This set is non-empty because $0 \in S$ and it is bounded by 1. Let c be the least upper bound of S. We contend that the point P defined by

$$P = x + c(y - x)$$

belongs to the boundary of A. If not, then we must consider two cases.
Case 1. If P is in the interior of A, then some open ball centered at P is contained in A. So for some small $\delta > 0$, the point $c + \delta \in S$ because the ball is convex. This contradicts the fact that c is the least upper bound for S.
Case 2. If P is in the interior of the complement of A, arguing as in case 1, we find that for some small $\delta > 0$, $c - \delta$ is an upper bound for S which again contradicts the definition of c.

Exercise XX.1.9 *Let A be an admissible set and let S be a rectangle. Prove that precisely one of the following possibilities holds: S is contained in the interior of A, S intersects the boundary of A, S is contained in the complement of the closure of A.*

Solution. If S is contained in the interior of A, then clearly, S does not intersect the boundary of A and S is not contained in the complement of the closure of A. Suppose S is not contained in the interior of A, so S intersects the complement of the interior of A. We consider two cases.
Case 1. Suppose S also intersects the interior of A, then by Exercises 7 and 8 we see that S intersects the boundary of A and clearly S is not contained in the complement of the closure of A.
Case 2. If S does not intersect the interior of A, then either S intersects the closure of A in which case S intersects the boundary of A, or S is contained in the complement of the closure of A.

Exercise XX.1.10 *Let A be an admissible set in $\mathbf{R}^n$, contained in some rectangle R. Show that*

$$\mathrm{Vol}(A) = \mathrm{lub}_P \sum_{S \subset A} v(S),$$

the least upper bound being taken over all partitions of R, and the sum taken over all subrectangles S of P such that $S \subset A$. Also prove: Given ϵ, there exists δ such that if size $P < \delta$, then

$$\left| \mathrm{Vol}(A) - \sum_{S \subset A} v(S) \right| < \epsilon,$$

the sum being taken over all subrectangles S of P contained in A. Finally, prove that

$$\mathrm{Vol}(A) = \mathrm{glb}_P \sum_{S \cap A \text{ not empty}} v(S),$$

the sum now being taken over all subrectangles S of the partition P having a non-empty intersection with A.

Solution. Let $f = 1_A$. Then by definition, $\mathrm{Vol}(A) = I_A(1) = I_R(f)$. Given a partition P of R, the lower sum is given by

$$L(P, f) = \sum_{S \in P} \inf_S (f) v(S).$$

If S is not contained in A, then $\inf_S(f) = 0$ and is S if contained in A, then $\inf_S(f) = 1$, consequently

$$L(P, f) = \sum_{S \in P, S \subset A} v(S).$$

By definition of the integral, we conclude that

$$\text{Vol}(A) = \text{lub}_P \sum_{S \subset A} v(S).$$

Since f is admissible, there exists $\delta > 0$ such that if P is a partition of size $< \delta$, then

$$U(P, f) - L(P, f) < \epsilon.$$

But

$$L(P, f) \leq \text{Vol}(A) \leq U(P, f),$$

thus

$$\left| \text{Vol}(A) - \sum_{S \subset A} v(S) \right| < \epsilon,$$

as was to be shown.

If a rectangle S intersects A, then $\sup_S(f) = 1$ and if S does not intersect A, then $\sup_S(f) = 0$, so

$$U(P, f) = \sum_{S \cap A \text{ not empty}} v(S).$$

By definition of $\text{Vol}(A)$ it follows that

$$\text{Vol}(A) = \text{glb}_P \sum_{S \cap A \text{ not empty}} v(S).$$

Exercise XX.1.11 *Let R be a rectangle and f an integrable function on R. Suppose that for each rectangle S in R we are given a number $I_S^* f$ satisfying the following conditions:*
(i) If P is a partition of R, then

$$I_R^* f = \sum_S I_S^* f.$$

(ii) If there are numbers m and M such that on a rectangle S we have

$$m \leq f(x) \leq M \quad \text{for all } x \in S,$$

then

$$mv(S) \leq I_S^* f \leq M v(s).$$

Show that $I_R^* f = I_R f$.

Solution. Let S be a subrectangle of a partition P of R. Then for all $x \in S$ we have

$$\inf_S(f) \leq f(x) \leq \sup_S(f),$$

so

$$\inf_S(f)v(S) \leq I_S^* f \leq \sup_S(f)v(S).$$

Thus

$$L(P, f) \leq \sum_{S \in P} I_S^* f \leq U(P, f),$$

and therefore $I_R^* f = I_R f$.

Exercise XX.1.12 *Let U be an open set in $\mathbf{R}^n$ and let $P \in U$. Let g be a continuous function on U. Let V_r be the volume of the ball of radius r. Let $B(P, r)$ be the ball of radius r centered at P. Prove that*

$$g(P) = \lim_{r \to 0} \frac{1}{V_r} \int_{B(P,r)} g.$$

Solution. Given $\epsilon > 0$, there exists δ such that if $X \in B(P, r)$ with $r < \delta$, then

$$g(P) - \epsilon \leq g(X) \leq g(P) + \epsilon,$$

so if $r < \delta$, then

$$(g(P) - \epsilon) \int_{B(P,r)} 1 \leq \int_{B(P,r)} g \leq (g(P) + \epsilon) \int_{B(P,r)} 1.$$

If $r > 0$, then $V_r > 0$ because we can always inscribe a small non-degenerate rectangle in the open ball $B(P, r)$, thus for all $r < \delta$ we have

$$\left| g(P) - \frac{1}{V_r} \int_{B(P,r)} g \right| < \epsilon.$$

So the limit

$$\lim_{r \to 0} \frac{1}{V_r} \int_{B(P,r)} g$$

exists and is equal to $g(P)$.

XX.2   Criteria for Admissibility

Exercise XX.2.1 *Let g be a continuous function defined on an interval $[a, b]$. Show that the graph of g is negligible.*

Solution. Given $\epsilon > 0$, choose n such that $|g(x) - g(y)| < \epsilon$ whenever $|x - y| \leq (b - a)/n$. For $k = 0, \dots, n$ let $a_k = a + \frac{k}{n}(b - a)$ and let $I_j = [a_j, a_{j+1}]$ for $j = 0, \dots, n - 1$. Then for each $j = 0, \dots, n - 1$ consider the rectangle

$$R_j = I_j \times [g(a_j) - \epsilon, g(a_j) + \epsilon].$$

By definition, the graph of g is the set of all pairs of points $(x, g(x))$ in $[a, b] \times \mathbf{R}$, so we see that

$$\text{graph}(g) \subset \bigcup_j R_j$$

and since $v(R_j) = 2\epsilon(b - a)/n$ we have

$$\sum_j v(R_j) = 2\epsilon(b - a),$$

which proves that the graph of g is negligible.

Exercise XX.2.2 *Let g_1, g_2 be continuous functions on $[a, b]$ and assume that $g_1 \leq g_2$. Let A be the set of points (x, y) such that $a \leq x \leq b$ and $g_1(x) \leq y \leq g_2(x)$. Show that A is admissible.*

Solution. The set A is clearly bounded. The map defined by $x \mapsto (x, g_1(x))$ is continuous, and so is the map $x \mapsto (x, g_2(x))$ so we see that the boundary of A is contained in the union of the graphs of g_1 and g_2 and the two vertical segments given by $[(a, g_1(a)), (a, g_2(a))]$ and $[(b, g_1(b)), (b, g_2(b))]$. The union of negligible sets is negligible, so A is admissible.

Exercise XX.2.3 *Let U be open in $\mathbf{R}^n$ and let $f : U \to \mathbf{R}^n$ be a map of class C^1. Let R be a closed cube contained in U, and let A be the subset of U consisting of all x such that*

$$\text{Det } f'(x) = 0.$$

Show that $f(A \cap R)$ is negligible. [Hint: Partition the cube into N^n subcubes each of side s/N where s is the side of R, and estimate the diameter of each $f(A \cap S)$ for each subcube S of the partition.]

Solution. Let $\epsilon > 0$. Since R is compact and f is of class C^1, the map f' is uniformly continuous on R, so there exists N such that if we subdivide R in N^n subcubes, then for each subcube S we have $|f'(y) - f'(x)| < \epsilon$ whenever $x, y \in S$. Given $x \in S$ let $g(y) = f(y) - f'(x)y$. Then $g'(y) = f'(y) - f'(x)$ for all $y \in S$, so by the mean value theorem we see that $|g(x) - g(y)| \leq \epsilon|x - y|$ for all $x, y \in S$. Hence

$$|f(y) - f(x) - f'(x)(y - x)| < \epsilon|x - y| < \epsilon C_1 \frac{s}{N} \quad \text{for all } y \in S,$$

where C_1 is a positive constant depending only on n. If $A \cap S$ is non-empty, we can take $x \in S$. In this case, the above inequality shows that if V is the image of $f'(x)$, then $\{f(y) : y \in S\}$ lies within $\epsilon C_1 s/N$ of $f(x) + V$. But since $x \in A \cap S$, the dimension of V is $\leq n - 1$. On the other hand, there exists a number M such that

$$|f(y) - f(x)| \leq M|x - y| \quad \text{for all } x, y \in R,$$

hence $|f(y) - f(x)| \leq C_2 s/N$ where C_2 is a positive constant. This implies that $\{f(y) : y \in S\}$ is contained in a cylinder of height $\epsilon C_1 s/N$ whose base is an $(n-1)$-dimensional sphere of radius $\leq C_2 s/N$. The volume of this cylinder is $\leq \epsilon C_3 (s/N)^n$ where C_3 is a positive constant. There are at most N^n subcubes S which intersect A so $f(S \cap A)$ lies in a set of volume $\leq \epsilon C_3 s^n$. This concludes the exercise.

XX.3   Repeated Integrals

Exercise XX.3.1 *Let f be defined on the square S consisting of all points (x, y) such that $0 \leq x \leq 1$ and $0 \leq y \leq 1$. Let f be the function on S such that*

$$f(x, y) = \begin{cases} 1 & \text{if } x \text{ is irrational,} \\ y^3 & \text{if } x \text{ is rational.} \end{cases}$$

(a) Show that

$$\int_0^1 \left[\int_0^1 f(x, y) dy \right] dx$$

does not exist.

(b) Show that the integral $I_S(f)$ does not exist.

Solution. (a) If x is irrational, then

$$\int_0^1 f(x, y) dy = 1.$$

If x is rational, then we have

$$\int_0^1 f(x, y) = \int_0^1 y^3 dy = \left[\frac{y^4}{4} \right]_0^1 = \frac{1}{4}.$$

So if $F(x) = \int_0^1 f(x, y) dy$ we found that

$$F(x) = \begin{cases} 1 & \text{if } x \text{ is irrational,} \\ 1/4 & \text{if } x \text{ is rational.} \end{cases}$$

This implies that every upper Riemann sum of F is equal to 1 and every lower Riemann sum is equal to $1/4$, and therefore the repeated integral

$$\int_0^1 \left[\int_0^1 f(x, y) dy \right] dx$$

does not exist.

(b) We can partition $[0, 1]$ in n equal subdivisions by letting $a_k = k/n$ for $k = 0, \ldots, n$. We call such a partition an n-partition of $[0, 1]$ and we denote it by P_n.

Let P be a partition of S. Then the upper sum with respect to this partition is equal to 1. It suffices to show that

$$L(P_n \times P_n, f) \le \frac{1}{4},$$

for all n, because there exists a partition P' and an integer m such that $L(P, f) - L(P', f)$ is small and $P_m \times P_m$ is a refinement of P'. The volume of a rectangle in $P_n \times P_n$ is $1/n^2$, so adding the contribution of each row of rectangles we find that

$$L(P_n \times P_n, f) = \frac{1}{n^2} \left[n \left(\frac{0}{n} \right)^3 + n \left(\frac{1}{n} \right)^3 + \cdots + n \left(\frac{n-1}{n} \right)^3 \right]$$

$$= \frac{1}{n^4} [1^3 + \cdots + (n-1)^3]$$

$$= \frac{1}{n^4} \left[\frac{n(n-1)}{2} \right]^2,$$

and the inequality drops out. Thus f is not integrable.

XX.4   Change of Variables

Exercise XX.4.1 *Let A be an admissible set, symmetric about the origin (that means: if $x \in A$, then $-x \in A$). Let f be an admissible function on A such that*

$$f(-x) = -f(x).$$

Show that

$$\int_A f = 0.$$

Solution. Let φ be the function defined by $\varphi(x) = -x$. Then

$$\int_{\varphi(A)} f = \int_A (f \circ \varphi) |\Delta_\varphi|.$$

But $\varphi(A) = A$, $|\Delta_\varphi| = 1$, and $f \circ \varphi(x) = -f(x)$, so

$$\int_A f = -\int_A f$$

which proves that $\int_A f = 0$.

Exercise XX.4.2 *Let $T: \mathbf{R}^n \to \mathbf{R}^n$ be an invertible linear map, and let B be a ball centered at the origin in $\mathbf{R}^n$. Show that*

$$\int_{T^{-1}(B)} e^{-\langle Ty, Ty \rangle}\, dy = \int_B e^{-\langle x, x \rangle}\, dx |\det T^{-1}|.$$

(The symbol $\langle, \rangle$ denotes the ordinary dot product in $\mathbf{R}^n$.) Taking the limit as the ball's radius goes to infinity, one gets

$$\int_{\mathbf{R}^n} e^{-\langle Ty, Ty \rangle}\, dy = \int_{\mathbf{R}^n} e^{-\langle x, x \rangle}\, dx |\det T^{-1}|$$
$$= \pi^{n/2} |\det T^{-1}|.$$

Solution. Let $f = T^{-1}$ in the change of variable formula so that

$$\int_{T^{-1}(B)} e^{-\langle Ty, Ty \rangle}\, dy = \int_B e^{-\langle TT^{-1}x, TT^{-1}x \rangle} |\det T^{-1}|\, dx$$
$$= \int_B e^{-\langle x, x \rangle}\, dx |\det T^{-1}|.$$

Exercise XX.4.3 *Let $B_n(r)$ be the closed ball of radius r in $\mathbf{R}^n$, centered at the origin, with respect to the euclidean norm. Find its volume $V_n(r)$. [Hint: First, note that*

$$V_n(r) = r^n V_n(1).$$

We may assume $n \geq 2$. The ball $B_n(1)$ consists of all $(x_1, \ldots, x_n)$ such that

$$x_1^2 + \cdots + x_n^2 \leq 1.$$

Put $(x_1, x_2) = (x, y)$ and let g be the characteristic function on $B_n(1)$. Then

$$V_n(1) = \int_{-1}^1 \int_{-1}^1 \left[\int_{R_{n-2}} g(x, y, x_3, \ldots, x_n)\, dx_3 \cdots dx_n \right] dx\, dy,$$

where R_{n-2} is a rectangle of radius 1 centered at the origin in $(n-2)$-space. If $x^2 + y^2 > 1$, then $g(x, y, x_3, \ldots, x_n) = 0$. Let D be the disc of radius 1 in $\mathbf{R}^2$. If $x^2 + y^2 \leq 1$, then $g(x, y, x_3, \ldots, x_n)$, viewed as a function of $(x_3, \ldots, x_n)$, is the characteristic function of the ball

$$B_{n-2}(\sqrt{1 - x^2 - y^2}).$$

Hence the inner integral is equal to

$$\int_{R_{n-2}} g(x, y, x_3, \ldots, x_n)\, dx_3 \cdots dx_n = (1 - x^2 - y^2)^{(n-2)/2} V_{n-2}(1)$$

so that

$$V_n(1) = V_{n-2}(1) \int_D (1 - x^2 - y^2)^{(n-2)/2} dxdy.$$

Using polar coordinates, the last integral is easily evaluated, and we find

$$V_{2n}(1) = \frac{\pi^n}{n!} \quad and \quad V_{2n-1}(1) = \frac{2^n \pi^{n-1}}{1 \cdot 3 \cdot 5 \cdots (2n-1)}.$$

Suppose that Γ is a function such that $\Gamma(x+1) = x\Gamma(x), \Gamma(1) = 1$, and $\Gamma(1/2) = \sqrt{\pi}$. Show that

$$V_n(1) = \frac{\pi^{n/2}}{\Gamma(1 + n/2)}.]$$

Solution. We compute the integral $\int_D (1 - x^2 - y^2)^{(n-2)/2} dxdy$. Switching to polar coordinates, we see that

$$\int_D (1 - x^2 - y^2)^{(n-2)/2} dxdy = \int_0^1 \int_0^{2\pi} r(1 - r^2)^{(n-2)/2} d\theta dr$$

$$= 2\pi \left[\frac{-1}{n}(1 - r^2)^{n/2} \right]_0^1 = \frac{2\pi}{n}.$$

So we get the desired formulas for $V_n(1)$ by induction. Now we prove the formula $V_n(1) = \pi^{n/2}/\Gamma(1 + n/2)$. In Exercise 13, §3, of Chapter XIII, we have proved the formulas

$$\Gamma(n) = (n-1)! \quad and \quad \Gamma(1/2 + n) = \frac{1 \cdot 3 \cdot 5 \cdots (2n-1)}{2^n} \pi^{1/2}.$$

If n is even, write $n = 2p$, so that $\Gamma(1 + n/2) = \Gamma(1 + p) = p!$, therefore the quotient $\pi^{n/2}/\Gamma(1 + n/2)$ is equal to $\pi^p/p!$, whence

$$\pi^{n/2}/\Gamma(1 + n/2) = \pi^p/p! = V_{2p}(1) = V_n(1).$$

If n is odd, then write $n = 2p - 1$, so that

$$\Gamma(1 + n/2) = \Gamma(1/2 + p) = \frac{1 \cdot 3 \cdot 5 \cdots (2p-1)}{2^p} \pi^{1/2},$$

and therefore

$$\pi^{n/2}/\Gamma(1 + n/2) = \pi^{p-1} \frac{2^p}{1 \cdot 3 \cdot 5 \cdots (2p-1)} = V_{2p-1}(1) = V_n(1)$$

which concludes the proof.

Exercise XX.4.4 *Determine the volume of the region in $\mathbf{R}^n$ defined by the inequality*

$$|x_1| + \cdots + |x_n| \leq r.$$

Solution. We argue as in Exercise 3. Let $V_n(r)$ be the volume of the given region. Then $V_n(r) = r^n V_n(1)$, so if g is the characteristic function of the region described by $|x_1| + \cdots + |x_n| \leq 1$, we have

$$V_n(1) = \int_{-1}^{1} \left[\int_{R_{n-1}} g(x_1, \ldots, x_n) dx_2 \cdots dx_n \right] dx_1$$

and therefore g viewed as a function on $x_2, \ldots, x_n$ is the characteristic function of the region defined by $|x_2| + \cdots + |x_n| \leq 1 - |x_1|$, thus

$$V_n(1) = V_{n-1}(1) \int_{-1}^{1} (1 - |x_1|)^{n-1} dx_1.$$

Integrating from -1 to 0 and then from 0 to 1 we find that the last integral is equal to

$$\int_{-1}^{1} (1 - |x_1|)^{n-1} dx_1 = \frac{2}{n}.$$

Hence

$$V_n(1) = \frac{2^{n-1}}{n!} V_1(1) = \frac{2^n}{n!}.$$

Exercise XX.4.5 *Determine the volume of the region in* $\mathbf{R}^{2n} = \mathbf{R}^2 \times \cdots \times \mathbf{R}^2$ *defined by*

$$|z_1| + \cdots + |z_n| \leq r,$$

where $z_i = (x_i, y_i)$ *and* $|z_i| = \sqrt{x_i^2 + y_i^2}$ *is the euclidean norm in* $\mathbf{R}^2$.

Solution. We argue as in Exercises 3 and 4. Let $V_{2n}(r)$ be the volume of the given region. Then $V_{2n}(r) = r^{2n} V_{2n}(1)$, and we have

$$V_{2n}(1) = \int_{|z_1| \leq 1} \left[\int_{R_{n-2}} g(z_1, \ldots, z_n) dz_2 \cdots dz_n \right] dz_1,$$

where g is the characteristic function of the region described by $|z_1| + \cdots + |z_n| \leq 1$ and where $dz_i = dx_i dy_i$. Consequently

$$V_{2n}(1) = V_{2n-2}(1) \int_{|z_1| \leq 1} (1 - |z_1|)^{2n-2} dz_1.$$

This last integral is equal to

$$\int_{D} (1 - \sqrt{x^2 + y^2})^{2n-2} dx dy = \int_{0}^{1} \int_{0}^{2\pi} r(1 - r)^{2n-2} d\theta dr$$

$$= 2\pi \int_{0}^{1} r(1 - r)^{2n-2} dr.$$

Integrating by parts we find that

$$\int_0^1 r(1-r)^{2n-2}dr = \frac{1}{2n(2n-1)},$$

and since $V_2(1) = \pi$ we obtain

$$V_{2n}(1) = \frac{(2\pi)^n}{(2n)!}.$$

Exercise XX.4.6 (Spherical Coordinates) *(a) Define $f : \mathbf{R}^3 \to \mathbf{R}^3$ by*

$$\begin{aligned}
x_1 &= r\cos\theta_1, \\
x_2 &= r\sin\theta_1\cos\theta_2, \\
x_3 &= r\sin\theta_1\sin\theta_2.
\end{aligned}$$

Show that

$$\Delta_f(r,\theta_1,\dots,\theta_{n-1}) = r^2\sin\theta_1.$$

Show that f is invertible on the open set

$$0 < r, \quad 0 < \theta_1 < \pi, \quad 0 < \theta_2 < 2\pi,$$

and that the image under f of this rectangle is the open set obtained from $\mathbf{R}^3$ by deleting the set of points $(x, y, 0)$ with $y \geq 0$, and x is arbitrary.

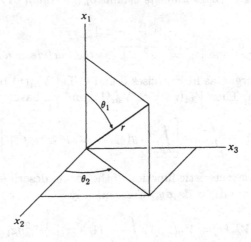

Let $S(r_1)$ be the closed rectangle of points (r, θ_1, θ_2) satisfying

$$0 \leq r \leq r_1, \quad 0 \leq \theta_1 \leq \pi, \quad 0 \leq \theta_2 \leq 2\pi.$$

Show that the image of $S(r_1)$ is the closed ball of radius r_1 centered at the origin in $\mathbf{R}^3$.

(b) Let g be a continuous function of one variable, defined for $r \geq 0$. Let

$$G(x_1, x_2, x_3) = g(\sqrt{x_1^2 + x_2^2 + x_3^2}).$$

Let $B(r_1)$ denote the closed ball of radius r_1. Show that

$$\int_{B(r_1)} G = W_3 \int_0^{r_1} g(r)r^2 dr,$$

where $W_3 = 3V_3$, and V_3 is the volume of the three-dimensional ball of radius 1 in $\mathbf{R}^3$.

(c) The n-dimensional generalization of the spherical coordinates is given by the following formulas:

$$\begin{aligned}
x_1 &= r\cos\theta_1, \\
x_2 &= r\sin\theta_1\sin\theta_2,
\end{aligned}$$

$$\cdots$$

$$\begin{aligned}
x_{n-1} &= r\sin\theta_1\sin\theta_2\cdots\sin\theta_{n-2}\cos\theta_{n-1}, \\
x_n &= r\sin\theta_1\sin\theta_2\cdots\sin\theta_{n-2}\sin\theta_{n-1}.
\end{aligned}$$

We take $0 < r$, $0 < \theta_i < \pi$ for $i = 1,\ldots,n-2$ and $0 < \theta_{n-1} < 2\pi$. The Jacobian determinant is then given by

$$\begin{aligned}
\Delta_f(r,\theta_1,\ldots,\theta_{n-1}) &= r^{n-1}\sin^{n-2}\theta_1\sin^{n-3}\theta_2\cdots\sin\theta_{n-2} \\
&= r^{n-1}J(\theta).
\end{aligned}$$

Then one has the n-dimensional analogue of $dxdy = rdrd\theta$, namely

$$dx_1\cdots dx_n = r^{n-1}J(\theta)drd\theta_1\cdots d\theta_{n-1} \quad \text{abbreviated} \quad r^{n-1}drd\mu(\theta).$$

Assuming this formula, define the $(n-1)$-dimensional area of the sphere to be

$$W_n = A(\mathbf{S}^{n-1}) = \int d\mu(\theta),$$

where the multiple integral on the right is over the intervals prescribed above for $\theta = (\theta_1,\ldots,\theta_{n-1})$. Prove that

$$A(\mathbf{S}^{n-1}) = nV_n,$$

where V_n is the n-dimensional volume of the n-ball of radius 1. This generalizes the formula $W_3 = 3V_3$ carried out in 3-space.

Solution. (a) The Jacobian of f is given by

$$J_f = \begin{pmatrix} \cos\theta_1 & -r\sin\theta_1 & 0 \\ \sin\theta_1\cos\theta_2 & r\cos\theta_1\cos\theta_2 & -r\sin\theta_1\sin\theta_2 \\ \sin\theta_1\sin\theta_2 & r\cos\theta_1\sin\theta_2 & r\sin\theta_1\cos\theta_2 \end{pmatrix},$$

so we have $\Delta_f = r^2\sin\theta_1$.

Let S be the set obtained from $\mathbf{R}^3$ by deleting the set of points $(x_1, x_2, 0)$ with $x_2 \geq 0$ and x_1 arbitrary. Suppose that $x_3 = 0$, then $\theta_2 = \pi$, so we have $x_2 < 0$, thus $f(R) \subset S$. Conversely, if $(x_1, x_2, x_3) \in S$, let $r = \sqrt{x_1^2 + x_2^2 + x_3^2}$, and let θ_1 be such that $0 < \theta_1 < \pi$ and $r \cos \theta_1 = x_1$. Then polar coordinates in the plane (x_2, x_3) imply that there exists $\rho > 0$ and $0 < \theta_2 < 2\pi$ such that $x_2 = \rho \cos \theta_2$ and $x_3 = \rho \sin \theta_2$. Since $x_1^2 + x_2^2 + x_3^2 = r^2$ we must have $\rho^2 = r^2 \sin^2 \theta_1$. This shows that f is invertible on the rectangle and its inverse is C^1 because the Jacobian of f is everywhere non-zero.

Since $x_1^2 + x_2^2 + x_3^2 = r^2$ we see that (x_1, x_2, x_3) belongs to $B(r_1)$. Conversely, if $(x_1, x_2, x_3) \in B(r_1)$, then we have $x_1^2 + x_2^2 + x_3^2 \le r_1^2$ and since we set $r = \sqrt{x_1^2 + x_2^2 + x_3^2}$ we conclude that the image of $S(r_1)$ is $B(r_1)$.

(b) We use the change of variable formula along with spherical coordinates, so that

$$\int_{B(R_1)} G = \int_0^{r_1} \int_0^{\pi} \int_0^{2\pi} g(r) r^2 \sin \theta_1 \, d\theta_2 d\theta_1 dr,$$

hence

$$\int_{B(R_1)} G = 2\pi \left(\int_0^{r_1} g(r) r^2 dr \right) \left(\int_0^{\pi} \sin \theta_1 d\theta_1 \right) = 4\pi \int_0^{r_1} g(r) r^2 dr.$$

But $V_3 = 4\pi/3$ so the formula drops out.

(c) The n-dimensional volume of the n-ball B^n is defined to be

$$V_n = \int_{B^n} dx_1 \cdots dx_n,$$

so switching to spherical coordinates we find that

$$V_n = \int_0^1 r^{n-1} dr \int d\mu(\theta) = A(\mathbf{S}^{n-1}) \int_0^1 r^{n-1} dr = \frac{A(\mathbf{S}^{n-1})}{n}.$$

Exercise XX.4.7 *Let $T : \mathbf{R}^n \to \mathbf{R}^n$ be a linear map whose determinant is equal to 1 or -1. Let A be an admissible set. Show that*

$$\text{Vol}(TA) = \text{Vol}(A).$$

(Examples of such maps are the so-called unitary maps, i.e. those T for which $\langle Tx, Tx \rangle = \langle x, x \rangle$ for all $x \in \mathbf{R}^n$.)

Solution. The linear map is invertible so we can apply the change of variable formula

$$\text{Vol}(TA) = \int_{TA} 1_{TA} = \int_A 1_A |\det T| = \int_A 1_A = \text{Vol}(A).$$

Exercise XX.4.8 *(a) Let A be the subset of $\mathbf{R}^2$ consisting of all points*

$$t_1 e_1 + t_2 e_2$$

with $0 \le t_1$ and $t_1 + t_2 \le 1$. *(This is just a triangle.) Find the area of A by integration.*
(b) Let v_1, v_2 be linearly independent vectors in $\mathbf{R}^2$. Find the area of the set of points $t_1 v_1 + t_2 v_2$ with $0 \le t_i$ and $t_1 + t_2 \le 1$, in terms of $\mathrm{Det}(v_1, v_2)$.

Solution. (a) The vectors e_1 and e_2 are the unit vectors, so using repeated integration we find

$$\mathrm{Vol}(A) = \int_0^1 \int_0^{1-x} dy dx = \int_0^1 1 - x dx = \frac{1}{2}.$$

(b) Let T be the linear transformation such that $Te_1 = v_1$ and $Te_2 = v_2$. Using the notation of (a) we see that we want to compute the area of TA. Thus

$$\mathrm{Vol}(TA) = |\mathrm{Det}(v_1, v_2)|\mathrm{Vol}(A) = \frac{|\mathrm{Det}(v_1, v_2)|}{2}.$$

Exercise XX.4.9 *Let $v_1, \ldots, v_n$ be linearly independent vectors in $\mathbf{R}^n$. Find the volume of the solid consisting of all points*

$$t_1 v_1 + \cdots + t_n v_n$$

with $0 \le t_i$ and $t_1 + \cdots + t_n \le 1$.

Solution. Let S be the solid described in the text and let T be the invertible linear transformation such that $Te_i = v_i$ for all i. Let A be the subset of $\mathbf{R}^n$ consisting of all points $t_1 e_1 + \cdots + t_n e_n$ with $0 \le t_i$ and $t_1 + \cdots + t_n \le 1$. Then we have

$$\mathrm{Vol}(S) = |\det(T)|\mathrm{Vol}(A).$$

We contend that $\mathrm{Vol}(A) = 1/n!$. This result is a consequence of Exercise 4. Indeed, in the expression $|x_1| + \cdots + |x_n|$ each x_i can be either positive or negative, so the region defined by $|x_1| + \cdots + |x_n| \le 1$ is the union of 2^n disjoint regions and the volume of each of these subregions is equal to $\mathrm{Vol}(A)$. The formula obtained in Exercise 4 proves our contention.

Exercise XX.4.10 *Let B_a be the closed ball of radius $a > 0$, centered at the origin. In n-space, let $X = (x_1, \ldots, x_n)$ and let $r = |X|$, where $|\cdot|$ is the euclidean norm. Take*

$$0 < a < 1,$$

and let A_a be the annulus consisting of all points X with $a \le |X| \le 1$. Both in the case $n = 2$ and $n = 3$ (i.e. in the plane and in 3-space), compute the integral

$$I_a = \int_{A_a} \frac{1}{|X|} dx_1 \cdots dx_n.$$

Show that this integral has a limit as $a \to 0$. Thus, contrary to what happens in 1-space, the function $f(X) = 1/|X|$ can be integrated in a neighborhood of

0. [Hint: Use polar or spherical coordinates. Actually, using n-dimensional spherical coordinates, the result also holds in n-space.] Show further that in 3-space, the function $1/|X|^2$ can be similarly integrated near 0.

Solution. In the case $n = 2$, we use polar coordinates together with a change of variable to obtain

$$\int_{A_a} \frac{1}{|X|} dx_1 dx_2 = \int_{A_a} \frac{1}{\sqrt{x^2 + y^2}} dx dy = \int_a^1 \int_0^{2\pi} \frac{1}{r} r d\theta dr = 2\pi(1 - a),$$

so

$$\lim_{a \to 0} \int_{A_a} \frac{1}{|X|} dx_1 dx_2 = 2\pi.$$

In the case $n = 3$ we use spherical coordinates and a change of variables to get

$$\int_{A_a} \frac{1}{\sqrt{x_1^2 + x_2^2 + x_3^2}} dx_1 dx_2 dx_3 = \int_a^1 \int_0^\pi \int_0^{2\pi} r \sin\theta_1 d\theta_2 d\theta_1 dr$$

$$= 4\pi \frac{1 - a^2}{2} = 2\pi(1 - a^2),$$

hence

$$\lim_{a \to 0} \int_{A_a} \frac{1}{|X|} dx_1 dx_2 dx_3 = 2\pi.$$

Spherical coordinates and the change of variable formula imply that

$$\int_{A_a} \frac{1}{|X|^2} dx_1 dx_2 dx_3 = \int_a^1 \int_0^\pi \int_0^{2\pi} \sin\theta_1 d\theta_2 d\theta_1 dr = 4\pi(1 - a),$$

so

$$\lim_{a \to 0} \int_{A_a} \frac{1}{|X|^2} dx_1 dx_2 dx_3 = 4\pi.$$

Exercise XX.4.11 Let B be the region in the first quadrant of $\mathbf{R}^2$ bounded by the curves $xy = 1, xy = 3, x^2 - y^2 = 1$ and $x^2 - y^2 = 4$. Find the value of the integral

$$\iint_B (x^2 + y^2) dx dy$$

by making the substitution $u = x^2 - y^2$ and $v = xy$. Explain how you are applying the change of variables formula.

Solution. Let A be the rectangle defined by

$$A = \{(u, v) \in \mathbf{R}^2 : 1 \leq u \leq 4 \text{ and } 1 \leq v \leq 3\}.$$

Let ψ be the function defined by $\psi(x, y) = (u, x)$ where $u = x^2 - y^2$ and $v = xy$. Then $\psi : B \to A$ is C^1 invertible, so by the change of variables formula

$$\int_A 1_A = \int_{\psi(B)} 1_{\psi(B)} = \int_B 1_B |\Delta_\psi|.$$

But

$$J_\psi = \begin{pmatrix} 2x & -2y \\ y & x \end{pmatrix},$$

so we have $|\Delta_\psi| = 2(x^2 + y^2)$. Therefore

$$\frac{1}{2} \iint_A 1_A = \iint_B x^2 + y^2 \, dxdy,$$

and since

$$\iint_A 1_A = (3-1)(4-1) = 6$$

we find that

$$\iint_B x^2 + y^2 \, dxdy = 3.$$

Exercise XX.4.12 *Prove that*

$$\iint_A e^{-(x^2+y^2)} \, dxdy = ae^{-a^2} \int_0^\infty \frac{e^{-u^2}}{a^2 + u^2} \, du,$$

where A denotes the half plane $x \geq a > 0$. [Hint: Use the transformation

$$x^2 + y^2 = u^2 + a^2 \quad and \quad y = vx.]$$

Solution. We represent a point in the right half plane $x \geq a$ by the intersection of a circle centered at the origin and a line passing through the origin. The variable u determines the radius of the circle and v determines the slope of the line. For a fixed u we see that v ranges between $-u/a$ and u/a.

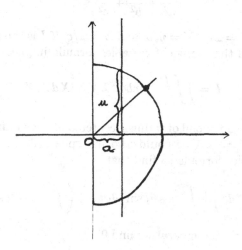

Since $x^2 + y^2 = a^2 + u^2$ and $y = vx$ we find that

$$x = \left(\frac{u^2 + a^2}{1 + v^2}\right)^{1/2} \quad \text{and} \quad y = v\left(\frac{u^2 + a^2}{1 + v^2}\right)^{1/2}.$$

So if $g(u, v) = (x, y)$ some elementary computations show that the determinant of the Jacobian is

$$\Delta = \frac{u}{1 + v^2}.$$

If I is the integral we want to compute, the change of variables formula implies

$$I = \int_0^\infty \int_{-u/a}^{u/a} e^{-(u^2 + a^2)} \frac{u}{1 + v^2} \, dv \, du.$$

However,

$$\int_{-u/a}^{u/a} \frac{dv}{1 + v^2} = \arctan(u/a)$$

so

$$I = e^{-a^2} \int_0^\infty 2u e^{-u^2} \arctan(u/a) \, du.$$

Integrate by parts using the fact that $e^{-u^2} \arctan(u/a) \, du \to 0$ as $u \to \infty$ and $\arctan(0) = 0$ to find the desired expression for I.

Exercise XX.4.13 *Find the integral*

$$\iiint xyz \, dx \, dy \, dz$$

taken over the ellipsoid

$$\frac{x^2}{a^2} + \frac{y^2}{b^2} + \frac{z^2}{c^2} \leq 1.$$

Solution. Let $X = x/a, Y = y/b$, and $Z = z/c$. If I is the integral we want to compute, then the change of variables formula implies

$$I = \iiint_{B(1)} (abc)^2 XYZ \, dX \, dY \, dZ,$$

where $B(1)$ is the open ball of radius 1 centered at the origin. Then Exercise 1 implies that $I = 0$. One could also use spherical coordinates and the change of variables formula to find that

$$I = \left(\int_0^1 r^5 \, dr\right) \left(\int_0^\pi \cos\theta_1 \sin^3\theta_1 \, d\theta_1\right) \left(\int_0^{2\pi} \cos\theta_2 \sin\theta_2 \, d\theta_2\right).$$

Evaluating the middle integral we find 0.

Exercise XX.4.14 *Let f be in the Schwartz space on $\mathbf{R}^n$. Define a normalization of the* **Fourier transform** *by*

$$f^{\vee}(y) = \int_{\mathbf{R}^n} f(x)e^{-2\pi i x \cdot y} dx.$$

Prove that the function $h(x) = e^{-\pi x^2}$ is self dual, that is $h^{\vee} = h$.

Solution. We first investigate the case $n = 1$. In the chapter on the Fourier integral, it was shown that

$$\frac{1}{\sqrt{2\pi}} \int e^{-x^2/2} e^{-ixy} dx = e^{-y^2/2}.$$

Putting $y = \sqrt{2\pi}v$ and changing variables $x = \sqrt{2\pi}u$ in the above formula we get

$$\frac{1}{\sqrt{2\pi}} \sqrt{2\pi} \int e^{-\pi u^2} e^{-2\pi i u v} du = e^{-\pi v^2}.$$

This takes care of the case $n = 1$. In the general case, we have

$$\int e^{-\pi x^2} e^{-2\pi i x \cdot y} dx = \int \cdots \int e^{-\pi(x_1^2 + \cdots + x_n^2)} e^{-2\pi(x_1 y_1 + \cdots + x_n y_n)} dx_1 \cdots dx_n$$

$$= \int \cdots \int \prod_{k=1}^{n} e^{-\pi x_k^2} e^{-2\pi x_k y_k} dx_1 \cdots dx_n$$

$$= \prod_{k=1}^{n} \int e^{-\pi x_k^2} e^{-2\pi x_k y_k} dx_k$$

$$= \prod_{k=1}^{n} e^{-\pi y_k^2} = e^{-\pi y^2}.$$

Exercise XX.4.15 *Let B be an $n \times n$ non-singular real matrix. Define $(f \circ B)(x) = f(Bx)$. Prove that the dual of $f \circ B$ is given by*

$$(f \circ B)^{\vee}(y) = \frac{1}{\|B\|} f^{\vee}({}^t B^{-1} y),$$

where $\|B\|$ is the absolute value of the determinant of B.

Solution. We use the change of variables formula

$$(f \circ B)^{\vee}(y) = \int f(Bx)e^{-2\pi i x \cdot y} dx = \frac{1}{\|B\|} \int f(x)e^{-2\pi i B^{-1} x \cdot y} dx$$

$$= \frac{1}{\|B\|} \int f(x)e^{-2\pi i x \cdot {}^t B^{-1} y} dx = \frac{1}{\|B\|} f^{\vee}({}^t B^{-1} y).$$

Exercise XX.4.16 *For $b \in \mathbf{R}^n$ define $f_b(x) = f(x - b)$. Prove that*

$$(f_b)^{\vee}(y) = e^{-2\pi i b \cdot y} f^{\vee}(y).$$

Solution. The change of variables formula implies

$$(f_b)^\vee(y) = \int f(x-b)e^{-2\pi i x \cdot y}dx = \int f(x)e^{-2\pi i(x+b)\cdot y}dx = e^{-2\pi i b \cdot y}f^\vee(y).$$

XX.5   Vector Fields on Spheres

Exercise XX.5.1 *Prove the statements depending on inverse mapping theorems which have been left as exercises in the proof of the section.*

Solution. Consider the annulus

$$A = \{x \in \mathbf{R}^3 : 1/2 \leq |X| \leq 3/2\}.$$

We extend E to A by $E(rU) = rE(U)$ for any unit vector U and any number r such that $1/2 \leq r \leq 3/2$. Since E is continuously differentiable, there exists a positive constant $c > 0$ which satisfies the Lipschitz condition on A

$$|E(X) - E(Y)| \leq c|X - Y| \quad \text{for all } X, Y \in A.$$

Choose t so small that $0 < t < 1/3$ and $0 < t < c^{-1}$. Let $Y \in S^3$ and consider the map $g_Y(X) = Y - tE(X)$ defined on A. Since $|tE(X)| < 1/2$ we have

$$1 - \frac{1}{2} \leq |Y - tE(X)| \leq 1 + \frac{1}{2},$$

so g_Y is a map of the complete metric space S into itself. Moreover, g_Y is a contraction because

$$|g_Y(X_1) - g_Y(X_2)| = t|E(X_1) - E(X_2)| \leq tc|X_1 - X_2|.$$

By the shrinking lemma, g_Y has a unique fixed point, so there exists a unique X in A such that $X + tE(X) = Y$. Conclude the exercise by multiplying X and Y by $\sqrt{1+t^2}$.

XXI
Differential Forms

XXI.1 Definitions

Exercise XXI.1.1 *Show that $ddf = 0$ for any function f, and also for a 1-form.*

Solution. Let f be a function. Then we have

$$df = D_1 f \, dx_1 + \cdots + D_n f \, dx_n = \sum_{i=1}^{n} D_i f \, dx_i,$$

so that

$$ddf = \sum_{j=1}^{n} \sum_{i=1}^{n} D_j D_i f \, dx_j \wedge dx_i.$$

But since $dx_j \wedge dx_i = -dx_i \wedge dx_j$ and $dx_i \wedge dx_i = 0$ we see that

$$ddf = \sum_{1 \leq s < k \leq n} (D_s D_k f - D_k D_s f) \, dx_s \wedge dx_k,$$

and since we assume f to be C^∞, the partials commute and $ddf = 0$.

Now suppose that ω is a 1-form. It suffices to prove that $dd\omega = 0$ when ω is decomposable. We write $\omega = g \, dx_k$ so that

$$d\omega = dg \wedge dx_k = \left(\sum_{i=1}^{n} f_i \, dx_i \right) \wedge dx_k,$$

where $f_i = D_i g$. An argument similar to the one given above shows that

$$ddw = \sum_{i=1}^{n} df_i \wedge dx_i \wedge dx_k = 0.$$

Exercise XXI.1.2 *Show that $ddw = 0$ for any differential form w.*

Solution. We prove the result by induction on the degree of the form. Suppose $dd\varphi = 0$ for all m-forms where $m \leq p-1$. Let w be a decomposable p-form and write $w = \eta \wedge \psi$ where ψ is of degree 1. Then applying the formula of Theorem 1.1 we obtain

$$dw = d\eta \wedge \psi + (-1)^r \eta \wedge d\psi,$$

where $r = \deg \eta$. Applying the same formula again we find

$$ddw = dd\eta \wedge \psi + (-1)^{r+1} d\eta \wedge d\psi + (-1)^r d\eta \wedge d\psi + (-1)^s \eta \wedge dd\psi.$$

The induction hypotheses and Exercise 1 imply $dd\eta = dd\psi = 0$ so $ddw = 0$.

Exercise XXI.1.3 *In 3-space, express dw in standard form for each one of the following w:*
(a) $w = x\, dx + y\, dz$.   (b) $w = xy\, dy + x\, dz$.
(c) $w = (\sin x)\, dy + dz$.   (d) $w = e^y\, dx + y\, dy + e^{xy}\, dz$.

Solution. (a) $dw = dy \wedge dz$.
(b) $dw = y\, dx \wedge dy + dx \wedge dz$.
(c) $dw = (\cos x)\, dx \wedge dy$.
(d) $dw = e^y\, dy \wedge dx + ye^{xy}\, dx \wedge dz + xe^{xy}\, dy \wedge dz$.

Exercise XXI.1.4 *Find the standard expression for dw in the following cases:*
(a) $w = x^2 y\, dy - xy^2\, dx$.   (b) $w = e^{xy}\, dx \wedge dz$.
(c) $w = f(x, y)\, dx$ where f is a function.

Solution. (a) $dw = 4xy\, dx \wedge dy$.
(b) $dw = xe^{xy}\, dy \wedge dx \wedge dz$.
(c) $dw = \frac{\partial f}{\partial y}\, dy \wedge dx$.

Exercise XXI.1.5 *(a) Express dw in standard form if*

$$w = x\, dy \wedge dz + y\, dz \wedge dx + z\, dx \wedge dy.$$

(b) Let f, g, h be functions, and let

$$w = f\, dy \wedge dz + g\, dz \wedge dx + h\, dx \wedge dy.$$

Find the standard form for dw.

Solution. (a) We simply have

$$dw = 3\, dx \wedge dy \wedge dz.$$

(b) In this case we find

$$dw = \left(\frac{\partial f}{\partial x} + \frac{\partial f}{\partial y} + \frac{\partial f}{\partial z}\right) dx \wedge dy \wedge dz.$$

Exercise XXI.1.6 *In n-space, find an $(n-1)$-form ω such that*

$$d\omega = dx_1 \wedge \cdots \wedge dx_n.$$

Solution. Take for example

$$\omega = x_1 dx_2 \wedge \cdots \wedge dx_n.$$

Exercise XXI.1.7 *Let ω be a form of odd degree on U, and let f be a function such that $f(x) \neq 0$ for all $x \in U$, and such that $d(f\omega) = 0$. Show that $\omega \wedge d\omega = 0$.*

Solution. Let r be the degree of the form ω. Then

$$\omega \wedge \omega = (-1)^{r^2} \omega \wedge \omega = -\omega \wedge \omega,$$

thus $\omega \wedge \omega = 0$. This implies that

$$0 = d(f\omega \wedge \omega) = d(f\omega) \wedge \omega - f\omega \wedge d\omega.$$

By assumption, we know that $d(f\omega) = 0$ so $f\omega \wedge d\omega = 0$ and since $f(x) \neq 0$ for all $x \in U$ we conclude that $\omega \wedge d\omega = 0$.

Exercise XXI.1.8 *A form ω on U is said to be **exact** is there exists a form ψ such that $\omega = d\psi$. If ω_1, ω_2 are exact, show that $\omega_1 \wedge \omega_2$ is exact.*

Solution. Assume that $\omega_1 = d\psi_1$ and $\omega_2 = d\psi_2$. Then

$$d(\psi_1 \wedge \psi_2) = d\psi_1 \wedge d\psi_2 + (-1)^r \psi_1 \wedge dd\psi_2 = d\psi_1 \wedge d\psi_2 = \omega_1 \wedge \omega_2,$$

so the form $\omega_1 \wedge \omega_2$ is exact.

Exercise XXI.1.9 *Show that the form*

$$\omega(x,y,z) = \frac{1}{r^3}(x\, dy \wedge dz + y\, dz \wedge dz + z\, dx \wedge dy)$$

is closed but not exact. As usual, $r^2 = x^2 + y^2 + z^2$ and the form is defined on the complement of the origin in $\mathbf{R}^3$.

Solution. To prove that the form ω is closed we must show that $d\omega = 0$. The only term that survives after taking the exterior derivative is $dx \wedge dy \wedge dz$, so we must show that its coefficient is 0. We see that

$$\frac{\partial}{\partial x}\left(\frac{x}{(x^2 + y^2 + z^2)^{3/2}}\right) = \frac{-2x^2 + y^2 + z^2}{(x^2 + y^2 + z^2)^{5/2}},$$

so by symmetry, we see that $d\omega$ is equal to

$$\frac{-2x^2 + y^2 + z^2 - 2y^2 + x^2 + z^2 - 2z^2 + x^2 + y^2}{(x^2 + y^2 + z^2)^{5/2}} \, dx \wedge dy \wedge dz = 0,$$

as was to be shown. Integrating this form over the unit 2-sphere in $\mathbf{R}^3$ we find 4π so ω is not exact.

XXI.3   Inverse Image of a Form

Exercise XXI.3.1 *Let the polar coordinate map be given by*

$$(x, y) = f(r, \theta) = (r\cos\theta, r\sin\theta).$$

Give the standard form for $f^*(dx)$, $f^*(dy)$, *and* $f^*(dx \wedge dy)$.

Solution. We have

$$f^*(dx) = \cos\theta dr - r\sin\theta d\theta,$$
$$f^*(dy) = \sin\theta dr + r\cos\theta d\theta,$$
$$f^*(dx \wedge dy) = f^*(dx) \wedge f^*(dy) = rdr \wedge d\theta.$$

Note that $\Delta_f = r$.

Exercise XXI.3.2 *Let the spherical coordinate map be given by*

$$(x_1, x_2, x_3) = f(r, \theta_1, \theta_2) = (r\cos\theta_1, r\sin\theta_1\cos\theta_2, r\sin\theta_1\sin\theta_2).$$

Give the standard form for $f^*(dx_1)$, $f^*(dx_2)$, $f^*(dx_3)$, $f^*(dx_1 \wedge dx_2)$, $f^*(dx_1 \wedge dx_3)$, $f^*(dx_2 \wedge dx_3)$, *and* $f^*(dx_1 \wedge dx_2 \wedge dx_3)$.

Solution. The computations show that

$$f^*(dx_1) = \cos\theta_1 dr - r\sin\theta_1 d\theta_1,$$
$$f^*(dx_2) = \sin\theta_1\cos\theta_2 dr + r\cos\theta_1\cos\theta_2 d\theta_1 - r\sin\theta_1\sin\theta_2 d\theta_2,$$
$$f^*(dx_3) = \sin\theta_1\sin\theta_2 dr + r\cos\theta_1\sin\theta_2 d\theta_1 + r\sin\theta_1\cos\theta_2 d\theta_2.$$

Also, $f^*(dx_1 \wedge dx_2)$ is equal to

$$(r\cos\theta_2)dr \wedge d\theta_1 + (-r\cos\theta_1\sin\theta_1\sin\theta_2)dr \wedge d\theta_2 + (r^2\sin^2\theta_1\sin\theta_2)d\theta_1 \wedge \theta_2$$

and $f^*(dx_1 \wedge dx_3)$ is equal to

$$(r \sin \theta_2)dr \wedge d\theta_1 + (r \sin \theta_1 \cos \theta_1 \cos \theta_2)dr \wedge d\theta_2 + (-r^2 \sin^2 \theta_1 \cos \theta_2)d\theta_1 \wedge \theta_2$$

and $f^*(dx_2 \wedge dx_3)$ is equal to

$$(r \sin^2 \theta_1)dr \wedge d\theta_1 + (r^2 \sin \theta_1 \cos \theta_1)d\theta_1 \wedge \theta_2$$

and $f^*(dx_1 \wedge dx_2 \wedge dx_3)$ is equal to

$$r^2 \sin \theta_1 dr \wedge d\theta_1 \wedge d\theta_2.$$

XXI.4   Stokes' Formula for Simplices

Exercise XXI.4.1 *Instead of using rectangles, one can use triangles in Stokes' theorem. Develop this parallel theory as follows. Let $v_0, \ldots, v_k$ be elements of $\mathbf{R}^n$ such that $v_i - v_0$ $(i = 1, \ldots, k)$ are linearly independent. We define the **triangle** spanned by $v_0, \ldots, v_k$ to consist of all points*

$$t_0 v_0 + \cdots + t_k v_k$$

with real t_i such that $0 \le t_i$ and $t_0 + \cdots + t_k = 1$.
We denote this triangle by T, or $T(v_0, \ldots, v_k)$.
(a) Let $w_i = v_i - v_0$ for $i = 1, \ldots, k$. Let S be the set of points

$$s_1 w_1 + \cdots + s_k w_k$$

with $s_i \ge 0$ and $s_1 + \cdots + s_k \le 1$. Show that $T(v_0, \ldots, v_k)$ is the translation of S by v_0.
Define the oriented boundary of the triangle T to be the chain

$$\partial^0 T = \sum_{j=0}^{k} (-1)^j T(v_0, \ldots, \widehat{v_j}, \ldots, v_k).$$

(b) Assume that $k = n$, and that T is contained in an open set U of $\mathbf{R}^n$. Let ω be an $(n-1)$-form on U. In analogy to Stokes' theorem for rectangles, show that

$$\int_T d\omega = \int_{\partial^0 T} \omega.$$

Solution. (a) Since

$$s_1 w_1 + \cdots + s_k w_k + v_0 = (1 - s_1 - \cdots - s_k)v_0 + s_1 v_1 + \cdots + s_k v_k,$$

we see at once that $T(v_0, \ldots, v_k)$ is the translation of S by v_0.
(b) The proof is carried out in Chapter 10 of Rudin's *Principles of Mathematical Analysis*.

The subsequent exercises do not depend on anything fancy, and occur in $\mathbf{R}^2$. Essentially, you don't need to know anything from this chapter.

Exercise XXI.4.2 *Let A be the region of $\mathbf{R}^2$ bounded by the inequalities*

$$a \le x \le b$$

and

$$g_1(x) \le y \le g_2(x)$$

where g_1, g_2 are continuous functions on $[a, b]$. Let C be the path consisting of the boundary of this region, oriented counterclockwise, as on the following picture:

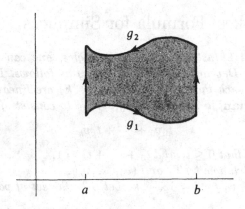

Show that if P is a continuous function of two variables on A, then

$$\int_C P dx = \iint_A -\frac{\partial P}{\partial y} dy dx.$$

Prove a similar statement for regions defined by similar inequalities but with respect to y. This yields **Green's theorem** *in special cases. The general case of Green's theorem is that if A is the interior of a closed piecewise C^1 path C oriented counterclockwise and ω is a 1-form then*

$$\int_C \omega = \iint_A d\omega.$$

Solution. The boundary of the region A can be written as

$$C = C_1 + L_2 - C_2 - L_1$$

where C_1, C_2, L_1, and L_2 are the curves shown on the figure. We have the following parametrization

$$
\begin{aligned}
C_1(t) &= (t, g_1(t)), \quad a \le t \le b, \\
C_2(t) &= (t, g_2(t)), \quad a \le t \le b, \\
L_1(t) &= (a, t), \quad g_1(a) \le t \le g_2(a), \\
L_2(t) &= (b, t), \quad g_1(b) \le t \le g_2(b).
\end{aligned}
$$

We have

$$
\iint_A -\frac{\partial P}{\partial y} dy dx = \int_a^b \int_{g_1(x)}^{g_2(x)} -\frac{\partial P}{\partial y} dy dx
$$

$$
= \int_a^b \left(-P(x,y) \big]_{g_1(x)}^{g_2(x)} \right) dx
$$

$$
= -\int_a^b P(x, g_2(x)) dx + \int_a^b P(x, g_1(x)) dx.
$$

But

$$
\int_{C_i} P dx = \int_a^b C_i^*(P dx) = \int_a^b P(t, g_i(t)) dt
$$

consequently

$$
\iint_A -\frac{\partial P}{\partial y} dy dx = \int_{C_1} P dx - \int_{C_2} P dx.
$$

It is sufficient to show that $\int_{L_1} P dx = \int_{L_2} P dx = 0$, which follows from the fact that the derivative of a constant is 0. This proves Green's theorem in this case.

In the subsequent exercises, you may assume Green's theorem.

Exercise XXI.4.3 *Assume that the function f satisfies Laplace's equation*

$$
\frac{\partial^2 f}{\partial x^2} + \frac{\partial^2 f}{\partial y^2} = 0,
$$

on a region A which is the interior of a curve C, oriented counterclockwise. Show that

$$
\int_C \frac{\partial f}{\partial y} dx - \frac{\partial f}{\partial x} dy = 0.
$$

Solution. Taking the exterior derivative of the form $\frac{\partial f}{\partial y} dx - \frac{\partial f}{\partial x} dy$ we get

$$
d \left(\frac{\partial f}{\partial y} dx - \frac{\partial f}{\partial x} dy \right) = \frac{\partial^2 f}{\partial x^2} dy \wedge dx - \frac{\partial^2 f}{\partial y^2} dx \wedge dy = 0.
$$

Applying Green's theorem yields

$$
\int_C \frac{\partial f}{\partial y} dx - \frac{\partial f}{\partial x} dy = -\iint_A \frac{\partial^2 f}{\partial x^2} + \frac{\partial^2 f}{\partial y^2} dx dy = 0.
$$

Exercise XXI.4.4 *If $F = (Q, P)$ is a vector field, we recall that its divergence is defined to be div $F = \partial Q/\partial x + \partial P/\partial y$. If C is a curve, we say that C is parametrized by arc length if $\|C'(s)\| = 1$ (we then use s as the parameter). Let*

$$
C(s) = (g_1(s), g_2(s))
$$

be parametrized by arc length. Define the unit normal vector at s to be the vector

$$N(s) = (g_2'(s), -g_1'(s)).$$

Verify that this is a unit vector. Show that if F is a vector field on a region A, which is the interior of a closed curve C, oriented counterclockwise, and parametrized by arc length, then

$$\iint_A (\text{div } F) dy dx = \int_C F \cdot N ds.$$

*If C is not parametrized by arc length, we define the **unit normal vector** by*

$$\mathbf{n}(t) = \frac{N(t)}{|N(t)|},$$

*where $|N(t)|$ is the euclidean norm. For any function f we define the **normal derivative** (the directional derivative in the normal direction) to be*

$$D_\mathbf{n} f = (\text{grad } f) \cdot \mathbf{n}.$$

So for any value of the parameter t, we have

$$(D_\mathbf{n} f)(t) = \text{grad } f(C(t)) \cdot \mathbf{n}(t).$$

Solution. Since $\|C'(s)\| = 1$ this means that

$$(g_1')^2(s) + (g_2')^2(s) = 1$$

which in turn implies that $\|N(s)\| = 1$. Applying Green's theorem to the vector field $(-P, Q)$ we obtain

$$\iint_A (\text{div } F) dy dx = \int_C -P dx + Q dy = \int_C F \cdot N ds.$$

Exercise XXI.4.5 *Prove Green's formulas for a region A bounded by a simple closed curve C, always assuming Green's theorem.*
(a) $\iint_A [(\text{grad } f) \cdot (\text{grad } g) + g\Delta f] dx dy = \int_C g D_\mathbf{n} f ds.$
(b) $\iint_A (g\Delta f - f\Delta g) dx dy = \int_C (g D_\mathbf{n} f - f D_\mathbf{n} g) ds.$

Solution. (a) We write

$$g \text{ grad } f = (g D_1 f, g D_2 f),$$

so that

$$\text{div}(g \text{ grad } f) = \sum_{i=1}^{2} D_i g f D_i f + g D_i D_i f.$$

By definition, $\Delta f = D_1 D_1 f + D_2 D_2 f$, so applying the divergence theorem to $g$grad f we obtain

$$\iint_A [(\text{grad } f) \cdot (\text{grad } g) + g\Delta f] dx dy = \int_C g D_n f ds.$$

(b) By symmetry we also have

$$\iint_A [(\text{grad } g) \cdot (\text{grad } f) + f\Delta g] dx dy = \int_C f D_n g ds.$$

Taking the difference of the above two equations we obtain Green's second formula.

Exercise XXI.4.6 *Let* $C \colon [a, b] \to U$ *be a* C^1-*curve in an open set* U *of the plane. If* f *is a function on* U *(assumed to be differentiable as needed), we define*

$$\int_C f = \int_a^b f(C(t)) \|C'(t)\| dt$$

$$= \int_a^b f(C(t)) \sqrt{\left(\frac{dx}{dt}\right)^2 + \left(\frac{dy}{dt}\right)^2} \, dt.$$

For $r > 0$, *let* $x = r \cos \theta$ *and* $y = r \sin \theta$. *Let* φ *be the function of* r *defined by*

$$\varphi(r) = \frac{1}{2\pi r} \int_{C_r} f = \frac{1}{2\pi r} \int_0^{2\pi} f(r \cos \theta, r \sin \theta) r d\theta,$$

where C_r *is the circle of radius* r, *parametrized as above. Assume that* f *satisfies Laplace's equation*

$$\frac{\partial^2 f}{\partial x^2} + \frac{\partial^2 f}{\partial y^2} = 0.$$

Show that $\varphi(r)$ *does not depend on* r *and in fact*

$$f(0, 0) = \frac{1}{2\pi r} \int_{C_r} f.$$

[Hint: First take $\varphi'(r)$ *and differentiate under the integral, with respect to* r. *Let* D_r *be the disc of radius* r *which is the interior of* C_r. *Using Exercise 4, you will find that*

$$\varphi'(r) = \frac{1}{2\pi r} \iint_{D_r} \text{div grad } f(x, y) dy dx = \frac{1}{2\pi r} \iint_{D_r} \left(\frac{\partial^2 f}{\partial x^2} + \frac{\partial^2 f}{\partial y^2}\right) dy dx = 0.$$

Taking the limit as $r \to 0$, *proves the desired assertion.]*

Solution. Differentiating under the integral sign we get

$$\varphi'(r) = \frac{1}{2\pi} \int_0^{2\pi} \frac{\partial f}{\partial x} \cos \theta + \frac{\partial f}{\partial y} \sin \theta d\theta.$$

Applying Stokes' theorem to the form $\omega = -(\partial f/\partial y)dx + (\partial f/\partial x)dy$ or Exercise 4 to the vector field $(D_1 f, D_2 f)$, we see that

$$\iint_A \frac{\partial^2 f}{\partial x^2} + \frac{\partial^2 f}{\partial y^2} dx dy = r \int_0^{2\pi} \frac{\partial f}{\partial x} \cos\theta + \frac{\partial f}{\partial y} \sin\theta d\theta,$$

and therefore $\varphi'(r) = 0$, thus φ is constant. Taking the limit as $r \to 0$ in the expression for φ we obtain $f(0,0)$ so that $\varphi(r) = f(0,0)$, which proves the desired assertion.